Dynamics and Control of Electrical Drives

Piotr Wach

Dynamics and Control of Electrical Drives

 Springer

Author

Prof. Piotr Wach
Politechnika Opolska
Institute of Electromechanical Systems and Applied Informatics
Mikołajczyka 5 str.
45-271 Opole
Poland
E-mail: p.wach@po.opole.pl

ISBN 978-3-642-44183-7   ISBN 978-3-642-20222-3 (eBook)

DOI 10.1007/978-3-642-20222-3

© 2011 Springer-Verlag Berlin Heidelberg

Typeset & Cover Design: Scientific Publishing Services Pvt. Ltd., Chennai, India.

Printed on acid-free paper

9 8 7 6 5 4 3 2 1

springer.com

To my dear wife Irena

Acknowledgments

The author would like to express warm gratitude to all who contributed in various ways so that this book finally has been completed in its form.

Firstly, I would like to express remembrance and pay tribute to Prof. Arkadiusz Puchała (Academy of Mining and Metallurgy in Cracow), who was my principal tutor and supervisor of my Ph.D. thesis. However, our contacts terminated early, when he died quite young in 1973.

Then I would like to thank warmly my colleagues from the Institute of Electromechanical Systems and Applied Informatics, who encouraged me and created favorable conditions and nice, friendly atmosphere to work and research: Prof. Krystyna Macek-Kamińska – Director of the Institute, Prof. Marian Łukaniszyn – our present Faculty Dean as well as Prof. Jerzy Hickiewicz and Prof. Sławomir Szymaniec – partners in several research undertakings.

Then I would like to thank Dr. Krzysztof Tomczewski, Dr. Ryszard Beniak, Dr. Andrzej Witkowski, Dr. Krzysztof Wróbel my former Ph.D students, work with whom gave me a lot of experience, exchange of ideas and excellent opportunity to discuss.

Finally I would like to thank my dear son Szymon Wach for his very good – I am sure, translation of the book into English and Mr Eugeniusz Głowienkowski for preparation of those technical drawings that were not produced automatically by MAPLE™, as an outcome of computer simulations.

Piotr Wach

Notation Index

$a = 2\pi/3$	- phase shift between 3-phase symmetrical sine curves
$\mathbf{a} = \dot{\mathbf{v}} = \ddot{\mathbf{r}}$	- acceleration vector
$\delta A, \delta A_m, \delta A_e$	- virtual work, its mechanical and electrical component
$\mathbf{A}$	- vector potential of a magnetic field
$\mathbf{A}_2, \mathbf{A}_3$	- skew symmetric matrices: 2- and 3- dimentional respectively
$\mathbf{B}$	- magnetic induction vector
C	- electrical capacity
D	- viscous damping factor
e_k	- electromotive force (EMF) induced in k-th winding
E	- total energy of a system
$f_k(\ldots)$	- analytical notation of holonomic constraints function
$\mathbf{F}, F_i$	- vector of external forces, i-th component of this vector
f_L, f_s, f_r	- frequency of voltage (current): feeding line, stator, rotor
$\mathbf{g}$	- acceleration vector of earth gravitation force
g	- number of branches of electric network
h	- number of holonomic constraints
$i = \dot{Q}$	- electric current as a derivative of electrical charge
i_f, i_a	- excitation current, armature current
$\underline{I}$	- symbolic value of sinusoidal current
$\mathbf{I}$	- matrix of inertia of a rigid body
$\mathbf{i}_s = [i_{s1} \ i_{s2} \ i_{s3}]^T$	- vector of a 3-phase stator currents
$\mathbf{i}_{s12} = [i_{s1} \ i_{s2}]^T$	- vector of a 3-phase stator currents in a star connected system
$\mathbf{i}_{r13} = [i_{r1} \ i_{r3}]^T$	- vector of a 3-phase rotor currents in a star connected system
$\mathbf{i}_r = [i_{r1} \ i_{r2} \ldots i_{rm}]^T$	- vector of a m-phase rotor currents
$\mathbf{i}_{s0uv}, \mathbf{i}_{r0uv}$	- vectors of transformed stator, rotor currents in $0, u, v$ axes
$\mathbf{i}_{suv}, \mathbf{i}_{ruv}$	- vectors of transformed stator, rotor currents to u, v axes

Q_k	- electric charge of the k-th element characterized by electrical capacity
$\dot{Q}_k = i_k$	- electric current in k-th winding as a derivative of respective charge
$\mathbf{Q}, \dot{\mathbf{Q}}$	- vector of electrical charges of a system, vector of electric currents
q	- number of pulses of power electronic converter
$q_k, \dot{q}_k, \ddot{q}_k$	- k-th generalized coordinate, - velocity, - acceleration respectively
$\mathbf{q}, \dot{\mathbf{q}}, \ddot{\mathbf{q}}$	- vectors of generalized coordinates, velocities, accelerations of a system
$\delta\mathbf{q}$	- vector of virtual displacements for generalized coordinates
δq_k	- virtual displacement for k-th generalized coordinate
R_s, R_r	- phase winding resistance for stator and rotor respectively
$\mathbf{r}_i, \mathbf{r}$	- radius-vector pointing i-th particle, radius-vector for whole system in Cartesian coordinates
$\delta\mathbf{r}$	- vector of virtual displacements of a system in Cartesian coordinates
$\mathbf{R}$	- vector of reaction forces of constraints in a system
s	- slip of an induction motor rotor motion in respect to magnetic field
s	- total number of degrees of freedom
s_m, s_e	- number of mechanical, electrical degrees of freedom of a system
S	- action function of a system
T, T_{el}, T_{me}	- kinetic energy, its electrical and mechanical component
T', T'_{el}, T'_{me}	- kinetic co-energy, its electrical and mechanical component
T_e, T_l	- electromagnetic torque of a motor, load torque
T_b, T_{st}, T_n	- break torque, starting torque, rated torque of an induction motor
T_p	- period of a single pulsation sequence in PWM control method
T_f	- friction force
$\mathbf{T}_s, \mathbf{T}_r$	- orthogonal matrices of transformation for stator and rotor variables
u, u_k	- electric voltage, voltage supplied to k-th winding
u_s, u_r	- stator and rotor voltages respectively
U_n, U_{sn}	- rated voltage, stator rated voltage
U_{12}, U_{23}, U_{31}	- phase to phase voltages in 3-phase electrical system
$u_{s12}, u_{s23}, u_{s31}$	- stator's phase to phase voltages in 3-phase system

U_{sph} — phase voltage of stator's winding

U_L — feeding line voltage

$U_{d0} = \dfrac{3\sqrt{2}}{\pi} U_L$ — average value of rectified voltage in a 6 pulse 3-phase system

$\underline{U}$ — symbolic value of sinusoidal voltage

U, U_{me}, U_{el} — potential energy of a system, its mechanical and electrical component

$\mathbf{U}_s, \mathbf{U}_{sph} = [u_{s1}, u_{s2}, u_{s3}]^T$ — stator's voltage vector, stator's phase voltages vector

$\mathbf{U}_r, \mathbf{U}_{rph} = [u_{r1}, u_{r2}, \ldots, u_{rm}]^T$ — vector of rotor voltages, vector of phase voltages of rotor windings

$\mathbf{u}_{s0uv}$ — vector of stator voltages transformed to $0, u, v$ system of axes

$\mathbf{u}_{suv}$ — vector of stator voltages transformed to u, v system of axes

$\mathbf{u}_{sxy}$ — vector of stator voltages transformed to x, y system of axes

$\mathbf{u}_{sxy\rho}$ — vector of stator voltages transformed to x_ρ, y_ρ field oriented axes

$\mathbf{v} = \dot{\mathbf{r}}$ — vector of velocities of a system

w — number of nodes of an electric network

X — reactance of a winding

X_s, X_r, X_m — reactance of a stator, rotor and magnetizing one respectively

$\mathbf{X} = (\chi_1, \ldots \chi_n)$ — vector of coordinates in a primary coordinate system

Z_s, Z_r — number of stator's, rotor's teeth of SRM machine

α_k, α_l — angles determining axis position of windings

$\alpha_{on}, \alpha_{off}$ — switch on and switch off control angles of SRM machine

γ — phase shift angle

$\delta\xi_k$ — virtual displacement of k-th Cartesian coordinate in unified coordinate system

$\Xi = (\xi_1, \ldots \xi_n)$ — vector of Cartesian coordinates in unified system of coordinates ξ

η — energy efficiency factor of a system

θ_r — rotation angle

$\dot{\theta}_r = \Omega_r$ — rotational speed of a rotor

ρ — number of a magnetic field harmonic

ρ — field orientation angle of x_ρ, y_ρ axes (vector control)

σ — leakage coefficient of windings

ϕ — scalar potential of electromagnetic field

$\varphi_j(...)$ — analytical notation of nonholonomic constraints

φ_T — field angle (DTC)

ψ_k — flux linkage of k-th winging

ψ_f — excitation flux

$\Psi_{sph} = [\psi_{s1}, \psi_{s2}, \psi_{s3}]^T$ — flux linkage vector of stator windings

Ψ_{rph} — flux linkage vector of rotor windings

Ψ_{suv}, Ψ_{ruv} — u,v transformed flux linkage vectors of stator and rotor windings

Ψ_{s0uv}, Ψ_{s0uv} — $0,u,v$ transformed flux linkage vectors of stator and rotor windings

Ψ_{sdq}, Ψ_{rdq} — d,q transformed flux linkage vectors of stator and rotor windings

Ψ_{sxy}, Ψ_{rxy} — x,y transformed flux linkage vectors of stator and rotor windings

$\Psi_{s\alpha\beta}, \Psi_{r\alpha\beta}$ — α,β transformed flux linkage vectors of stator and rotor windings

$\omega_L = 2\pi f_L$ — AC supply line pulsation

ω_s, ω_r — pulsation of stator, rotor sinusoidal voltages, currents

ω_c — reference pulsation (angular speed) in transformed $0,u,v$ system

$\omega_e = p\Omega_r$ — electrical angular speed of a rotor

$\omega_f = \Omega_0 = \omega_s / p$ — synchronous speed of rotating magnetic field, idle run speed

$\boldsymbol{\omega}$ — vector of angular velocity of rigid body

$\Omega_0 = \omega_s / p$ — ideal idle run speed of an induction motor's rotor

Ω_r — angular speed of a rotor

Contents

Chapter 1
Introduction

Abstract. First Chapter is an introductory one and it generally presents the scope of this book, methodology used and identifies potential readers. It also develops interrelations between modern electric drives, power electronics, mechatronics and application of control methods as the book to some degree covers all these fields. The content is concentrated around electromechanical energy conversion based on Lagrange's method and its clear and subsequent application to control of electric drives with induction machines, brushless DC motors and SRM machines. It does not cover stepper motors and synchronous PM drives. All computer simulation results are outcome of original mathematical models and based on them computations carried out with use of MAPLE™ mathematical package.

Electrical drives form a continuously developing branch of science and technology, which dates back from mid-19th century and plays an increasingly important role in industry and common everyday applications. This is so because every day we have to do with dozens of household appliances, office and transportation equipment, all of which contain electrical drives also known as actuators. In the same manner, industry and transport to a large extent rely on the application of electrical drive for the purposes of effective and precise operation. The electrical drives have taken over and still take on a large share of the physical efforts that were previously undertaken by humans as well as perform the type of work that was very needed but could not be performed due to physical or other limitations. This important role taken on by the electrical drive is continuously expanding and the tasks performed by the drives are becoming more and more sophisticated and versatile [2,11,12,13,20]. Electrical drives tend to replace other devices and means of doing physical work as a result of their numerous advantages. These include a common accessibility of electrical supply, energy efficiency and improvements in terms of the control devices, which secures the essential quality of work and fit in the ecological requirements that apply to all new technology in a modern society [1]. The up-to-date electrical drive has become more and more intelligent, which means fulfillment of the increasingly complex requirements regarding the shaping of the trajectories of motion, reliable operation in case of interference and in the instance of deficiency or lack of measurement information associated with the executed control tasks [4,8,16,21]. This also means that a large number of components are involved in information gathering and processing, whose role is to ensure the proper operation, diagnostics and protection of the drive. The reasons for such intensive development of electrical drives are numerous and the basic reason for that is associated with the need of intelligent, effective, reliable and un-disturbed execution of mechanical work. Such drives are designed in a very wide range of electrical capacity as electromechanical devices with the power rating from [mW] to [MW]. Additionally, the processes are accompanied by multi-parameter motion control and the primary role can be attributed to speed regulation along with the required levels of force or torque produced by the drive. The present capabilities of fulfilling such complex requirements mostly result from the development of two technological branches, which made huge progress at the end of the 20th century and the continuous following developments. One of such areas involves the branch of technological materials used for the production of electrical machines and servomotors. What is meant here is the progress in the technology of manufacturing and accessibility of inexpensive permanent magnets, in particu-lar the ones containing rare earth elements such as samarium (Sr) and neodymium (Nd). In addition, progress in terms of insulation materials, their service lives and small losses for high frequencies of electric field strength, which result from con-trol involving the switching of the supply voltage. Moreover, considerable pro-gress has occurred in terms of the properties of ferromagnetic materials, which are constantly indispensable for electromechanical conversion of energy. The other branch of technology which has enabled such considerable and quick development of electrical drive is the progress made in microelectronics and power electronics. As a result of the development of new integrated circuits microelectronics has made it possible to gather huge amount of information in a comfortable and inex-pensive way, accompanied by its fast processing, which in turn offers the applica-tion of complex methods of drive control. Moreover, up-to-date power electronics markets new current flow switches that allow the control over large electric power with high frequency thus enabling the system to execute complex control tasks. This occurs with very small losses of energy associated with switching, hence playing a decisive role in the applicability of such devices for high switching fre-quencies. The versatility and wide range of voltages and currents operating in the up-to-date semiconductor switches makes it possible to develop electric power converters able to adapt the output of the source to fulfill the parameters resulting from instantaneous requirements of the drive [5,7,23]. Among others this capabil-ity has led to the extensive application of sliding mode control in electrical drive which very often involves rapid switching of the control signal in order to follow the given trajectory of the drive motion [22].

Such extensive and effective possibility of the development of electrical drives, which results from the advancements in electronics and a rapid increase in the ap-plication range of the actuating devices, has given rise to the area of mechatronics.

Mechatronics can either be thought of as a separate scientific discipline or a relevant and modern division of the electrical drive particularly relating to elec-tronics, control and large requirements with regard to the dynamic parameters of the drive [19].

It is also possible to discuss this distinction in terms of the number of degrees of freedom of the device applied for the processing of information followed by electric power conversion into mechanical work. In a traditional electrical drive we have to do with a number of degrees of freedom for the electric state variables and a single one for the mechanical motion. In standard electrical machines the variables include: angle of the rotation of the rotor or translational motion in a linear motor. In mechatronics it is assumed that the number of the degrees of mechanical freedom can be larger, i.e. from the mechanical viewpoint the device executes a more complex capabilities than simply a rotation or translational motion. In addition, in mechatronics it is not necessary that the medium serving for the conversion of energy is the magnetic field. Mechatronic devices may operate under electric field, which is the case in electrostatic converters [14,15]. In conclusion, it can be stated that mechatronics has extended the set of traditional devices in the group of electrical drives to cover a wider choice of them as well as has supplemented extra methods and scope of research. However, it can be stated beyond doubt that mechatronics constitutes nowadays a separate scientific discipline which realizes its aim in an interdisciplinary manner while applying equally the findings of computer engineering, electronics and electromechanics in order to create a multi-dimensional trajectory of the mechanical motion. Such understanding of mechatronics brings us closer to another more general scientific discipline as robotics. What is important to note is that if a manipulator or a robot has electrical joint drives, in its electromechanical nature it constitutes a mechatronic device.

Concurrently, robotics has even more to it [18]. Not to enter the definitions and traditions in this discipline, what is generally meant is the autonomous nature of the robots in terms of its capability of recognition of its environment and scope of decision making, i.e. the application of artificial intelligence. By looking at a manipulator or a robot produced in accordance with up-to-date technology we start to realize its capabilities with regard to its orientation in space and organization of the imposed control tasks. However, one should also give merit to its speed, precision, repeatability and reliability of operation, all of which relate to mechatronics.

The reference to robotics in a book devoted to electrical drive results from the fact that in its part devoted to theory and in the presented examples a reference is made to the methods and solutions originating from robotics, the focus in which has often been on the motion in a multi dimensional mechanical systems with constraints [3].

The following paragraphs will be devoted to the presentation of the overview of the current book, which contains 4 chapters (besides the introduction) devoted to the issues of the up-to-date electrical drives and their control.

Chapter 2 covers the issues associated with the dynamics of mechanical and electromechanical systems. The subjects of the subsequent sections in this chapter focus on mechanical systems with a number of degrees of freedom as well as holonomic and non-holonomic constraints. The presented concept covers a physical system which is reduced to a set of material points and a system defined as a set of rigid bodies. A detailed method of the development of a mathematical models is introduced involving Lagrange's functions and equations departing from the principle of least action for a charged material particle in the magnetic field. Subsequently, this concept has been extended to cover macroscopic systems capable of accumulation of energy of the magnetic and electric fields in the form of kinetic and potential energy, respectively. The dissipation of energy and transformation of the dissipation coefficients into terms of the state variables are taken down to the negative term of the virtual work of the system. This is a way that is formally correct from the point of methodology of research. Additionally, it proves effective in the practice of the formulation of the equations of motion. The section devoted to electromechanical systems has been illustrated by numerous examples whose difficulty level is intermediate.

In general, the examples of the application of theory in the book are quite numerous and have been selected in a manner that should not pose excessive difficulty while maintaining them at a level that can serve for the purposes of illustrating specific characteristics of the applied method but are never selected to be trivial.

Chapter 3 focuses on induction machine drives. The presented mathematical models have been developed with the aid of Lagrange's method for electromechanical systems. The models transformed into orthogonal axes are presented in a classical manner along with the models of an induction machine for which the variables on one side, i.e. stator's or rotor's are untransformed and remain in the natural–phase coordinates. This plays an important role in the drive systems supplied from power electronic converters. The adequate and more detailed modeling of the converter system requires natural variables of the state, i.e. untransformed ones in order to more precisely realize the control of the drive. These models, i.e. models without the transformation of the variables on one side of the induction motors are presented in their applications in the further sections in this book. The presentation focuses on various aspects of their supply, regulation and control with the application of converters. The classical subject matters include presentation of DC braking, Scherbius drive, as well as the operation of a soft-starter. Concurrently, the up-to-date issues associated with induction machine drives cover two - level and three-level Voltage Source Inverters (VSI), Sinusoidal Pulse Width Modulation (SPWM), Space Vector Modulation (SVM), Discontinuous Space Vector Modulation (DSVM) and PWM Current Source Inverter (CSI) control. Further on, beside VC methods the Direct Torque Control (DTC) is presented in theory and in examples. The final section of Chapter 3 is devoted to the presentation of structural linearization of a model of induction motor drive along with several state observers applicable for the induction motor.

Chapter 4 is devoted to permanent magnet brushless DC motor drives and control of such drives. Firstly, the characteristics and properties of the up-to-date permanent magnets (PM) are presented together with simplified methods applied for their modeling. The example of a pendulum coil swinging over a stationary PM serves for the purposes of presenting the effect of simplifications in the model of the magnet on the trajectories of the motion of such an electromagnetic system. Further on, the transformed d-q model of a BLDC machine is derived along with an untransformed model in which the commutation occurs in accordance with the courses of the natural variables of the machine. The presentation of the mathematical model of BLDC does not cover the subject of nonholonomic constraints in this type of machines. In a classical DC machine with a mechanical commutator the occurrence of such constraints involving the dependence of the configuration of electrical circuits on the angle of the rotation of the rotor is quite evident. Concurrently, in an electronically commuted machine the supply of the particular windings of the armature is still relative to the angle of the rotation of the rotor; however, the introduction of nonholonomic constraints in the description is no longer necessary. Commutation occurs with the preservation of the fixed structure of electrical circuits of the electronic commutator coupled with the phase windings of the machine's armature and the switching of the current results from the change of the parameters of the impedance of the semiconductor switch in the function of the angle of rotation. This chapter focuses on the characteristics and dynamic courses illustrating the operation of BLDC motors with a comparison between the results of modeling drives with the aid of two-axial d-q transformation as well as untransformed model. On this basis it is possible to select a model for the simulation of the issues associated with the drive depending on the dimension of the whole system and the level of the detail of the output from modeling. The presented static characteristics and dynamic courses of BLDC drives focus on adequate characterizing the capabilities and operating parameters of such drives without control. Subsequently, research focuses on the control of BLDC drives and the presentation of the control using PID regulator, control with the given speed profile and the given profile of the position as well as inverse dynamics control. The illustrations in the form of dynamic courses are extensive and conducted for two different standard BLDC motors.

The final chapter, i.e. Chapter 5 is devoted to the presentation of switched reluctance motor (SRM) drives. Before the development of the mathematical model, magnetization characteristics of SRM motors are presented and the important role of non-linearity of characteristics in the conversion of energy by the reluctance motor is remarked. Subsequently, the presentation follows with the mathematical model accounting for the magnetic saturation reflected by magnetization characteristics with regard to the mutual position of the stator and rotor teeth. This is performed in a way that is original since the inductance characteristics that are relative to two variables are presented here in the form of a product of the function of the magnetic saturation and the function of the rotor's position angle. Such an approach has a number of advantages since it enables one to analyze the effect of particular parameters on the operation of the motor. The derived model of the SRM motor does not account for magnetic coupling between phase windings; however, from the examples of two standard SRM motors it was possible to indicate a little effect of such couplings on the characteristics and operation of the motors. In this manner such simplifications included in the mathematical model are justified. The further sections of this chapter focus on a number of issues regarding the dynamics and control of SRM drives. The presentation includes a solution to the problem of the pulse based determination of the starting sequence during starting SRM drive for the selected direction of the rotation of the motor, direct start-up with the limitation of the current as well as braking and discussion of the issue of very specific generator regime of operation. The presentation also covers the selection of the regulation parameters for SRM with the aim of gaining high energy efficiency and reducing torque ripple level. The section devoted to the control involves the presentation of sliding control applied for this drive type, current control as well as DTC control and the possibility of limiting the pulsations of the torque as a result of applying specific control modes. In addition, the presentation briefly covers sensor and sensorless control of SRM drive and the application of state observers while providing for the exclusion of the position sensor.

Following this brief overview of the content it is valuable that the reader notes that the book contains a large number of examples in the area of dynamics and control of specific drives, which is reflected by waveforms illustrating the specific issues that are presented in the figures. All examples as well as illustrations come from computer simulations performed on the basis of mathematical models developed throughout the book. This has been performed for standard examples of motors the detailed data and parameters of which are included in the particular sections of the book. Computer simulations and graphical illustrations gained on this basis were performed in MAPLE™ mathematical programming system, which has proved its particular applicability and flexibility in this type of modeling. All calculations were performed on the basis of programs originally developed by the author.

As one can see from this short overview, the scope of this book is limited and does not involve some types of drives used in practice, i.e. stepper motor drives and synchronic machine drive with permanent magnets. The missing types of drive have similar characteristics in terms of the principle of energy conversion and mathematical models to SRM motor drives and BLDC drives, respectively. However, the details of construction and operation are dissimilar and only a little effort can enable one to apply the corresponding models in this book in order to develop dedicated programs for computer simulations and research of the two missing drive types.

A final remark concerns the target group of this book, which in the author's opinion includes students of postgraduate courses and Ph.D. students along with engineers responsible for the design of electrical drives in more complex industrial systems.

References

[1] Almeida, A.T., Ferreira, F.J., Both, D.: Technical and Economical Considerations on the Applications of Variable-Speed Drives with Electric Motor System. IEEE Trans. Ind. Appl. 41, 188–198 (2005)
[2] Boldea, I., Nasar, A.: Electric Drives. CRC Press, Boca Raton (1999)
[3] Canudas de Wit, C., Siciliano, B., Bastin, G.: Theory of Robot Control. Springer, Berlin (1997)
[4] Dawson, D.M., Hu, J., Burg, T.C.: Nonlinear Control of Electric Machinery. Marcel Dekker, New York (1998)
[5] El-Hawary, M.E.: Principles of Electric Machines with Power Electronic Applications, 2nd edn. John Wiley & Sons Inc., New York (2002)
[6] Fitzgerald, A.E., Kingsley, C., Kusko, A.: Electric Machinery. McGraw-Hill, New York (1998)

[7] Heier, S.: Wind Energy Conversion Systems, 2nd edn. John Wiley & Sons, Chichester (2006)

[8] Isidiri, A.: Nonlinear Control Systems. Springer, New York, Part I-(1995), Part II-(1999)

[9] Kokotovic, P., Arcak, M.: Constructive nonlinear control: a historical perspective. Automatica 37, 637–662 (2001)

[10] Krause, P.C.: Analysis of Electrical Machinery. Mc Graw Hill, New York (1986)

[11] Krause, P.C., Wasynczuk, O., Sudhoff, S.D.: Analysis of Electric Machinery and Drive Systems, 2nd edn. John Wiley & Sons Inc., New York (2002)

[12] Krishnan, R.: Electric Motor Drives: Modeling, Analysis and Control. Prentice-Hall, Upper Saddle River (2002)

[13] Leonhard, W.: Control of Electrical Drives, 3rd edn. Springer, Berlin (2001)

[14] Li, G., Aluru, N.R.: Lagrangian approach for electrostatics analysis of deformable con-ductors. IEEE J. Microelectromech. Sys. 11, 245–251 (2002)

[15] Li, Y., Horowitz, R.: Mechatronics of electrostatic actuator for computer disc drive dual-stage servo systems. IEEE Trans. Mechatr. 6, 111–121 (2001)

[16] Mackenroth, U.: Robust Control Systems. Springer, Heidelberg (2004)

[17] Mohan, N.: Advanced Electric Drives. NMPERE, Minneapolis (2001)

[18] Sciavicco, L., Siciliano, B.: Modeling and Control of Robot Manipulators. Springer, London (2000)

[19] de Silva, C.W.: Mechatronics: An Integrative Approach. CRC Press, Boca Raton (2004)

[20] Slemon, G.R.: Electric Machines and Drives. Addison-Wesley, Reading (1992)

[21] Slotine, J.J., Li, W.: Applied Nonlinear Control. Prentice Hall, New Jersey (1991)

[22] Utkin, V.I.: Sliding Modes in Control and Optimization. Springer, Berlin (1992)

[23] Zhu, Z., Howe, D.: Electrical Machines and Drives for Electric, Hybrid and Fuel Cell Vehicles. Proc IEEE 95, 746–765 (2007)

Chapter 2
Dynamics of Electromechanical Systems

Abstract. Chapter is devoted to dynamics of mechanical and electromechanical systems. Sections dealing with mechanical systems concern holonomic and non-holonomic objects with multiple degrees of freedom. The concept of an object represented by a system of connected material points and the concept of a rigid body and connected bodies are presented. The Lagrange's method of dynamics formulation is thoroughly covered, starting from d'Alembert's virtual work principle. Carefully selected examples are used to illustrate the method as well as the application of the theory. Electromechanical system's theory is also introduced on the basis of the Lagrange's equation method, but starting from the principle of least action for a electrically charged particle in a stationary electromagnetic field. Subsequently, the method is generalized for macroscopic systems whose operation is based on electric energy and magnetic co-energy conversion. Nonlinear systems are discussed and the concept of kinetic co-energy is explained. Energy dissipation is introduced as a negative term of the virtual work of the system, and transformation of dissipation coefficients to the terms of generalized coordinates are presented in accordance with Lagrange method. Finally a number of examples is presented concerning electromechanical systems with magnetic and electric field and also selected robotic structures.

2.1 Mechanical Systems

2.1.1 Basic Concepts

Discrete system - is a system whose position is defined by a countable number of variables. In opposition to discrete system a continuous system (or distributed parameters' system) is defined as a system with continuously changing variables along coordinates in space. Both these concepts are a kind of idealization of real material systems.

Particle – is an idealized object that is characterized only by one parameter – mass. To define its position in a three-dimensional space (3D) three variables are necessary. This idealization is acceptable for an object whose mass focuses closely around the center of the mass. In that case its kinetic energy relative the to linear (translational) motion is strongly dominant over the kinetic energy of the rotational motion. Besides, it is possible to consider large bodies as particles (material points) in specific circumstances, for example when they do not rotate or their rotation does not play significant role in a given consideration. This is the case in the examination of numerous astronomical problems of movement of stars and planets.

Rigid body – is a material object, for which one should take into account not only the total mass M, but also the distribution of that mass in space. It is important for the rotating objects while the kinetic energy of such movement plays an important role in a given dynamical problem. In an idealized way a rigid body can be considered as a set of particles, of which each has an specific mass m_i, such that $\Sigma m_i = M$. Formally, the body is rigid if the distances d_{ij} between particles i,j are constant. The physical parameters that characterize a rigid body from the mechanical point of view are the total mass M, and a symmetrical matrix of the dimension 3, called the matrix of inertia. This matrix accounts for moments of inertia on its diagonal and deviation moments, which characterize the distribution of m_i masses within the rigid body in a Cartesian coordinate system.

Constraints – are physical limitations on the motion of a system, which restrict the freedom of the motion of that system. The term system used here denotes a particle, set of connected particles, a rigid body or connected bodies as well as other mechanical structures. These limitations defined as constraints are diverse: they can restrict the position of a system, the velocity of a system as well as the kind of motion. They can be constant, time dependent or specific only within a limited sub-space. Formally, the constraints should be defined in an analytical form to enable their use in mathematical models and computer simulations of motion. Hence, they are denoted in the algebraic form as equations or inequalities.

Cartesian coordinate system - is the basic, commonly used coordinate system, which in a three dimensional space (3D) introduces three perpendicular straight axes. On these axes it is possible to measure the actual position of a given particle in an unambiguous way using three real numbers.

Position of the particle P_i in that system is given by a three dimensional vector, so called radius-vector $\mathbf{r}_i$:

$$\mathbf{r}_i = \mathbf{r}_i(x_i, y_i, z_i) \tag{2.1}$$

Its coordinates are, respectively: x_i, y_i, z_i ; see Fig. 2.1.

In complex mechanical systems, consisting of a number of particles: $i=1,2,...,N$ the generalized position vector for the whole system is defined as follows:

$$\mathbf{r} = (\mathbf{r}_1, \mathbf{r}_2, ..., \mathbf{r}_N) = (x_1, y_1, z_1, x_2, y_2, z_2, ..., x_{3N}, y_N, z_{3N})$$

which is placed in an abstract $3N$ space. For a more convenient operation of this kind of notation of the system's position, especially in application in various summation formulae, a uniform Greek letter ξ_j is introduced:

$$\mathbf{r}_i = (\xi_{3i-2}, \xi_{3i-1}, \xi_{3i})$$

As a result of above, the position vector for the whole system of particles takes the following form:

$$\mathbf{r} = (\mathbf{r}_1, \mathbf{r}_2, \dots, \mathbf{r}_N) =$$
$$= \Xi = (\xi_1, \xi_2, \xi_3, \xi_4, \dots, \xi_{3N-2}, \xi_{3N-1}, \xi_{3N}) \tag{2.2}$$

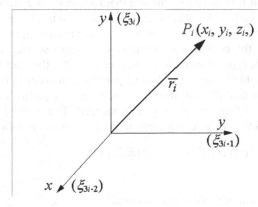

Fig. 2.1 Cartesian coordinates of a particle

Velocity and *acceleration* of a system
The formal definition of a velocity of a system is given below:

$$\mathbf{v}_i = \dot{\mathbf{r}}_i = \lim_{\Delta t \to 0} \frac{\mathbf{r}_i(t + \Delta t) - \mathbf{r}_i(t)}{\Delta t} \tag{2.3}$$

Because time t is a parameter of any motion, differentiation and calculation of derivatives in respect to time is a frequent operation in dynamics. Hence traditionally, the time derivative is briefly noted by a dot above a variable that is differentiated in respect to time, in the following form:

$$\mathbf{v}_i = \frac{d}{dt}\mathbf{r}_i = \dot{\mathbf{r}}_i \tag{2.4}$$

Acceleration is the time derivative of the velocity, which means it is the second derivative of the position in respect to time:

$$\mathbf{a}_i = \dot{\mathbf{v}}_i = \frac{d^2}{dt^2}\mathbf{r}_i = \ddot{\mathbf{r}}_i \tag{2.5}$$

According to the Newton's Second Law of Dynamics, which describes the relation between motion and its cause, i.e. the applied force (or torque), in the description of dynamics there is no need or place for higher order derivatives of the position of a body than ones of the second order – i.e. acceleration. This also means that in dynamics one has to do only with the position $\mathbf{r}$, velocity $\dot{\mathbf{r}}$ and the acceleration $\ddot{\mathbf{r}}$, time t as a parameter of motion, and forces (torques) as causes of motion.

2.1.2   Constraints, Classification of Constraints and Effects of Their Imposition

Constraints are physical limitations of motion that reduce the freedom of motion of a given system. The system denotes here a mechanical unit such as a particle, a set of connected particles, a rigid body or a set of connected rigid bodies. The limitations imposed by the constraints are various in nature so they may restrict the freedom of position, type of motion, as well as velocity; they act in a limited space and even are variable in time. For formal purposes, in order to perform analytical description of motion the constraints are denoted as equations or inequalities and a classification of the constraints is introduced. The general form of an analytical notation used to present constraints acting in a system is following:

$$f(\mathbf{r},\dot{\mathbf{r}},t) \; \Re \; 0 \quad \text{or} \quad f(\Xi,\dot{\Xi},t) \; \Re \; 0 \tag{2.6}$$

where:

f - is the analytical form of constraints function,
$\mathbf{r},\Xi$ - position vector of a system,
$\dot{\mathbf{r}},\dot{\Xi}$ - velocity vector of a system,
$\Re$ - the relation belonging to the set $\Re \in \{=,<,>\leq,\geq\}$

Stiff or *bilateral constraints* vs. *releasing* or *unilateral constraints*. This is a classification in respect to a relation $\Re$. Stiff constraints are expressed by the equality relation

$$f(\bullet,\bullet,\bullet) = 0 \qquad f(\bullet,\bullet,\bullet) \; \{<,>,\leq,\geq\}0 \tag{2.7}$$

while releasing constraints are ones that contain the relation of inequality.

Geometric vs. *kinematical* constraints. This classification accounts for the absence or presence of velocity in a relation of constraints. In case that the velocity is there the constraints are called kinematical

$$f(\bullet,\dot{\mathbf{r}},\bullet) \; \Re \; 0 \qquad f(\mathbf{r},\bullet) \; \Re \; 0 \tag{2.8}$$

and without explicit presence of velocity they are named geometric constraints.

Time depending (scleronomic) vs. *time independent* (reonomic) constraints. This is a division that takes into account the explicit presence of time in the relation of constraints:

$$f(\mathbf{r},\dot{\mathbf{r}}) \; \Re \; 0 \qquad f(\bullet,\bullet,t)\Re \; 0 \tag{2.9}$$

In that respect the first relation of (2.9) presents scleronomic constraints and the second one reonomic constraints.

Holonomic vs *nonholonomic* constraints. It is the basic classification of constraint types from the theory of dynamical systems point of view. The division of mechanical systems into holonomic and nonholonomic systems follows.

Holonomic constraints are all geometric constraints and those kinematical constraints that can be converted into geometric constraints by integration.

$f(\mathbf{r}, \bullet) \, \Re \, 0$ - geometric constraints

$f(\mathbf{r}, \dot{\mathbf{r}}, \bullet) \, \Re \, 0$ - those of kinematical constraints, for which exists: (2.10)

$F(\mathbf{r}, \bullet)$ - such that: $dF = f$

Nonholonomic constraints are all kinematical constraints that could not be integrated and hence cannot be converted into geometric ones. The formal division is clear, but it is more difficult to offer the physical explanation for this distinction. Simply speaking one can say that holonomic constraints restrict the position of a system and the velocity of that system in a uniform manner, while nonholonomic ones impose restrictions on the velocity without restricting the position. Consequently, one can say that nonholonomic constraints restrict the manner of motion without limiting the position in which such motion can result. For further applications, nonholonomic constraints will be denoted in the following form:

$$\varphi_j \left(\xi_1, \xi_2 \ldots, \xi_n, \dot{\xi}_1, \ldots, \dot{\xi}_n, t \right) = 0 \qquad j = 1, \ldots b \tag{2.11}$$

and in a specific case of the linear nonholonomic constraints:

$$\varphi_j = \sum_{i=1}^{n} h_{ij} \dot{\xi}_i + D_j \qquad j = 1, \ldots b \tag{2.12}$$

- where h_{ij} in the general case are functions of position coordinates and time.

To verify whether it is possible to integrate linear kinematical constraints it is sufficient to check if they are in the form of the Pfaff's differential equations with total differentials.

2.1.3 Examples of Constraints

Example 2.1. In a planar system (Fig.2.2) two steel balls are connected by a stiff rod. One has to define the analytical form for constraint notation and to define them in accordance with the presented classification.

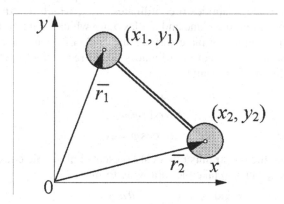

Fig. 2.2 System of two massive balls constrained by a stiff rod

a)   for balls connected with a stiff rod with the length l

$$|\mathbf{r}_1 - \mathbf{r}_2| = l \text{ hence:}$$

$$(x_1 - x_2)^2 + (y_1 - y_2)^2 - l^2 = 0 \qquad (2.13)$$

As a consequence, the examined case presents geometric, stiff, time independent constraints.

b)   for balls connected with a cord of length l whose thickness is negligible the equation (2.13) is replaced with an inequality in the form:

$$(x_1 - x_2)^2 + (y_1 - y_2)^2 - l^2 \leq 0 \qquad (2.14)$$

Hence the case represents releasing (unilateral) constraints. It is possible to differentiate the equation for constraints (2.13) with respect to time, hence the following form is obtained:

$$(x_1 - x_2)\dot{x}_1 - (x_1 - x_2)\dot{x}_2 + (y_1 - y_2)\dot{y}_1 - (y_1 - y_2)\dot{y}_2 = 0 \qquad (2.15)$$

This represents kinematic constraints resulting from geometric constraints (2.13), which can take also more general form:

$$f_1\dot{\xi}_1 + f_2\dot{\xi}_2 + f_3\dot{\xi}_3 + f_4\dot{\xi}_4 = 0$$

For the above equation the following condition is fulfilled:

$$\frac{\partial f_k}{\partial \xi_m} - \frac{\partial f_m}{\partial \xi_k} = 0 \qquad k, m = 1,2,3,4 \qquad (2.16)$$

which means that the conditions for the total differential are met. The equation (2.15) takes the form of the Pfaff's differential equation, which is quite self-evident due to its origin. As a result, constraints given by (2.15) are holonomic.

Example 2.2. A classical example of nonholonomic constraints can be illustrated by the slipless motion of a flat plate on a plane. The relations between the coordinates in this case are presented in Fig. 2.3. The description of the slipless motion of a plate on a surface Π applies 6 coordinates: x,y,z, which determine points of tangency of the plate and the plane and angles α,υ,φ which define: rotational angle of the plate, inclination of the plate surface and angle of intersection between the plate surface and the Cartesian coordinate system, respectively. This system is limited by the following constraints:

$$\begin{aligned} f_1 &: \quad z = 0 \\ f_2 &: \quad R\,d\alpha\sin\varphi = dx \\ f_3 &: \quad R\,d\alpha\cos\varphi = dy \end{aligned} \qquad (2.17)$$

The equation f_1 for the constraints obviously presents holonomic constraints, while the constraints f_2, f_3 can also take the following form:

$$R\dot{\alpha}\sin\varphi = \dot{x} \qquad\qquad R\dot{\alpha}\cos\varphi = \dot{y} \qquad (2.18)$$

which represents nonholonomic constraints, since angle φ constitutes a coordinate of the system in motion and does not form a value that is input or a function.

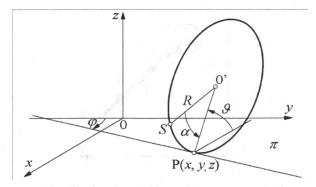

Fig. 2.3 Flat plate in slipless motion on a plane

Therefore, the two equations (2.18) cannot be integrated separately without prior establishment of a solution to the system of the equations of motion. One can note that the constraint equation f_1 enables one to eliminate variable z from the system of equations and; thus, it can be disregarded in the vector of system's position. At the same time, constraints equations f_2 and f_3 do not impose a limitation on the position of the system, while they constrain the motion, so that it is slipless. One can further observe that that the equations for nonholonomic constraints (2.18) can be transformed to take the form:

$$R^2\dot{\alpha}^2 = \dot{x}^2 + \dot{y}^2 = v^2 \tag{2.19}$$

which eliminates angle φ from constraint equation and denotes velocity of the motion of the tangency point P over a plane in which a plate rolls. This equation enables one to interpret nonholonomic constraints but does not offer grounds for their elimination. Equations for nonholonomic constraints are also encountered in electrical and electromechanical systems in such a form that the electrical node in which the branches of electrical circuits converge is movable and its position is relative to a mechanical variable. This type of nonholonomic constraints is encountered e.g. in electrical pantographs of rail vehicles and mechanical commutators in electrical machines involving sliding contact.

2.1.4 External Forces and Reaction Forces; d'Alembert Principle

2.1.4.1 External Forces and Reaction Forces

External forces are forces (torques) acting upon the components of a system. In this form they constitute the cause of motion in accordance with the Newton's second law. Reaction forces of constraints (Fig. 2.4) form the internal forces acting along the applied constrains and operate so that the system preserves the state which results from the imposed constraints. Hence, reaction forces of constraints do not constitute the cause of the motion but result in the preservation of the system in conformity with the constraints. In ideal circumstances the forces of constraint reactions do not exert any work associated with the motion of a system, which is applied in d'Alembert principle discussed later.

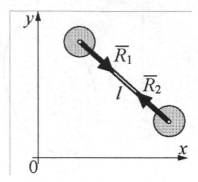

Fig. 2.4 Equilibrium between reaction forces of constraints $\mathbf{R}_1, \mathbf{R}_2$ resulting in constant distance l between balls in motion

2.1.4.2 Virtual Displacements

The introduction of the notion of virtual displacements, i.e. ones that are compatible with constraints is indispensable in analytical dynamics due to their role in elimination of constraint reaction forces occurring in constrain based systems [12,13,16]. The vector of virtual displacements is denoted analogically to the construction of the position vector (2.2)

$$\delta\mathbf{r} = (\delta\mathbf{r}_1, \delta\mathbf{r}_2, \delta\mathbf{r}_3, \ldots, \delta\mathbf{r}_N) \qquad (2.20)$$

where:

$$\delta\mathbf{r}_i = (\delta x_i, \delta y_i, \delta z_i) = (\delta\xi_{3i-2}, \delta\xi_{3i-1}, \delta\xi_{3i})$$

The vector of virtual displacements is constructed by the increments of variables which fulfill the following conditions:

    $1°$ - possess infinitesimal value

    $2°$ - are compatible with constraints

    $3°$ - their displacements occur within fixed a moment of time

These conditions also mean that virtual displacements are also referred to as infinitesimal displacements, i.e. small testing displacements which occur consistently with applied constraints without accounting for their duration. As a result, it is possible to compare work exerted by a system for various vectors of virtual displacements. Virtual displacements do not necessarily have to overlap with sections of actual paths of motion but need to be consistent with potential paths from the kinematics perspective. From the statement of consistency between virtual displacements and constraints the following relation can be established:

$$f(\mathbf{r} + \delta\mathbf{r}) - f(\mathbf{r}) = 0$$

which upon resolving into Taylor series relative to $\delta\mathbf{r}$ and omission of higher powers $(\delta\mathbf{r})^2$, $(\delta\mathbf{r})^3, \ldots$ leads to the statement of the relation between virtual displacements for a given j-th equation of constraints

$$\sum_{i=1}^{N}\frac{\partial f_j}{\partial \mathbf{r}_i}\delta\mathbf{r}_i = 0 \quad \text{or} \quad \sum_{k=1}^{3N}\frac{\partial f_j}{\partial \xi_k}\delta\xi_k = 0 \qquad (2.21)$$

The relation (2.21) also means that any equation for holonomic constraints, f_j, $j=1,...,h$ enables one to find an expression for a particular virtual displacement by use of the remaining ones

$$\delta\xi_k = -\frac{1}{a_{j,k}}(a_{j,1}\delta\xi_1 + a_{j,2}\delta\xi_2 + ... + a_{j,k-1}\delta\xi_{k-1} + a_{j,k+1}\delta\xi_{k+1} + ... + a_{j,n}\delta\xi_n) \quad (2.22)$$

where:
$$a_{j,k} = \frac{\partial f_j}{\partial \xi_k}$$

2.1.4.3 Perfect Constraints

It is only possible to define perfect constraints in a system in which friction forces are either missing or in the case where the inherent friction forces can be considered as external forces. After this prerequisite is fulfilled, it is possible to define perfect constraints. Such constraints satisfy the condition that total work exerted on the virtual displacements is equal to zero:

$$\sum_{i=1}^{N}\mathbf{R}_i\delta\mathbf{r}_i = 0 \qquad (2.23)$$

An example of perfect constraints include a rigid connection of material points which is not subjected to tension or bending. Historically, the concept of perfect constraints originates from d'Alembert principle and forms a postulate confirmed by numerous examples.

2.1.4.4 d'Alembert Principle

It constitutes the first analytical statement of the motion of a system in which particles are constrained. In order to eliminate forces of constraint reactions the principle applies the notion of perfect constraints. For a material point (particle) with mass m the equation of motion directly results from Newton's second law of motion:

$$m\ddot{\mathbf{r}} = \mathbf{F}$$

For a system with N material points limited by constraints, the above equation can be restated for every material point to account for the resulting force of constraint reactions $\mathbf{R}$ beside the external force $\mathbf{F}$

$$m_i\ddot{\mathbf{r}}_i = \mathbf{F}_i + \mathbf{R}_i \qquad i = 1...N \qquad (2.24)$$

The unknown constraint reaction forces do not yield it possible to directly apply equations (2.24). After summation of the equations it is possible to eliminate constraint reaction forces on the basis of the notion of the perfect constraint (2.23)

$$\sum_{i=1}^{N}(m_i\ddot{\mathbf{r}}_i - \mathbf{F}_i - \mathbf{R}_i)\delta\mathbf{r}_i = 0$$

which gives:
$$\sum_{i=1}^{N}(m_i\ddot{\mathbf{r}}_i - \mathbf{F}_i)\delta\mathbf{r}_i = 0 \tag{2.25}$$

Virtual displacements are not separate entities but are related to one another by equations resulting from the constraints. Hence, d'Alembert principle is expressed by the system of equations:

$$\begin{cases} \sum_{i=1}^{N}(m_i\ddot{\mathbf{r}}_i - \mathbf{F}_i)\delta\mathbf{r}_i = 0 \\ \sum_{i=1}^{N}\frac{\partial f_j}{\partial \mathbf{r}_i}\delta\mathbf{r}_i = 0 \qquad j=1...h \end{cases} \tag{2.26}$$

This is a set of differential - algebraic equations on the basis of which it is possible to obtain equations of motion e.g. using Lagrange indefinite multiplier method. This can be performed as follows: each of h algebraic equations in (2.26) is multiplied by indefinite factor λ_j and summed up:

$$\sum_{j=1}^{h}\sum_{i=1}^{N}\lambda_j\frac{\partial f_j}{\partial \mathbf{r}_i}\delta\mathbf{r}_i = 0 \tag{2.27}$$

The expression in (2.27) is subsequently subtracted from the equation of motion, thus obtaining:

$$\sum_{i=1}^{N}(m_i\ddot{\mathbf{r}}_i - \mathbf{F}_i - \sum_{j=1}^{h}\lambda_j\frac{\partial f_j}{\partial \mathbf{r}_i})\delta\mathbf{r}_i = 0 \tag{2.28}$$

For the resulting sum of N parenthetical expressions multiplied by subsequent virtual displacements $\delta\mathbf{r}_i$, the following procedure is followed: for the first h expressions in parentheses $i=1,...,h$ the selection of multipliers λ_j should be such that the value of the expression in parenthesis is equal to zero. In consequence, for the remaining parenthetical expressions the virtual displacements $\delta\mathbf{r}_i$, $i=h+1,...,N$ are already independent, hence, the parenthetical expressions must be equal to zero. The final equations of motion take the form:

$$m_i\ddot{\mathbf{r}}_i = \mathbf{F}_i + \sum_{j=1}^{h}\lambda_j\frac{\partial f_j}{\partial \mathbf{r}_i} = 0 \qquad i=h+1...N \tag{2.29}$$

The term thereof: $\sum_{j=1}^{h} \lambda_j \dfrac{\partial f_j}{\partial \mathbf{r}_i}$ constitutes the reaction force of nonholonomic

constraints for equations (2.24). In a similar manner it is possible to extend d'Alembert principle to cover systems limited by nonholonomic constraints (2.11-2.12). As a result the following expression is obtained:

$$m_i \ddot{\mathbf{r}}_i = \mathbf{F}_i + \sum_{j=1}^{h} \lambda_j \frac{\partial f_j}{\partial \mathbf{r}_i} + \sum_{l=1}^{nh} \mu_l \frac{\partial \varphi_l}{\partial \dot{\mathbf{r}}_i} = 0 \qquad i = h+1 \ldots N \qquad (2.30)$$

where: φ_l - nonholonomic functions of constrain type (2.12)

   μ_l - indefinite multipliers for nonholonomic constraints

d'Alembert principle leads to the statement of a system of equations with constraints; however, this procedure is time-consuming and quite burdensome since the obtained forms of equations are extensive and complex due to the selection of coordinates of motion that is far from optimum. This can be demonstrated in a simple presentation.

Example 2.3. In a planar system presented in Fig. 2.5 a set of two balls of mass m_1 and m_2 are connected by a stiff rod. They are put in motion under the effect of external forces $\mathbf{F}_1$ and $\mathbf{F}_2$, in which gravity pull and friction force are already accounted for. The equation of motion are subsequently stated in accordance with d'Alembert principle.

   Solution: the single equation of constraints stated in accordance with (2.13) takes the form:

$$f_1 : \qquad (x_1 - x_2)^2 + (y_1 - y_2)^2 - l^2 = 0$$

The vector of Cartesian coordinates for this system is as follows:

$$\mathbf{r} = (\mathbf{r}_1, \mathbf{r}_2) = \Xi(\xi_1, \xi_2, \xi_3, \xi_4)$$

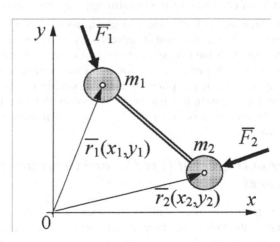

Fig. 2.5 Set of two balls connected by a stiff rod

The system of differential - algebraic equations written in accordance with (2.29)
takes the form:

$$
\left.\begin{aligned}
m_1\ddot{x}_1 &= F_{1x} + 2\lambda_1(x_1 - x_2) \\
m_1\ddot{y}_1 &= F_{1y} + 2\lambda_1(y_1 - y_2) \\
m_2\ddot{x}_2 &= F_{2x} + 2\lambda_1(-x_1 + x_2) \\
m_2\ddot{y}_2 &= F_{2y} + 2\lambda_1(-y_1 + y_2) \\
(x_1 - x_2)^2 &+ (y_1 - y_2)^2 - l^2 = 0
\end{aligned}\right\}
\tag{2.31}
$$

The equation of holonomic constraints eliminates one of the Cartesian coordinates
since it is an dependent variable in the description of dynamics. At this points, let
us assume that it is y_2, hence:

$$
\lambda_1 = \frac{F_{2y} - m_2\ddot{y}_2}{2(y_1 - y_2)} \quad ; \quad y_2 = y_1 \mp \sqrt{l^2 - (x_1 - x_2)^2}
$$

By introducing these variables into (2.31) we obtain

$$
\left.\begin{aligned}
m_1\ddot{x}_1 + \alpha m_2\ddot{y}_2 &= F_{1x} + \alpha F_{2y} \\
m_1\ddot{y}_1 + m_2\ddot{y}_2 &= F_{1y} + F_{2y} \\
m_2\ddot{x}_2 - \alpha m_2\ddot{y}_2 &= F_{2x} - \alpha F_{2y} \\
y_2 = y_1 \mp \sqrt{l^2 - (x_1 - x_2)^2}
\end{aligned}\right\}
\tag{2.32}
$$

where:

$$
\alpha = \alpha(\mathbf{r}) = \frac{x_1 - x_2}{y_1 - y_2}
$$

The resulting system of equations of motion still requires the elimination of y_2, $\ddot{y}_2$,
which is only an algebraic problem. This set of equations is very complex in its
analytical notation despite the fact that it presents a very simple mechanical system. This is associated with the necessity of application of Cartesian coordinates,
which is not the most adequate choice for the case of equations containing
constraints, in particular from the point of view of simple notation of dynamic
equations. A favorable option in this respect is offered by the introduction of generalized coordinates and expression of the equations of motion in the form of
Lagrange's equations.

2.1.5 Number of Degrees of Freedom and Generalized Coordinates

The most general definition states that the number of degrees of freedom in a
system is made up of the number of independent virtual displacements $\delta\xi$ (2.20).
For a holonomic system it also represents the number of coordinates (variables)

necessary and sufficient in order to define the position of a system. In accordance with this description every equation for holonomic constraints reduces the number of degrees of freedom by one (see 2.21, 2.22). This can be defined by the relation

$$s = n - h \tag{2.33}$$

where:

s - the number of degrees of freedom of a holonomic system
n - the number of coordinates necessary for description of the position of an unconstrained system
h - the number of holonomic constraints

Under such assumptions regarding the number of degrees of freedom, the equation of nonholonomic constraints (2.11, 2.12) also leads to the reduction of the degrees of freedom despite the fact that the position of the system is not limited. This also means that the number of degrees of freedom of nonholonomic systems is lower than the number of coordinates necessary for the description of the position of such a system. For the time being we shall, however, focus on holonomic systems.

Generalized coordinates form the vector of $\mathbf{q} = (q_1, q_2, \ldots, q_s)$, and the components of this vector include any variables that fulfill three pre-requisites:

1° the number s of generalized coordinates is equal to the number of degrees of freedom

2° generalized coordinates are selected in such a manner that they are compatible with constraints present in the system, i.e. they fulfill the condition of identity with the equations of constraints

$$f_j(\mathbf{r}(\mathbf{q}))) = 0 \tag{2.34}$$

3° generalized coordinates need to be linearly independent, which means that the selection of them has to enable one to uniformly express Cartesian coordinates $\mathbf{r} = \mathbf{r}(\mathbf{q})$, alternatively $\Xi = \Xi(\mathbf{q})$, or coordinates of the primary description $\mathbf{X} = \mathbf{X}(\mathbf{q})$, which gives

$$\xi_j = \xi_j(q_1, \ldots, q_s) \quad \text{or} \quad \chi_j = \chi_j(q_1, \ldots, q_s) \tag{2.35}$$

Formally it means that the functional Jacobian matrix

$$\left[\frac{\partial \xi_i}{\partial q_k} \right] \quad \text{or else} \quad \left[\frac{\partial \chi_i}{\partial q_k} \right] \tag{2.36}$$

- is of s order in the entire area of the variation of coordinates.

The second of the equations (2.35) defines the so called *primary description coordinates*, which form an alternative to the Cartesian coordinate system, as they involve an arbitrary set of variables for the description of the position of a system, without an imposed limitation on the number of coordinates used in such a description. The practical selection of generalized coordinates can be performed in a number of ways and tends to be much easier than it is implied from the study of formal requirements (2.34-2.36). Among Cartesian, polar, spherical or other variables used in the description of a physical model of a system (which means all coordinates of the primary description) it is necessary to select such s of independent variables which are compatible with constraints and offer a comfortable source for the description of the position of a system. For the case of holonomic constraints the geometry of the constraints often suggests the selection of such variables. After an appropriate selection of the variables the resulting equations are succinct and short, while for other selection the resulting equations of motion might be complex and involve a lot of other components. However, the total number of equations of motion remains constant (or the total order of a system of equations), which amounts to s equations of the second order for a holonomic system. The appropriate selection of the generalized coordinates in such a manner that simple and short forms of equations ensue can be found later in the text.

Transformational formulae – are functional relations which express the relations between Cartesian coordinates of motion (**r**, Ξ) or coordinates of primary description (**X**) and the vector of generalized coordinates. Similar transformational formulae account for the relations between velocities, which can be gained for holonomic constraints by differentiation of relations regarding position with respect to time. The transformational formulae which are expressed by equations (2.35) for position could be completed by explicit relation to time for the purposes of the general consideration. Such instances are non-isolated systems, e.g.

$$\Xi = \Xi(q_1 q_2,...,q_s,t) \quad \text{or} \quad X = X(q_1,q_2,...q_s,t) \tag{2.37}$$

From these relations transformational formula for velocity ensues in the form

$$\dot{\xi}_i = \sum_{k=1}^{s} \frac{\partial \xi_i}{\partial q_k} \dot{q}_k + \frac{\partial \xi_i}{\partial t} \quad \text{and} \quad \dot{\chi}_i = \sum_{k=1}^{s} \frac{\partial \chi_i}{\partial q_k} \dot{q}_k + \frac{\partial \chi_i}{\partial t} \tag{2.38}$$

Similarly, during the calculation of the variation of variables (2.35), the result takes the form of virtual displacement of Cartesian coordinates (of the primary description) expressed in terms of virtual displacements (variations) of generalized coordinates

$$\delta \xi_i = \sum_{k=1}^{s} \frac{\partial \xi_i}{\partial q_k} \delta q_k \quad \text{or} \quad \delta \chi_i = \sum_{k=1}^{s} \frac{\partial \chi_i}{\partial q_k} \delta q_k \tag{2.39}$$

One can note that the transformational formulae for virtual displacements (2.39) are the same as the ones resulting from the calculation of total differential of variables for transformational formulae (2.35) not accounting for time. One also should note at this point that independence of virtual displacements for generalized coordinates comes as a consequence of the fulfillment of constraint equations by the generalized coordinates

$$\delta \mathbf{q} = (\delta q_1, \delta q_2,...,\delta q_s) \tag{2.40}$$

and hence they can assume arbitrary values with the role of indefinite multipliers.

2.1.6 Lagrange's Equations

Lagrange's equations are based on generalized coordinates $\mathbf{q}$ and apply their independence as well as independence of virtual displacements $\delta\mathbf{q}$. Lagrange's equations for a mechanical system can be derived from d'Alembert principle for holonomic systems by calculating variation from transformational formulae for Cartesian coordinates (2.35). After introduction of the transformation of virtual displacements into the first of equations (2.26)

$$\delta\mathbf{r}_i = \sum_{k=1}^{s} \frac{\partial \mathbf{r}_i}{\partial q_k} \delta q_k = 0 \qquad (2.41)$$

we obtain:

$$\sum_{k=1}^{s}\sum_{i=1}^{N}(m_i\ddot{\mathbf{r}}_i - \mathbf{F}_i)\frac{\partial \mathbf{r}_i}{\partial q_k}\delta q_k = 0 \qquad (2.42)$$

Concurrently, the second equation in d'Alembert principle (2.26) disappears due to the independence of variation δq_k, which formally means that generalized coordinates fulfill these constraints. As a result of the transformation of (2.42) we obtain:

$$\sum_{k=1}^{s}\left(\sum_{i=1}^{N}m_i\ddot{\mathbf{r}}_i\frac{\partial \mathbf{r}_i}{\partial q_k} - \sum_{i=1}^{N}\mathbf{F}_i\frac{\partial \mathbf{r}_i}{\partial q_k}\right)\delta q_k = 0 \qquad (2.43)$$

The second of the expressions in parenthesis denotes generalized force (not accounting for friction forces) acting along the generalized coordinate

$$\tilde{P}_k = \sum_{i=1}^{N}\mathbf{F}_i\frac{\partial \mathbf{r}_i}{\partial q_k} \quad \text{or} \quad \tilde{P}_k = \sum_{l=1}^{3N}F_l\frac{\partial \xi_l}{\partial q_k} \qquad (2.44)$$

This constitutes the total of projection of all external forces expressed in a Cartesian system towards the generalized coordinate. Since the transformational formulae are in the general case non-linear, partial derivatives in (2.44) give the formulae of force projection. Concurrently, the first component in the bracket in (2.43) can be transformed as follows

$$\sum_{i=1}^{N}m_i\ddot{\mathbf{r}}_i\frac{\partial \mathbf{r}_i}{\partial q_k} = \sum_{i=1}^{N}m_i\frac{d}{dt}\left(\dot{\mathbf{r}}_i\frac{\partial \mathbf{r}_i}{\partial q_k}\right) - \sum_{i=1}^{N}m_i\dot{\mathbf{r}}_i\frac{d}{dt}\left(\frac{\partial \mathbf{r}_i}{\partial q_k}\right) =$$

$$= \sum_{i=1}^{N}m_i\frac{d}{dt}\left(\dot{\mathbf{r}}_i\frac{\partial \dot{\mathbf{r}}_i}{\partial \dot{q}_k}\right) - \sum_{i=1}^{N}m_i\left(\dot{\mathbf{r}}_i\frac{\partial \dot{\mathbf{r}}_i}{\partial q_k}\right) = \qquad (2.45)$$

$$= \frac{d}{dt}\frac{\partial\left(\sum_i \frac{1}{2}m_i\dot{\mathbf{r}}_i^2\right)}{\partial \dot{q}_k} - \frac{\partial\left(\sum_i \frac{1}{2}m_i\dot{\mathbf{r}}_i^2\right)}{\partial q_k} = \frac{d}{dt}\frac{\partial T}{\partial \dot{q}_k} - \frac{\partial T}{\partial q_k}$$

The expressions in brackets denote kinetic energy of the system in the Cartesian coordinates

$$T = \sum_i \tfrac{1}{2} m_i \dot{\mathbf{r}}_i{}^2 \tag{2.46}$$

This energy associated with velocities of system masses has to be expressed in relation to generalized velocities and generalized coordinates, by employing transformational formulae (2.37-2.38). After these transformations the following is obtained

$$T = T(\dot{\mathbf{q}}, \mathbf{q}, t) \tag{2.47}$$

As a result from the initial equation (2.43) we get

$$\sum_{k=1}^{s} \left(\frac{d}{dt} \frac{\partial T(\dot{\mathbf{q}}, \mathbf{q}, t)}{\partial \dot{q}_k} - \frac{\partial T(\dot{\mathbf{q}}, \mathbf{q}, t)}{\partial q_k} - \tilde{P}_k \right) \delta q_k = 0, \quad k = 1, \ldots, s \tag{2.48}$$

which in the consideration of independence of virtual displacements leads to s separate equations in the form

$$\frac{d}{dt} \left(\frac{\partial T(\dot{\mathbf{q}}, \mathbf{q}, t)}{\partial \dot{q}_k} \right) - \frac{\partial T(\dot{\mathbf{q}}, \mathbf{q}, t)}{\partial q_k} = \tilde{P}_k \tag{2.49}$$

Potential forces (2.50) acting in the system, which are derivatives of a potential $U(\mathbf{q},t)$ with respect to position, can be easily integrated into equations of motion. In this case these forces are omitted in the consideration of external forces $\mathbf{F}$.

$$P_u = -\frac{\partial U(\mathbf{q},t)}{\partial q_k} \tag{2.50}$$

The result takes the following form:

$$\frac{d}{dt} \frac{\partial (T - U)}{\partial \dot{q}_k} - \frac{\partial (T - U)}{\partial q_k} = P_k$$

or finally

$$\frac{d}{dt} \left(\frac{\partial L}{\partial \dot{q}_k} \right) - \frac{\partial L}{\partial q_k} = P_k, \quad k = 1 \ldots s \tag{2.51}$$

which is called *Lagrange's equation* for an examined dynamic system. Generalized force P_k does not include the component resulting from potential external forces, which was already incorporated in the form of (2.50). The component $U(\mathbf{q},t)$ denoting the potential energy of the system has been incorporated also into the first expression of Lagrange's equation (2.51) for symmetry, despite the fact that its differentiation with respect to $\dot{q}_k$ returns the result zero. Function L defined as

$$L(\dot{\mathbf{q}}, \mathbf{q}, t) = T(\dot{\mathbf{q}}, \mathbf{q}, t) - U(\mathbf{q}, t) \tag{2.52}$$

is called Lagrange's function for a mechanical system. In the classical mechanics it determines the difference between the kinetic and potential energy of a system. Lagrange's equations are widely applied in the studies of dynamic properties and control of mechanical systems, as well as electromechanical ones, including servomechanisms, manipulators and robots. Besides, they form one of the two fundamental methods used for the statement of dynamic models. Lagrange's equations in the form (2.51) could be extended [12,16] to cover systems limited by nonholonomic constraints in the form (2.11, 2.12).

$$f(\xi_1,\xi_2,...,\xi_n,\dot{\xi}_1,\dot{\xi}_2,...,\dot{\xi}_n,t)=0 \tag{2.53}$$

If (2.53) represents holonomic constraints, it is possible to integrate it and the function takes the form $F(\xi_1,\xi_2,...,\xi_n,t)=0$ such that $\dot{F}=f$. It is notable that the conditions that these constraints impose on virtual displacements are identical

$$\sum_{i=1}^{3N}\frac{\partial F}{\partial \xi_i}\delta\xi_i=0 \quad \text{or} \quad \sum_{i=1}^{3N}\frac{\partial f}{\partial \dot{\xi}_i}\delta\xi_i=0 \tag{2.54}$$

since:

$$f=\dot{F}=\sum_{i=1}^{3N}\frac{\partial F}{\partial \xi_i}\dot{\xi}_i+\frac{\partial F}{\partial t}=0 \quad \text{hence} \quad \frac{\partial f}{\partial \dot{\xi}}=\frac{\partial F}{\partial \xi_i}$$

After the introduction of generalized coordinates these formulae are omitted as a result of the fulfillment of the equation of constraints. However, if kinematic constraints of the type (2.53) were not integrated, the introduction of generalized coordinates would not result in the omission of the equation type (2.53); in fact, it is then transformed into the equation of nonholonomic constraints in generalized coordinates:

$$\psi_j(q_1,q_2,...,q_c,\dot{q}_1,\dot{q}_2,...,\dot{q}_c,t)=0 \qquad j=1...nh \tag{2.55}$$

and in the case of linear ones: $\psi_j=\sum_{k=1}^{c}h_{jk}\dot{q}_k+d_k$

where: nh - is the number of nonholonomic constraints

                   $c=n-h$ - is the number of generalized coordinates

                   $s=n-h-nh$ - is the number of degrees of freedom

As a result of the expansion of the equations (2.55) into multiple variable Taylor series or by analogy to (2.54) one can demonstrate that these equations can be used to relate the virtual displacements of generalized coordinates δq_k

$$\sum_{k=1}^{c}\frac{\partial \psi_j}{\partial \dot{q}_k}\delta q_k=0 \quad \text{or} \quad \sum_{k=1}^{c}h_{jk}\delta q_k=0 \qquad j=1...nh \tag{2.56}$$

The present virtual displacements are not independent as it was the case for holonomic constraints and, hence, the value of the particular parentheses (2.48) need not be equal to zero, which has given the equations of motion (2.49) in case of holonomic constraints. In these circumstances, as well as in the derivation of equations of motion on the basis of d'Alembert principle (2.28) we will apply the method of indefinite multipliers. All nh components in relation (2.56) will be multiplied by successive indefinite multipliers μ_j and subsequently added to equations (2.51). As a result, we obtain c expressions in parenthesis in (2.57).

$$\sum_{l=1}^{c}\left(\frac{d}{dt}\frac{\partial L(\dot{\mathbf{q}},\mathbf{q},t)}{\partial \dot{q}_l}-\frac{\partial L(\dot{\mathbf{q}},\mathbf{q},t)}{\partial q_l}-P_l-\sum_{j=1}^{nh}\mu_j\frac{\partial \psi_j}{\partial \dot{q}_l}\right)\delta q_l=0 \qquad (2.57)$$

For example, in the first nh components of the sum in (2.57) the selection of μ_j is made so that the value of expressions in parenthesis is equal to zero. The corresponding dependent variations are δq_j, $j=1,\ldots,nh$. In that case the remaining variations of generalized variables are already independent and the corresponding parenthesis of the sum (2.57) have to be equal to zero. As a result, the equations of motion take the form

$$\frac{d}{dt}\frac{\partial L(\dot{\mathbf{q}},\mathbf{q},t)}{\partial \dot{q}_k}-\frac{\partial L(\dot{\mathbf{q}},\mathbf{q},t)}{\partial q_k}=P_k+\sum_{j=1}^{nh}\mu_j\frac{\partial \psi_j}{\partial \dot{q}_k} \qquad (2.58)$$

$$k=1\ldots s, \qquad j=1\ldots nh$$

The resulting system of equations consists of s Lagrange's equations in the form (2.58) and nh equations of nonholonomic constraints in generalized coordinates (2.55), in which the unknown include s in q_k variables and nh in μ_j multipliers. In equation (2.58) the potential forces in the form (2.50) have already been separated from generalized forces $\tilde{P}_k$ and integrated into Lagrange's function L, while the non potential components of generalized forces P_k are preserved. For the case of nonholonomic linear constraints the equations of motion for a system with nonholonomic constraints, accounting for (2.56), take the following form:

$$\frac{d}{dt}\frac{\partial L(\dot{\mathbf{q}},\mathbf{q},t)}{\partial \dot{q}_k}-\frac{\partial L(\dot{\mathbf{q}},\mathbf{q},t)}{\partial q_k}=P_k+\sum_{j=1}^{nh}\mu_j h_{jk} \qquad (2.59)$$

$$k=1\ldots s, \qquad j=1\ldots nh$$

2.1.7 Potential Mechanical Energy

Potential mechanical energy, i.e. accumulated energy regardless of velocity, can be stored in two ways: in gravitational field and in the form of elastic tension. In the gravitational field the potential energy is accumulated during the displacement of mass towards increasing potential. This energy equal to the work exerted during this displacement is:

$$U = \int_a^b \mathbf{F}d\mathbf{l} = \int_a^b m\,grad\varphi\,d\mathbf{l} = m(\varphi(b) - \varphi(a)) \tag{2.60}$$

For the case of constant gravitational field, i.e. one with constant vector of gravitational acceleration $\mathbf{g}$, potential energy is expressed as

$$U = m(\varphi(b) - \varphi(a)) = m\mathbf{g}(\mathbf{h}(b) - \mathbf{h}(a)) = m\mathbf{g}\mathbf{h} = mgh \tag{2.61}$$

where $\mathbf{h}$ is the vector called the height of a point above the reference level measured in the parallel direction to the vector of gravitational acceleration $\mathbf{g}$. For a solid of mass M the height of the center of mass S is associated with the potential energy of the solid as a total.

Potential energy in elastic element is associated with the work accumulated during its elastic strain. The symbol of this component is a spring (Fig. 2.6).

$$U = \int_a^b \mathbf{F}(x)dx \tag{2.62}$$

For a spring with linear characteristics the force that a spring exerts is proportional to deformation (extension, compression, torsion) in relation to the stationary state of the spring denoted as x_0. The elastic modulus of an spring k [N/m] is defined as the slope of its stress-strain curve in the elastic deformation region. As a consequence

$$\mathbf{F} = k\Delta\mathbf{x} \quad \text{hence} \quad U = \int_{x_0}^x k(\tilde{\mathbf{x}} - \mathbf{x}_0)d\tilde{\mathbf{x}} = \tfrac{1}{2}k(x - x_0)^2 \tag{2.63}$$

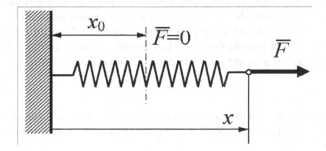

Fig. 2.6 Accumulation of potential energy in an elastic element

For the case of parallel connection between springs the total extension of all springs is identical and the force affecting the springs is the sum of the forces needed for the elastic strain, thus

$$\mathbf{F} = \mathbf{F}_1 + \mathbf{F}_2 + \ldots \mathbf{F}_n = k_1\Delta\mathbf{x} + k_2\Delta\mathbf{x} + \ldots + k_n\Delta\mathbf{x} = (k_1 + k_2 + \ldots + k_n)\Delta\mathbf{x}$$

and the resultant elastic strain is

$$k = \frac{F}{\Delta x} = k_1 + k_2 + \ldots + k_n \tag{2.64}$$

For the case of serial connection of springs, we have:

$$\mathbf{F} = \mathbf{F}_1 = \mathbf{F}_2 = \ldots = \mathbf{F}_n \qquad \Delta \mathbf{x} = \Delta \mathbf{x}_1 + \Delta \mathbf{x}_2 + \ldots + \Delta \mathbf{x}_n$$

$$\mathbf{F}_1 = k_1 \Delta \mathbf{x}_1 \quad \mathbf{F}_2 = k_2 \Delta \mathbf{x}_2 \quad \ldots \quad \mathbf{F}_n = k_n \Delta \mathbf{x}_n$$

hence:

$$k^{-1} = \frac{1}{\mathbf{F}} \sum_i \Delta \mathbf{x}_i = k_1^{-1} + k_2^{-1} + \ldots + k_n^{-1} \tag{2.65}$$

2.1.8 Generalized Forces, Exchange of Energy with Environment

Generalized forces resulting from the transformation of Cartesian forces acting on a system (2.44) are presented in detail earlier in the book i.e. during the derivation of Lagrange's equations. In this section they will be extended to cover friction forces, which is required as a result of consideration of perfect constraints using d'Alembert principle. The exchange of energy with the environment occurs as a result of work exerted on the system by external forces and friction forces, the effect of the latter is always negative, i.e. results in dissipation of energy of a system. Potential forces may be integrated into Lagrange's function by consideration of potential energy (2.61-2.63), while the right hand side of Lagrange's equations is reduced only to consideration of active forces. However, potential energy is also capable of playing the role of exchange of energy with the environment, if it is explicitly relative to time $U(\mathbf{q},t)$ - this is the case in non-isolated systems. We shall, however, take into consideration friction forces, which occur in virtually any system, have often nonlinear characteristics and tend to require notation in a complex way. The most basic system used for the notation of friction forces associated with motion is based on an assumption that they are proportional to the velocity. This case is denoted with the term viscous friction, which takes the form

$$\mathbf{F}_t = -D\dot{\xi} \tag{2.66}$$

The exchange of energy with the environment due to external forces is calculated on the basis of virtual work δA exerted by these forces on the particular virtual displacements:

$$\delta A = \sum_{i=1}^{3N} (F_i - D_i \dot{\xi}_i) \delta \xi_i \tag{2.67}$$

The transformation of virtual work into generalized coordinates applies transformational formulae of velocity (2.38) and virtual displacements (2.39) under the assumption that the reference system is inertial and, as a result, $\partial \xi_i / \partial t = 0$.

Hence:

$$\delta A = \sum_{k=1}^{s} \left(\sum_{i=1}^{3N} F_i \frac{\partial \xi_i}{\partial q_k} - \sum_{m=1}^{s} \sum_{i=1}^{3N} D_i \frac{\partial \xi_i}{\partial q_k} \frac{\partial \xi_i}{\partial q_m} \dot{q}_m \right) \delta q_k \tag{2.68}$$

The first component of virtual work (2.68) is already familiar to us and known as generalized force (2.44) acting in the direction of k-th generalized coordinate, while the second component denotes transformed friction forces. Furthermore, it is possible to identify the transformed friction coefficient

$$D_{km} = \sum_{i=1}^{3N} D_i \frac{\partial \xi_i}{\partial q_k} \frac{\partial \xi_i}{\partial q_m} \tag{2.69}$$

Generalized force accounting for friction for k-th generalized coordinate could be noted as:

$$P_k = \tilde{P}_k - \sum_{m=1}^{s} D_{km} \dot{q}_m \tag{2.70}$$

The virtual work of a dynamic system (2.68) can also be written in the form of a matrix using Jacobian matrix of the transformation of a coordinate system

$$\delta A = (\delta \mathbf{q})^T \mathbf{J}^T (\mathbf{F} - \mathbf{DJ\dot{q}}) = (\delta \mathbf{q})^T \mathbf{P} = (\mathbf{F} - \mathbf{DJ\dot{q}})^T \mathbf{J} \, \delta \mathbf{q} = \mathbf{P}^T \, \delta \mathbf{q} \tag{2.71}$$

where:

$$\mathbf{J} = \left[\frac{\partial \xi_i}{\partial q_k} \right] = \begin{bmatrix} \dfrac{\partial \xi_1}{\partial q_1} & \cdots & \dfrac{\partial \xi_1}{\partial q_s} \\ \vdots & \ddots & \\ \dfrac{\partial \xi_n}{\partial q_1} & \cdots & \dfrac{\partial \xi_n}{\partial q_s} \end{bmatrix}$$ - is the Jacobian matrix of transformation

$$i = 1,\ldots,3N \quad k = 1,\ldots,s$$

$\mathbf{D} = diag[D_1 \quad D_2 \quad \cdots \quad D_n]$ - diagonal matrix of coefficients of viscous friction

$\mathbf{F} = [F_1 \quad F_2 \quad \cdots \quad F_n]^T$ - vector of external forces in Cartesian coordinates.

As a result of the statement of virtual work in the form (2.71) we obtain the vector of generalized forces accounting for viscous friction with components from (2.68). It is particularly complex to transform friction forces in which there are co-efficients of mutual friction (2.69), or in the vector form of expression $\mathbf{J}^T \mathbf{DJ\dot{q}}$ in relation (2.71). In numerous practical cases friction coefficient can be derived directly for generalized variables $\mathbf{q}$ without reference to transformation of losses resulting from friction expressed in Cartesian coordinates.

2.1.9 Examples of Application of Lagrange's Equations Method for Complex Systems with Particles

Example 2.4. We shall one more time consider a system of two balls connected with a stiff rod of length *l* (*Example 2.3*, Fig. 2.5). The equations of motion for this system obtained from d'Alembert principle in the Cartesian coordinates (2.32) are very complex and of little use for computing the trajectory of this system. At present for the development of a mathematical model we shall use the method of Lagrange's equations with a different selection of coordinates (Fig. 2.7).

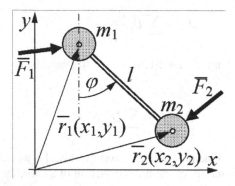

Fig. 2.7 Planar system containing two masses connected with a stiff rod

For this system the vector of position takes the form

$$\mathbf{r} = (\mathbf{r}_1, \mathbf{r}_2) = (x_1, y_1, x_2, y_2) = \Xi = (\xi_1, \xi_2, \xi_3, \xi_4),$$

which means that the number of coordinates in the description of unconstrained system amounts to $n = 4$. In this system we have one equation of holonomic constraints:

$$f_1: \quad |\mathbf{r}_1 - \mathbf{r}_2| = l \quad \text{or} \quad (x_1 - x_2)^2 + (y_1 - y_2)^2 - l^2 = 0 \tag{2.72}$$

Hence, the number of the degrees of freedom is: $s = n - h = 4 - 1 = 3$.

Cartesian coordinates of particle of mass m_1 and angle φ are assumed as the generalized coordinates:

$$\mathbf{q} = (q_1, q_2, q_3) = (x_1, y_1, \varphi) \tag{2.73}$$

Transformational formulae of the type (2.35) are following:

$$\xi_1 = q_1 \qquad \xi_2 = q_2$$
$$\xi_3 = x_2 = x_1 + l \sin \varphi = q_1 + l \sin q_3$$
$$\xi_4 = y_2 = y_1 - l \cos \varphi = q_2 - l \cos q_3$$

and for velocities, respectively:

$$\dot{\xi}_1 = \dot{q}_1 \qquad \dot{\xi}_2 = \dot{q}_2$$
$$\dot{\xi}_3 = \dot{x}_2 = \dot{x}_1 + l\dot{\varphi}\cos\varphi = \dot{q}_1 + l\dot{q}_3\cos q_3 \qquad (2.74)$$
$$\dot{\xi}_4 = \dot{y}_2 = \dot{y}_1 + l\dot{\varphi}\sin\varphi = \dot{q}_2 + l\dot{q}_3\sin q_3$$

Generalized coordinates are compatible with constraints, which can be verified by direct substitution of (2.74) into constraints equation (2.72); this also derives from the independence of virtual displacements of generalized coordinates considered as infinitesimal displacements. The kinetic energy of this system takes the following form

$$T = \tfrac{1}{2}m_1(\dot{x}_1^2 + \dot{y}_1^2) + \tfrac{1}{2}m_2(\dot{x}_2^2 + \dot{y}_2^2),$$

which after introduction of generalized velocities from (2.74) leads to the expression

$$T = \tfrac{1}{2}(m_1 + m_2)(\dot{x}_1^2 + \dot{y}_1^2) + \tfrac{1}{2}m_2 l^2\dot{\varphi}^2 + m_2 l\dot{\varphi}(\dot{x}_1\cos\varphi + \dot{y}_1\sin\varphi) \qquad (2.75)$$

$$U = 0 \quad \text{hence} \quad L = T - U = T$$

The vector of generalized forces in the right hand side of Lagrange's equations will be determined from relation (2.71). The vector of external forces and matrix of damping coefficients are

$$\mathbf{F} = \begin{bmatrix} F_{x1} & F_{y1} & F_{x2} & F_{y2} \end{bmatrix}^T$$

$$\mathbf{D} = diag\begin{bmatrix} D_{x1} & D_{y1} & D_{x2} & D_{y2} \end{bmatrix}^T$$

whereas Jacobian matrix of transformation

$$\mathbf{J} = \left[\frac{\partial \xi_i}{\partial q_k}\right] = \begin{bmatrix} 1 & 0 & 0 \\ 0 & 1 & 0 \\ 1 & 0 & l\cos\varphi \\ 0 & 1 & l\sin\varphi \end{bmatrix} \qquad (2.76)$$

Let us additionally assume that: $D_{x1} = D_{y1} = D_1 \qquad D_{x2} = D_{y2} = D_2$.

In this case on the basis of (2.71) we shall obtain the vector of generalized forces:

$$\mathbf{P} = \begin{bmatrix} P_1 \\ P_2 \\ P_3 \end{bmatrix} = \begin{bmatrix} F_{x1} + F_{x2} - (D_1 + D_2)\dot{x}_1 - D_2 l\dot{\varphi}\cos\varphi \\ F_{y1} + F_{y2} - (D_1 + D_2)\dot{y}_1 - D_2 l\dot{\varphi}\sin\varphi \\ l(F_{x2}\cos\varphi + F_{y2}\sin\varphi) - D_2 l(\dot{x}_1\cos\varphi + \dot{y}_1\sin\varphi + \dot{\varphi}l) \end{bmatrix} \qquad (2.77)$$

Hence, using (2.51), we are able to state Lagrange's equations:

1°   for $q_1 = x_1$ $\dfrac{d}{dt}\left(\dfrac{\partial L}{\partial \dot{x}_1}\right) - \dfrac{\partial L}{\partial x_1} = P_1$

$$\frac{d}{dt}\big((m_1 + m_2)\dot{x}_1 + m_2\dot{\varphi}l\cos\varphi\big) = P_1$$

$$(m_1 + m_2)\ddot{x}_1 + m_2l(\ddot{\varphi}\cos\varphi - \dot{\varphi}^2\sin\varphi) =$$
$$= F_{x1} + F_{x2} - (D_1 + D_2)\dot{x}_1 - D_2l\dot{\varphi}\cos\varphi$$

2°   for $q_2 = y_1$ $\dfrac{d}{dt}\left(\dfrac{\partial L}{\partial \dot{y}_1}\right) - \dfrac{\partial L}{\partial y_1} = P_2$

$$\frac{d}{dt}\big((m_1 + m_2)\dot{y}_1 + m_2\dot{\varphi}l\sin\varphi\big) = P_2$$

$$(m_1 + m_2)\ddot{y}_1 + m_2l(\ddot{\varphi}\sin\varphi + \dot{\varphi}^2\cos\varphi) =$$
$$= F_{y1} + F_{y2} - (D_1 + D_2)\dot{y}_1 - D_2l\dot{\varphi}\sin\varphi$$

3° for $q_3 = \varphi$ $\dfrac{d}{dt}\left(\dfrac{\partial L}{\partial \dot{\varphi}}\right) - \dfrac{\partial L}{\partial \varphi} = P_3$

$$\frac{d}{dt}\big(m_2l^2\dot{\varphi} + m_2l(\dot{x}_1\cos\varphi + \dot{y}_1\sin\varphi)\big) - m_2l\dot{\varphi}(-\dot{x}_1\sin\varphi + \dot{y}_1\cos\varphi) = P_3$$

$$m_2l^2(\ddot{\varphi} + (\ddot{x}_1\cos\varphi + \ddot{y}_1\sin\varphi)/l) =$$
$$= l(F_{x2}\cos\varphi + F_{y2}\sin\varphi) - D_2l(\dot{x}_1\cos\varphi + \dot{y}_1\sin\varphi + l\dot{\varphi})$$

The above equations of motion can be restated more simply by introduction of $\mu = m_2/(m_1 + m_2)$:

1° for $q_1 = x_1$:

$$\ddot{x}_1 + \mu l(\ddot{\varphi}\cos\varphi - \dot{\varphi}^2\sin\varphi) =$$
$$= \mu/m_2\big(F_{x1} + F_{x2} - (D_1 + D_2)\dot{x}_1 - D_2l\dot{\varphi}\cos\varphi\big)$$

2° for $q_2 = y_1$

$$\ddot{y}_1 + \mu l(\ddot{\varphi}\sin\varphi + \dot{\varphi}^2\cos\varphi) =$$
$$= \mu/m_2\big(F_{y1} + F_{y2} - (D_1 + D_2)\dot{y}_1 - D_2l\dot{\varphi}\sin\varphi\big)$$ (2.78)

3°   for $q_3 = \varphi$

$$\ddot{\varphi} + (\ddot{x}_1\cos\varphi + \ddot{y}_1\sin\varphi)/l =$$
$$= \big((F_{x2}\cos\varphi + F_{y2}\sin\varphi) - D_2(\dot{x}_1\cos\varphi + \dot{y}_1\sin\varphi + l\dot{\varphi})\big)/(m_2l)$$

The resulting system of equations of motion (2.78) can be verified for specific cases, for instance by immobilizing a particle of mass m_1, then: $x_1 = y_1 = const$. In this case the first two equations of motion (2.78) enable one to calculate reaction forces at the fixation point:

$1°$ for $q_1 = x_1$: $F_{x1} = m_2l(\ddot{\varphi}\cos\varphi - \dot{\varphi}^2\sin\varphi) - F_{x2} + D_2l\dot{\varphi}\cos\varphi$

$2°$ for $q_2 = y_1$: $F_{y1} = m_2l(\ddot{\varphi}\sin\varphi + \dot{\varphi}^2\cos\varphi) - F_{y2} + D_2l\dot{\varphi}\sin\varphi$

The equation $3°$ after introduction of gravitational force $F_{y2} = -m_2g$ and under the assumption that $F_{x2} = 0$ constitutes the equation of damped motion for pendulum:

$$3° \text{ for } q_3 = \varphi: \quad \ddot{\varphi} = -\frac{g}{l}\sin\varphi - \frac{D_2}{m_2}\dot{\varphi} \tag{2.79}$$

The conducted interpretation of equations leads to the identification of the particular terms and indicates how useful the model could be despite the complex form of the equations. However, the equations are much more simple than the ones in Cartesian coordinates, as shown in the discussion of d'Alembert principle (2.32).

Example 2.5. A planar mechanical system is given whose physical model is presented in Fig. 2.8. The system consists of a pendulum of mass m_2 attached to mass m_1 sliding along a horizontal bar. The motion being limited by spring of stiffness k and free length of d_0, and coefficient of viscous friction D_1. The coefficient of viscous friction of the pendulum is equal to D_2. The two forces operating in the system include gravity force G and the force of lateral pressure Q. The task involves the development of a mathematical model of the system motion using Lagrange's equations.

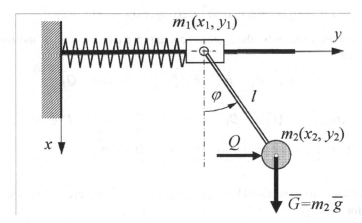

Fig. 2.8 Planar system of pendulum with mobile pivot

This system consists of two mobile masses, which in unconstrained state requires $n = 4$ coordinates in order to describe the position:

$$\mathbf{r} = (\mathbf{r}_1, \mathbf{r}_2) = (x_1, y_1, x_2, y_2) = (\xi_1, \xi_2, \xi_3, \xi_4)$$

In this case, there are two equations of holonomic constraints:

$$f_1: \quad x_1 = 0$$
$$f_2: \quad (x_1 - x_2)^2 + (y_1 - y_2)^2 - l^2 = 0$$

(2.80)

Hence, the number of degrees of freedom is

$$s = n - h = 4 - 2 = 2$$

The following variables are adopted as generalized coordinates:

$$\mathbf{q} = (q_1, q_2) = (y_1, \varphi)$$

(2.81)

which are independent and compatible with constraints. The transformational formulae (2.35) take the following form

$$\xi_1 = 0 \qquad \xi_3 = l\cos\varphi \qquad \dot{\xi}_3 = -l\dot{\varphi}\sin\varphi$$
$$\xi_2 = q_1 \qquad \xi_4 = y_1 + l\sin\varphi \qquad \dot{\xi}_4 = \dot{y}_1 + l\dot{\varphi}\cos\varphi$$

(2.82)

The kinetic energy of the system is

$$T = \tfrac{1}{2}m_1(\dot{x}_1^2 + \dot{y}_1^2) + \tfrac{1}{2}m_2(\dot{x}_2^2 + \dot{y}_2^2)$$

while the potential energy of the system is associated with the spring

$$U = \tfrac{1}{2}k(y_1 - d_0)^2$$

After the introduction of generalized coordinates (2.81) Lagrange's function takes the following form:

$$L = T - U =$$
$$= \tfrac{1}{2}(m_1 + m_2)\dot{y}_1^2 + \tfrac{1}{2}m_2(l^2\dot{\varphi}^2 + 2l\dot{\varphi}\dot{y}_1\cos\varphi) - \tfrac{1}{2}k(y_1 - d_0)^2$$

(2.83)

The virtual work exerted by the system accounts for external forces and forces of viscous friction

$$\mathbf{F} = \begin{bmatrix} F_{x1} & F_{y1} & F_{x2} & F_{y2} \end{bmatrix}^T = \begin{bmatrix} 0 & 0 & m_2 g & Q \end{bmatrix}^T$$

$$\delta A = (m_2 g \frac{\partial x_2}{\partial y_1} + Q\frac{\partial y_2}{\partial y_1} - D_1\dot{y}_1)\delta y_1 + (m_2 g \frac{\partial x_2}{\partial \varphi} + Q\frac{\partial y_2}{\partial \varphi} - D_2\dot{\varphi})\delta\varphi =$$
$$= \underbrace{(Q - D_1\dot{y}_1)}_{P_1}\delta y_1 + \underbrace{(-m_2 gl\sin\varphi + Ql\cos\varphi - D_2\dot{\varphi})}_{P_2}\delta\varphi$$

(2.84)

The equations of motion stated by aid of Lagrange's method are as follows:

1°   for $q_1 = y_1$

$$\frac{d}{dt}\left(\frac{\partial L}{\partial \dot{y}_1}\right) - \frac{\partial L}{\partial y_1} = P_1$$

$$\frac{d}{dt}\big((m_1 + m_2)\dot{y}_1 + m_2\dot{\varphi}l\cos\varphi\big) + k(y_1 - d_0) = P_1$$

2°   for $q_2 = \varphi$

$$\frac{d}{dt}\left(\frac{\partial L}{\partial \dot{\varphi}}\right) - \frac{\partial L}{\partial \varphi} = P_2$$

$$\frac{d}{dt}\left(m_2 l^2 \dot{\varphi} + m_2 l \, \dot{y}_1 \cos\varphi\right) + m_2 l \dot{\varphi} \dot{y}_1 \sin\varphi = P_2$$

After calculating time derivatives and introduction of generalized forces P_1, P_2, we obtain

1° $\ddot{y}_1 (m_1 + m_2) + m_2 l (\ddot{\varphi}\cos\varphi - \dot{\varphi}^2 \sin\varphi) + k(y_1 - d_0) = Q - D_1 \dot{y}_1$

2° $\ddot{y}_1 \cos\varphi + \ddot{\varphi} l + g \sin\varphi = (Ql\cos\varphi - D_2 \dot{\varphi})/m_2 l$ (2.85)

The above equations have been simulated in the mathematical package. The selected motion curves for input data and initial conditions given below are presented in Fig. 2.9 ... Fig. 2.12.

$$m_1 = 4.5\,[kg] \quad m_2 = 3.0\,[kg] \quad g = 9.81\,[m/s^2] \quad Q = -10.0\,[N]$$
$$k = 300.0\,[N/m] \quad l = 0.3\,[m] \quad d_0 = 0.5\,[m] \quad D_1 = D_2 = 5.0\,[Ns/m]$$
$$y_1(0) = 0.5\,[m] \quad \dot{y}_1(0) = 0.0\,[m] \quad \varphi(0) = 1.7\,[rad] \quad \dot{\varphi}(0) = 0.0\,[1/s]$$

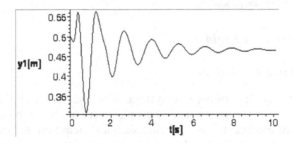

Fig. 2.9 Position y_l of the pendulum attachment point

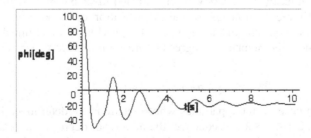

Fig. 2.10 Sway angle φ for the pendulum

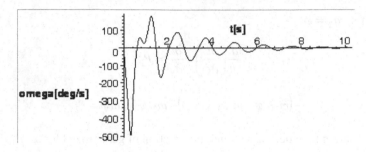

Fig. 2.11 Angular velocity $\dot{\varphi}$ of the pendulum swing

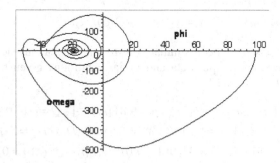

Fig. 2.12 Trajectory of the pendulum

2.1.10   Motion of a Rigid Body

2.1.10.1   Fundamental Notions

Rigid body – in a concept of discrete systems it is a set of material points (particles) whose distances remain constant. This idealization tends to be correct within a certain range of external forces affecting such a system both for flexible bodies with high rigidity and brittle ones. In order to determine the position of a body it is sufficient to determine three points on such a body that are not situated along a single straight line. Each successive point situated in such a body requires the determination of three coordinates for the identification of its position; however, it is kept at three constant distances from the previously determined points on this body. Hence, the number of degrees of freedom in accordance with (2.33) is equal to

$$s = n - h = 9 - 3 = 6 \qquad (2.86)$$

just as it is the case for three particles whose relations are determined by constant distances (Fig. 2.13). As a consequence, the number of degrees of freedom of such a body amounts to 6 (2.86); however, the description of the position of the body does not usually apply Cartesian coordinates related to three selected points on the body. The typical procedures followed in order to obtain the required 6 coordinates involves the determination of three Cartesian coordinates for a selected point on the body O_b, with a local system of coordinates $O_b x_b y_b z_b$ associated with this body as well as three angles for the description of the orientation of the system on the body in relation to the basic $Oxyz$ system. In brief, one can say that the first three coordinates determine the position of the body while the successive three define the orientation of the body in space. In order to establish the angles of the orientation of the body (or the local system $O_b x_b y_b z_b$ on the rigid body) a standard technique is to be applied, which involves the determination of Euler angles or navigation angles system RPY (roll, pitch, yaw) [1,6,20]. The angles determine the elementary revolute motion: in the first case with respect to current axes in the successive order of $xy'z''$, while in the latter case with respect to constant axes zyx.

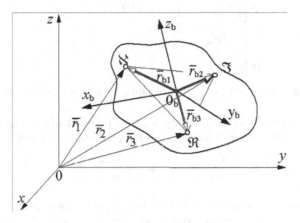

Fig. 2.13 Rigid body with local coordinates system and three points determining the position of the body

2.1.10.2 Motion of a Mass Point on a Rigid Body

The differential displacement of a point on a rigid dr_i could be made up of the displacement of the initial position of a local system on rigid $d\mathbf{R}_O$ and the differential vector of revolution $d\tau_i$ which accompanies the revolute motion of a rigid body. The differential vector of revolute motion (Fig. 2.14) is expressed as

$$|d\tau_i| = r_{bi}\sin\theta\,d\varphi \qquad d\tau_i = d\overline{\varphi}\times\mathbf{r}_{bi} \qquad (2.87)$$

Hence: $\qquad d\mathbf{r}_{bi} = d\mathbf{R}_O + d\tau_i \quad \text{or} \quad d\mathbf{r}_i = d\mathbf{R}_O + d\overline{\varphi}\times\mathbf{r}_{bi}$

As a result of dividing the differential displacements by the differential of time dt we obtain the velocity of the motion of an i-th point on a rigid body

$$\mathbf{v}_i = \mathbf{v}_O + \boldsymbol{\omega}\times\mathbf{r}_{bi} \qquad (2.88)$$

where: $\mathbf{v}_O$ is a linear velocity of O_b point with respect to the basic $Oxyz$ system and

$$\boldsymbol{\omega} = \frac{d\overline{\varphi}}{dt} \qquad (2.89)$$

is the angular speed of the rigid body. The axis determined by the direction of the differential angle $d\varphi$ of the revolute motion of a rigid, or angular speed ω, is called the instantaneous axis of rigid body's revolute motion.

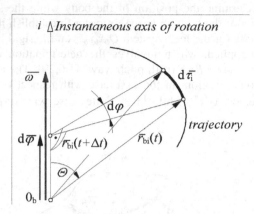

Fig. 2.14 Motion of a point on a rigid body resulting from its revolute motion

The center of mass S on a rigid body is a specific point which, can either belong to the rigid or not; however, it has to fulfill the following condition

$$M\mathbf{R}_S = \sum_i m_i \mathbf{r}_i \quad \text{or} \quad \mathbf{R}_S = \frac{1}{M}\sum_i m_i \mathbf{r}_i \tag{2.90}$$

The center of mass on a rigid body has some particular characteristics:

1° The potential energy U of a rigid could be expressed as the potential energy of the total mass M of rigid focused at the center of mass at point S. This regards a rigid body situated in gravitational field with constant vector $\mathbf{g}$ of gravitational acceleration (Fig. 2.15).

$$U = -\sum_i \mathbf{g}m_i\mathbf{h}_i = -\mathbf{g}\sum_i m_i\mathbf{h}_S - \mathbf{g}\underbrace{\sum_i m_i\mathbf{r}_{bi}}_{0} = -\mathbf{gh}_S\sum_i m_i = -\mathbf{gh}_S M = Mgh_S$$

Finally: $U = Mgh_S$ \hfill (2.91)

2° Center of mass S is also the center of the body's weight under constant gravitational field in the state of balance between torques due to gravity forces. Hence it represents the equilibrium state of a stationary rigid in gravitational field.

$$\mathbf{M} = \sum_i m_i\mathbf{r}_{bi} \times \mathbf{g} = \left(\sum_i m_i\mathbf{r}_{bi}\right)\times \mathbf{g} = 0 \tag{2.92}$$

3° After differentiation (2.90) with respect to time, we obtain

$$\mathbf{v}_S = \frac{1}{M}\sum_i m_i\mathbf{v}_i \tag{2.93}$$

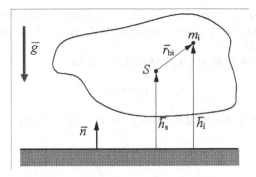

Fig. 2.15 Rigid body in gravitational field – potential energy

This means that the total momentum of a rigid body can be expressed as the product of solid's mass and velocity of the center of mass $\mathbf{v}_S$

$$\mathbf{p} = M\mathbf{v}_S \tag{2.94}$$

If the local system associated with a rigid body is situated in the center of mass, and $\mathbf{v}_s = 0$, which, consequently, gives $\mathbf{p} = 0$. This means that the center of mass is the point around which the internal momentum of a rigid is equal to zero.

2.1.10.3 Kinetic Energy of a Rigid Body

It is possible to express kinetic energy of a rigid body in a more synthetic manner than through the sum of kinetic energies of conventional mass points m_i with the total mass amounting to M, which are distributed in space but remain at constant

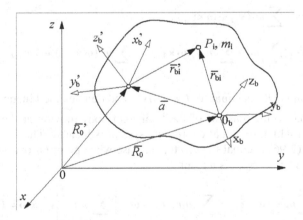

Fig. 2.16 Dependent and independent elements in relation to coordinate system situated on a rigid body

distances. However, one can note that angular speed ω refers to the rigid as a whole, which means that it is not relative to the location of the coordinate system on a rigid (Fig. 2.16).

For any point of mass m_i from (2.88) it follows that

$$\mathbf{v}_i = \mathbf{v}_O + \omega \times \mathbf{r}_{bi} \tag{i}$$

Using another system on a rigid with origin in O'_b, this velocity $\mathbf{v}_i$ can be noted as

$$\mathbf{v}_i = \mathbf{v}'_O + \omega' \times \mathbf{r}'_{bi} \tag{ii}$$

By substitution in (i) $\mathbf{r}_{bi} = \mathbf{a} + \mathbf{r}'_{bi}$ we obtain:

$$\mathbf{v}_i = \mathbf{v}_O + \omega \times \mathbf{a} + \omega \times \mathbf{r}'_{bi} \tag{iii}$$

The comparison of the results in (ii) with (iii) gives:

$$\mathbf{v}'_O = \mathbf{v}_O + \omega \times \mathbf{a} \qquad\qquad \omega = \omega' \tag{2.95}$$

Hence, it results that angular speed ω of rigid's revolute motion is the same along the entire area of the body and is not relative to the position of the local coordinate system associated with it.

The kinetic energy of the rigid is equal to

$$T = \tfrac{1}{2} \sum_i m_i \dot{\mathbf{r}}_i^2$$

Using (2.88) we obtain

$$\begin{aligned}
T &= \tfrac{1}{2} \sum_i m_i (\mathbf{v}_O + \omega \times \mathbf{r}_{bi})^2 = \\
&= \tfrac{1}{2} \mathbf{v}_O^2 \sum_i m_i + \sum_i m_i \mathbf{v}_O \omega \times \mathbf{r}_{bi} + \tfrac{1}{2} \sum_i m_i (\omega \times \mathbf{r}_{bi})(\omega \times \mathbf{r}_{bi})
\end{aligned} \tag{2.96}$$

The first term in (2.96) is equal to $T = \tfrac{1}{2} M \mathbf{v}_O^2$ and denotes the kinetic energy of a mass point equal to the mass M of the total rigid body moving at the velocity of the point O_b, in which the local coordinate system is located. The second term of the expression (2.96) is equal to zero for the case when point O_b overlaps with the center of mass S ($\mathbf{r}_{bi} = \mathbf{r}_{Si}$). As a result:

$$T_2 = \sum_i m_i \mathbf{v}_O \omega \times \mathbf{r}_{bi} = \sum_i m_i \mathbf{r}_{Si} \mathbf{v}_S \times \omega = \underbrace{\left(\sum_i m_i \mathbf{r}_{Si} \right)}_{0} \mathbf{v}_S \times \omega = 0 \tag{2.97}$$

Concurrently, the third term

$$T_3 = \frac{1}{2}\sum_i m_i\left((\omega_y z_i - \omega_z y_i)^2 + (\omega_z x_i - \omega_x z_i)^2 + (\omega_x y_i - \omega_y x_i)^2\right) =$$

$$= \frac{1}{2}\omega^T \underbrace{\begin{bmatrix} \sum_i m_i(y_i^2 + z_i^2) & -\sum_i m_i x_i y_i & -\sum_i m_i x_i z_i \\ -\sum_i m_i x_i y_i & \sum_i m_i(x_i^2 + z_i^2) & -\sum_i m_i y_i z_i \\ -\sum_i m_i x_i z_i & -\sum_i m_i y_i z_i & \sum_i m_i(x_i^2 + y_i^2) \end{bmatrix}}_{\mathbf{I}} \omega \qquad (2.98)$$

denotes the kinetic energy of the revolute motion around center of mass S with angular speed $\omega = [\, \omega_x \ \ \omega_y \ \ \omega_z \,]^T$. The elements present along the main diagonal of the matrix of inertia $\mathbf{I}$, i.e. I_x, I_y, I_z are called the moments of inertia of a rigid with respect to x,y,z axes, respectively. As a result

$$I_x = \sum_i m_i(y_i^2 + z_i^2) \qquad (2.99)$$

is the moment of inertia with respect to x axis and is calculated as the sum of the partial masses m_i multiplied by the square of the distance of the particular masses from x axis. Apart from the main axes there are the so called moments of deviation, e.g.

$$D_{xy} = \sum_i m_i x_i y_i \qquad (2.100)$$

The matrix of inertia $\mathbf{I}$ is symmetrical and positively determined

$$\mathbf{I} = \begin{bmatrix} I_x & -D_{xy} & -D_{xz} \\ -D_{xy} & I_y & -D_{yz} \\ -D_{xz} & -D_{yz} & I_z \end{bmatrix} \qquad (2.101)$$

The kinetic energy of a rigid body in free motion, under the assumption of a local coordinate system in the center of mass S, takes the form:

$$T = \frac{1}{2}Mv_S^2 + \frac{1}{2}\omega^T \mathbf{I}_S \omega \qquad (2.102)$$

where $\mathbf{I}_S$ denotes the matrix of inertia of a rigid body with respect to axes of Cartesian coordinates system intersecting at the rigid body's mass center S. For any irregularly shaped rigid body it is possible to select such directions of the axes of the Cartesian coordinate system $O_b x_b y_b z_b$ for which all moments of deviation are equal to zero; as a result, they disappear. For such axes the term principal axes of inertia is used

$$I' = \begin{bmatrix} I_{x'} & & \\ & I_{y'} & \\ & & I_{z'} \end{bmatrix}$$

(2.103)

In this case the kinetic energy of a rigid body associated with its revolute motion has only three terms:

$$T_3 = \tfrac{1}{2}(I_{x'}\omega_x^2 + I_{y'}\omega_y^2 + I_{z'}\omega_z^2)$$

(2.104)

For an irregularly shaped body the identification of the directions corresponding to the principal axes of inertia involves finding a solution to the characteristic equation for a square matrix $\mathbf{I}$ (2.101), i.e. the establishment of eigenvectors and eigenvalues using the methods of linear algebra. The term asymmetric top has been coned for such irregularly shaped rigid bodies for which the three principal moments of inertia $I_{x'}$, $I_{y'}$, $I_{z'}$ are different. The term symmetrical top is used with regard to a rigid body whose two principal moments of inertia are identical and the third is denoted with another value, while the term spherical top is used for a rigid whose all moments of inertia are equal. The identification of the center of mass and principal axes of rigid body's inertia is much simplified when the rigid displays the characteristics of symmetry (under the assumption of constant density and regular distribution of elementary masses). For the case of a body possessing an axis of symmetry the center of mass is situated along this axis and, hence, it represents one of the principal axes of inertia. For the case of a rigid body with a symmetry plane this plane contains two principal axes of inertia, while the third one is perpendicular to the intersection of the two axes. The final part of the present considerations regarding the motion of a rigid body will focus on the relation between the matrix of inertia $\mathbf{I}_S$ for the case when the initial point of the local system overlaps with the center of mass and the matrix of inertia $\mathbf{I}'$ calculated with respect to the coordinate system with parallel axes, while the initial point of the system is displaced from the center of mass by vector $\mathbf{a}$, such that

$$\mathbf{r}'_{bi} = \mathbf{r}_{Si} + \mathbf{a}, \text{ which gives:}$$

$$\mathbf{I}' = \mathbf{I}_S + M \begin{bmatrix} a^2 & -a_x a_y & -a_x a_z \\ -a_x a_y & a^2 & -a_y a_z \\ -a_x a_z & -a_y a_z & a^2 \end{bmatrix}$$

(2.105)

In this case the expression used for the description of the total energy of a rigid body, according to (2.97) includes

$$T_2 = \sum_i m_i (\mathbf{r}_{Si} + \mathbf{a}) \mathbf{v}_O \times \boldsymbol{\omega} = M\mathbf{a}(\mathbf{v}_O \times \boldsymbol{\omega})$$

(2.106)

and the total kinetic energy accounts for three components:

$$T = \tfrac{1}{2} M v_O^2 + M\mathbf{a}(\mathbf{v}_O \times \boldsymbol{\omega}) + \tfrac{1}{2} \boldsymbol{\omega}^T \mathbf{I}' \boldsymbol{\omega}$$

(2.107)

2.1.10.4 Motion of a Rigid Body around an Axis

Free motion of a rigid body or a system of rigid bodies is represented in a number of engineering issues in aviation, ballistics, space research, etc.; however, the most numerous group of engineering problems is concerned with the motion of a rigid body around a constant axis. This is the case in rotating machines, wheel bearings in a car, industrial manipulators and robots. In such cases the direction of the vector of angular speed ω is in conformity with the axis of revolute motion determined by the line of the bearings and reaction forces in bearings secure the maintenance of the constant direction of angular speed vector. The expression accounting for kinetic energy in motion around a constant axis is considerably simplified as it only involves one term of T_3 (2.104) representing a single component of angular speed

$$T = \tfrac{1}{2}\omega^2 I_O \tag{2.108}$$

where I_O is the moment of inertia of a rigid body with respect to the current axis of revolute motion. For the case when the axis of revolute motion does not intersect with the center of mass S the moment of inertia I_O can be determined from the relation (2.105), while angular speed has just a single component

$$I_O = I_S + Ma^2 \tag{2.109}$$

where: I_S - moment of inertia for an axis intersecting the center of mass S, parallel to the axis of the rigid body's revolute motion

a - distance between the two axes.

The same result can be derived from relation (2.107) under the assumption of the coordinate system in the center of mass S, so that

$$T = \tfrac{1}{2}M\omega^2 a^2 + \tfrac{1}{2}\omega^2 I_S = \tfrac{1}{2}\omega^2 (I_S + Ma^2) \tag{2.110}$$

This result is known in literature under the term König's theorem.

2.1.10.5 Rigid Body's Imbalance in Motion around a Constant Axis

In the considerations of the motion of a rigid body around a constant axis we have to do with problems associated with static and dynamic imbalance. The case of static imbalance (Fig. 2.17) is dealt with in the case when the axis of the rigid body's revolution does not cross its center of mass S, but is parallel to one of the principal axes of inertia.

In this case centrifugal force rotating along the center of mass is exerted on the bearings

$$\mathbf{F} = \frac{Mv^2}{r} = \frac{\mathbf{r}}{|\mathbf{r}|}\frac{M\omega^2 r^2}{r} = \frac{\mathbf{r}}{|\mathbf{r}|}M\omega^2 r \tag{2.111}$$

and it imposes two parallel forces on the bearings.

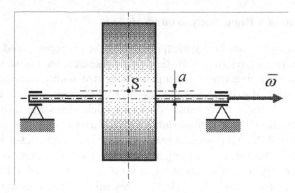

Fig. 2.17 State of static imbalance of a rigid body in motion around a constant axis

In a more complex case of dynamic imbalance the axis of rigid body's revolution overlaps its center of mass, but is not parallel to the principal axis of inertia. To simplify things, one can say that the bearings' axes of revolution are askew in relation to the axis of the rigid body' symmetry. However, it can be strictly established that dynamic imbalance is the case when angular momentum **L** is askew in relation to the vector of angular speed ω. In order to calculate the effects of dynamic imbalance it is necessary to apply the notion of angular momentum. For a system of particles the angular momentum is defined as

$$\mathbf{L} = \sum_i \mathbf{r}_i \times m_i \mathbf{v}_i \tag{2.112}$$

using the expression which determines the velocity for a particle in a rigid body (2.88) and limiting the scope to revolute motion we obtain:

$$\mathbf{L} = \mathbf{I}\,\omega \tag{2.113}$$

where **I** is the matrix of rigid body's inertia (2.101), and ω is the vector of angular speed expressed in the same axes as the matrix of inertia. It is notable that any change in the angular momentum for a system, not accounting for energy losses, is associated with the necessity of applying moment of force in the form

$$\mathbf{M} = \frac{d\mathbf{L}}{dt} \tag{2.114}$$

In a system accounting for losses the applied moment of force overcomes the moment of frictional resistance and potentially the moment associated with the exerted work; however, it is only its surplus or deficit that affects a change in the momentum in the system. The case of dynamic imbalance for a rigid with circular section, which has been slightly exaggerated, is presented in Fig. 2.18. For this case the angular momentum **L** does not overlap with the vector of angular speed ω due to various values of the moment of inertia in this disk for the principal axes of inertia.

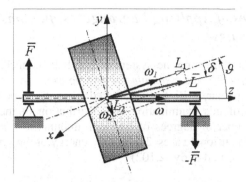

Fig. 2.18 Dynamic imbalance of a rotating disk

The angular speed ω of a disk is distributed along the principal axes of inertia is

$$\omega_1 = \omega\cos\vartheta \qquad \omega_2 = \omega\sin\vartheta$$

As a consequence, the components of the angular momentum **L** can be determined as

$$L_1 = I_1\omega_1 \qquad L_2 = I_2\omega_2 \qquad I_1 > I_2$$

In this case moment **M** is associated with maintenance of a steady direction of angular speed ω and results in a pair of reaction forces in bearings. It is expressed as

$$\mathbf{M} = \frac{d\mathbf{L}}{dt} = \mathbf{L}\times\omega \tag{2.115}$$

The value of this moment is

$$|\mathbf{M}| = \omega L \sin\varphi = \omega L \sin(\vartheta - \delta)$$

By application of relations

$$\cos\delta = \frac{L_1}{L} \qquad \sin\delta = \frac{L_2}{L}$$

we finally obtain:

$$M = \tfrac{1}{2}\omega^2(I_1 - I_2)\sin 2\vartheta \tag{2.116}$$

From formula in (2.116) it stems that the maximum value of this moment, which tends to orthogonalize orientation of a disk, occurs for the skewness angle $\vartheta = 45°$, which is the greater the bigger the difference between the moments of inertia for the principal axes. Certainly the moment resulting from centrifugal forces is relative to the square of angular speed. This moment acts on the bearings with forces whose directions are opposite (a pair of forces). The forces rotate as a results of the revolution of the disk (Fig. 2.18), according to vector product (2.115).

2.1.11 Examples of Applying Lagrange's Equations for Motion of Rigid Bodies

Example 2.6. For the third time the system of two balls connected by a stiff rod, referred to in *Example 2.1* and *Example 2.4* will be examined. This time the system despite being planar, will be considered as a rigid body. Due to two-dimensional nature of this examination and distribution of the masses along a straight line the number of degrees of freedom amounts to $s = 3$ in contrast to $s = 6$ in the three dimensional cases. The kinetic energy of the system is expressed in a way specific for a rigid body (2.102)

$$T = \tfrac{1}{2}(m_1 + m_2)(\dot{x}_S^2 + \dot{y}_S^2) + \tfrac{1}{2}I_S\dot{\vartheta}^2 \qquad (2.117)$$

Under the assumption of generalized coordinates $\mathbf{q} = (x_s, y_s, \vartheta)$, including the position of the center of mass and angle of rigid body's revolution, the kinetic energy expressed in (2.117) is not associated with the necessity of transforming variables. In the circumstances of the lack of elements which accumulate potential energy (gravitational forces are accounted for in the external forces), Lagrange's function for the system is equal to the kinetic energy (2.117):

$$L = T - U = T$$

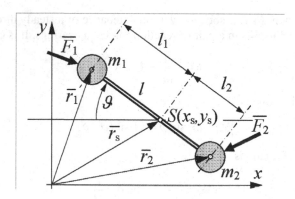

Fig. 2.19 System of two balls connected by stiff rod considered as a rigid body

Concurrently, transformational formulae are needed in order to determine generalized force on the basis of data from external forces expressed in Cartesian coordinates

$$\mathbf{F} = [F_1 \quad F_2 \quad F_3 \quad F_4]^T = [F_{x1} \quad F_{y1} \quad F_{x2} \quad F_{y2}]^T$$

For this purpose we have to determine the position of the center of mass S along the rod which connects the two balls

$$\left. \begin{array}{l} m_1 l_1 = m_2 l_2 \\ l = l_1 + l_2 \end{array} \right\} \qquad l_1 = \frac{m_2}{m_1 + m_2}l \qquad l_2 = \frac{m_1}{m_1 + m_2}l$$

and the moment of inertia for the system

$$I_S = m_1 l_1^2 + m_2 l_2^2 = \frac{m_1 m_2}{m_1 + m_2} l^2 \tag{2.118}$$

so that the transformation of coordinates (2.35) in this case takes the form

$$\begin{matrix} \xi_1 = x_1 = x_S - l_1 \cos \vartheta & \xi_3 = x_2 = x_S + l_2 \cos \vartheta \\ \xi_2 = y_1 = y_S + l_1 \sin \vartheta & \xi_4 = y_2 = y_S - l_2 \sin \vartheta \end{matrix} \tag{2.119}$$

Subsequently, following from (2.44)

$$Q_{xS} = \sum_{i=1}^{4} F_i \frac{\partial \xi_i}{\partial x_S} = F_{x1} + F_{x2}$$

$$Q_{yS} = \sum_{i=1}^{4} F_i \frac{\partial \xi_i}{\partial y_S} = F_{y1} + F_{y2}$$

$$Q_\vartheta = \sum_{i=1}^{4} F_i \frac{\partial \xi_i}{\partial \vartheta} = (F_{x1} l_1 - F_{x2} l_2) \sin \vartheta + (F_{y1} l_1 - F_{y2} l_2) \cos \vartheta$$

we obtain equations of motion in the following form

For $q_1 = x_S$:

$$\frac{d}{dt} \frac{\partial L}{\partial \dot{x}_S} - \frac{\partial L}{\partial x_S} = Q_{xS} \quad \Rightarrow \quad (m_1 + m_2) \ddot{x}_S = F_{x1} + F_{x2}$$

for $q_2 = y_S$:

$$\frac{d}{dt} \frac{\partial L}{\partial \dot{y}_S} - \frac{\partial L}{\partial y_S} = Q_{yS} \quad \Rightarrow \quad (m_1 + m_2) \ddot{y}_S = F_{y1} + F_{y2} \tag{2.120}$$

for $q_3 = \vartheta$:

$$\frac{d}{dt} \frac{\partial L}{\partial \dot{\vartheta}} - \frac{\partial L}{\partial \vartheta} = Q_\vartheta \Rightarrow$$

$$\Rightarrow \quad I_S \ddot{\vartheta} = (F_{x1} l_1 - F_{x2} l_2) \sin \vartheta + (F_{y1} l_1 - F_{y2} l_2) \cos \vartheta$$

This is the simplest form of the equations of motion among the three examined models (2.32, 2.78, 2.120) since generalized coordinates are selected in such a way that the form of notation of kinetic energy that is most succinct.

Example 2.7. It involves a model of dynamics for a spherical manipulator. The kinematic model of the manipulator, containing variables and parameters, is presented in Figs. 2.20 and 2.21. This manipulator is shown in its nominal position, which means that the values of coordinates in the joints are equal to zero.

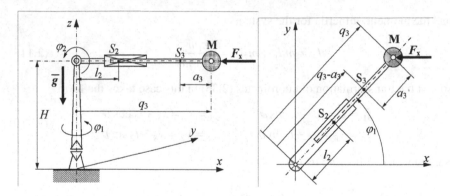

Fig. 2.20 Kinematic model of a spherical manipulator

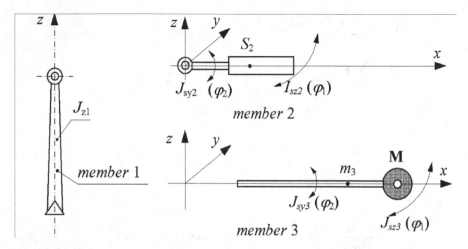

Fig. 2.21 Moments of inertia in particular members of manipulator in Fig. 2.20

The first two joints in this manipulator are revolute joints, while the third one is a translational joint. The successive joints of the local motion of the manipulator and the end effector in this model are represented by mass M at the end of the final member. Fig. 2.21 presents the particular joints and contains a representation of the moments of inertia in respect to the axes along which the manipulator's motion occurs. The moments of inertia J_{Sy2}, J_{Sz2}, J_{Sy3}, J_{Sz3} are determined for the axes crossing the centers of mass S_2 and S_3. Hence, the kinetic energy of the manipulator can be determined as the sum of four terms which correspond to the four members: $T = T_1 + T_2 + T_3 + T_M$, which can be expanded to

$$T = \tfrac{1}{2} J_{z1} \dot{\varphi}_1^2 + \tfrac{1}{2} J_{Sz2} \dot{\varphi}_1^2 \cos^2 \varphi_2 + \tfrac{1}{2} m_2 (\dot{x}_2^2 + \dot{y}_2^2 + \dot{z}_2^2) + \tfrac{1}{2} J_{Sy2} \dot{\varphi}_2^2 +$$
$$+ \tfrac{1}{2} J_{Sz3} \dot{\varphi}_1^2 \cos^2 \varphi_2 + \tfrac{1}{2} J_{Sy3} \dot{\varphi}_2^2 + \tfrac{1}{2} m_3 (\dot{x}_3^2 + \dot{y}_3^2 + \dot{z}_3^2) + \qquad (2.121)$$
$$+ \tfrac{1}{2} M (\dot{x}_M^2 + \dot{y}_M^2 + \dot{z}_M^2)$$

where: $\mathbf{r}_{S2} = (x_2, y_2, z_2)$; $\mathbf{r}_{S3} = (x_3, y_3, z_3)$; $\mathbf{r}_M = (x_M, y_M, z_M)$

are the respective Cartesian coordinates of the center of mass S_2, S_3 and the mass M. This system has 3 degrees of freedom and the generalized coordinates are

$$q = (q_1, q_2, q_3) = (\varphi_1, \varphi_2, q_3) \tag{2.122}$$

which denote angular variables in the first two joints and a linear variable in a translational joint. Prior to the presentation of transformational formulae it would be useful to introduce the convention of abbreviations used for the notation of trigonometric functions which are commonly used in robotics, namely:

$$\cos\varphi_1 \Rightarrow c1 \; ; \; \sin\varphi_2 = s2, \quad \text{etc.} \tag{2.123}$$

which tends to considerably shorten the notation associated with transformation of coordinates. The transformational formulae for the coordinates of center of mass S_2, S_3 and M, which apply the notation as in (2.123) are the following:

$$
\begin{aligned}
x_2 &= l_2 c1c2 & x_3 &= (q_3 - a_3)c1c2 & x_M &= q_3 c1c2 \\
y_2 &= l_2 s1c2 & y_3 &= (q_3 - a_3)s1c2 & y_M &= q_3 s1c2 \\
z_2 &= H + l_2 s2 & z_3 &= H + (q_3 - a_3)s2 & z_M &= H + q_3 s2
\end{aligned}
\tag{2.124}
$$

and for speed, respectively:

$$
\begin{aligned}
\dot{x}_2 &= -l_2 f_x & \dot{x}_3 &= \dot{q}_3 c1c2 - (q_3 - a_3)f_x & \dot{x}_M &= \dot{q}_3 c1c2 - q_3 f_x \\
\dot{y}_2 &= l_2 f_y & \dot{y}_3 &= \dot{q}_3 s1c2 + (q_3 - a_3)f_y & \dot{y}_M &= \dot{q}_3 s1c2 + q_3 f_y \\
\dot{z}_2 &= l_2 \dot{\varphi}_2 c2 & \dot{z}_3 &= \dot{q}_3 s2 + (q_3 - a_3)\dot{\varphi}_2 c2 & \dot{z}_M &= \dot{q}_3 s2 + q_3 \dot{\varphi}_2 c2
\end{aligned}
\tag{2.125}
$$

where:
$$
\begin{aligned}
f_x &= f_x(\varphi_1, \varphi_2, \dot{\varphi}_1, \dot{\varphi}_2) = \dot{\varphi}_1 s1c2 + \dot{\varphi}_2 c1s2 \\
f_y &= f_y(\varphi_1, \varphi_2, \dot{\varphi}_1, \dot{\varphi}_2) = \dot{\varphi}_1 c1c2 - \dot{\varphi}_2 s1s2
\end{aligned}
$$

The potential energy in this system, associated with gravity forces acting on members 2 and 3 of the manipulator and on mass M, is

$$U = m_2 g(H + l_2 s2) + m_3 g(H + (q_3 - a_3)s2) + Mg(H + q_3 s2)$$

which can be transformed to take the form

$$U = const + gs2(m_2 l_2 + m_3(q_3 - a_3) + Mq_3) \tag{2.126}$$

Lagrange's function for the manipulator accounting for (2.121), (2.125) and (2.126) takes the form:

$$
\begin{aligned}
L = T - U &= \tfrac{1}{2}\dot{\varphi}_1^2 \left\{ J_{z1} + (c2)^2(J_{O2z} + J_{S3z} + m_3(q_3 - a_3)^2 + Mq_3^2) \right\} + \\
&\quad \tfrac{1}{2}\dot{\varphi}_2^2 \left\{ J_{O2y} + J_{S3y} + m_3(q_3 - a_3)^2 + Mq_3^2 \right\} + \\
&\quad \tfrac{1}{2}\dot{q}_3^2(m_3 + M) - gs2(m_2 l_2 + m_3(q_3 - a_3) + Mq_3)
\end{aligned}
\tag{2.127}
$$

where: $J_{O2z} = m_2 l_2^2 + J_{S2z}$ $J_{O2y} = m_2 l_2^2 + J_{S2y}$

The exchange of energy with the environment surrounding the manipulator occurs as a result of virtual work relative to external force $\mathbf{F}=-F_x$, the driving torques M_1, M_2 and force F_3 originating from drives on the joints and friction forces. First, we will calculate the generalized forces for the particular generalized coordinates resulting from the external force. They take the form

$$Q_{\varphi 1} = -F_x \frac{\partial x_M}{\partial \varphi_1} = F_x q_3 s1c2$$

$$Q_{\varphi 2} = -F_x \frac{\partial x_M}{\partial \varphi_2} = F_x q_3 c1s2 \qquad (2.128)$$

$$Q_3 = -F_x \frac{\partial x_M}{\partial q_3} = -F_x c1c2$$

while the overall virtual work of the system is:

$$\delta A = Q_{\varphi 1}\delta\varphi_1 + Q_{\varphi 2}\delta\varphi_2 + Q_3\delta q_3 = (M_1 + F_x q_3 s1c2 - D_1\dot{\varphi}_1)\delta\varphi_1 +$$
$$(M_2 + F_x q_3 c1s2 - D_2\dot{\varphi}_2)\delta\varphi_2 + (F_3 - F_x c1c2 - D_3\dot{q}_3)\delta q_3 \qquad (2.129)$$

Familiar with the expressions (2.127), (2.129) we are able to state the equations of motion:

1° for $q_1 = \varphi_1$:

$$\frac{d}{dt}\frac{\partial L}{\partial \dot{\varphi}_1} - \frac{\partial L}{\partial \varphi_1} = Q_{\varphi 1}$$

and following the operations we obtain:

$$\ddot{\varphi}_1\left\{J_{z1} + (c2)^2(J_{O2z} + J_{S3z} + m_3(q_3 - a_3)^2 + Mq_3^2)\right\} =$$
$$+ \dot{\varphi}_1\dot{\varphi}_2 s(2\varphi_2)(J_{O2z} + J_{S3z} + m_3(q_3 - a_3)^2 + Mq_3^2) -$$
$$- 2\dot{\varphi}_1\dot{q}_3(c2)^2(m_3(q_3 - a_3) + Mq_3) + M_1 + F_x q_3 s1c2 - D_1\dot{\varphi}_1$$

As a result:

$$\ddot{\varphi}_1\left\{J_{z1} + (c2)^2 J_z(q_3)\right\} = \dot{\varphi}_1\dot{\varphi}_2 s(2\varphi_2)J_z(q_3) -$$
$$- 2\dot{\varphi}_1\dot{q}_3(c2)^2 M_3(q_3) + M_1 + F_x q_3 s1c2 - D_1\dot{\varphi}_1$$

2° for $q_2 = \varphi_2$:

$$\frac{d}{dt}\frac{\partial L}{\partial \dot{\varphi}_2} - \frac{\partial L}{\partial \varphi_2} = Q_{\varphi 2}$$

$$\ddot{\varphi}_2\left\{J_y(q_3)\right\} = -2\dot{\varphi}_2\dot{q}_3 M_3(q_3) - \tfrac{1}{2}\dot{\varphi}_1^2 s(2\varphi_2)J_z(q_3) -$$
$$gc2(m_2 l_2 + M_3(q_3)) + M_2 + F_x q_3 c1s2 - D_2\dot{\varphi}_2 \qquad (2.130)$$

3° for linear variable q_3 :

$$\frac{d}{dt}\frac{\partial L}{\partial \dot{q}_3} - \frac{\partial L}{\partial q_3} = Q_{q3}$$

$$\ddot{q}_3(m_3 + M) = M_3(q_3)(\dot{\varphi}_1^2 (c2)^2 + \dot{\varphi}_2^2) -$$
$$- gs2(m_3 + M) + F_3 - F_x c1c2 - D_3\dot{q}_3$$

where:
$$J_z(q_3) = J_{O2z} + J_{S3z} + m_3(q_3 - a_3)^2 + Mq_3^2$$
$$J_y(q_3) = J_{O2y} + J_{S3y} + m_3(q_3 - a_3)^2 + Mq_3^2$$
$$M_3(q_3) = m_3(q_3 - a_3) + Mq_3$$

The resulting equations are quite complex despite the fact that the manipulator has only three degrees of freedom. The model's complexity is becoming greater following an increase in the number of degrees of freedom and with the greater number of revolute joints in the kinematic chain of the manipulator. Hence, in order to avoid laborious transformations and calculations in the formalized course of model's statement using Lagrange's equations it is possible to program the model in a programming language handling symbolic operations [1,11] and input the data regarding the kinematics and parameters of the examined object. The resultant equations of manipulator motion can be verified for a number of particular instances by determining the values of selected variables and interpreting the terms in equations whose sum is not equal to zero. Let us assume for instance that the angular speed of the revolution of the vertical axis of the manipulator is $\dot{\varphi}_1 = const$, while the third joint – the translational one is immobile: $q_3 = const$ and for simplification purposes zero value of force **F**. In this case the equation of motion for the second member takes the form

$$\ddot{\varphi}_2\{J_y(q_3)\} =$$
$$= -\tfrac{1}{2}\dot{\varphi}_1^2 s(2\varphi_2)J_z(q_3) - gc2(m_2 l_2 + M_3(q_3)) + M_2 - D_2\dot{\varphi}_2 \qquad (2.131)$$

It can be concluded that the arm of the manipulator is lowered under the effect of gravity force, the moment of centrifugal force relative to $\dot{\varphi}_1^2$ tries to preserve the arm in the horizontal position ($\varphi_2 = 0$) and that this moment is the highest for angle $\varphi_2 = 45°$, which is quite logical. An active role is played by the driving moment M_2 and the passive role is attributed to the moment of friction. The members of the equation have physical interpretation and the correct sign denoting the sense of a moment. From the first of the equations (2.130) it is possible to calculate moment M_1 needed in order to maintain constant angular speed $\dot{\varphi}_1 = const$ of manipulator's motion

$$M_1 = -\dot{\varphi}_1\dot{\varphi}_2 s(2\varphi_2)J_z(q_3) + D_1\dot{\varphi}_1 \qquad (2.132)$$

This equation serves in order to determine the torque required to overcome the frictional resistance and changing the angular momentum associated with the motion of the second member with the velocity equal to $\dot{\varphi}_2$. The final equation for variable Q_3 makes it possible to determine the value of required force $\mathbf{F}_3$, necessary at the point where the translational joint is fixed, in order to make it immobile, which is

$$\mathbf{F}_3 = -\mathbf{M}_3(q_3)(\dot{\varphi}_1^2(c2)^2 + \varphi_2^2) + gs2(M + m_3) \qquad (2.133)$$

The case dealt with here involves reaction to the centrifugal force relative to both angular speeds in revolute joints and to the gravity force. The derived equations of manipulator's motion (2.130) could well be applicable for the selection of drive in manipulator's joints and in discussion of issues associated with its control.

2.1.12 General Properties of Lagrange's Equations

2.1.12.1 Laws of Conservation

An isolated system is a system which neither affects nor is affected by other external systems. For such a system Lagrange's function is not explicitly relative to time

$$L = L(\dot{\mathbf{q}}, \mathbf{q}) \qquad (2.134)$$

In an isolated system we don't have to do with external forces, neither in an active form or as frictional forces, hence, Lagrange's equation takes the following form:

$$\frac{d}{dt}\frac{\partial L(\dot{\mathbf{q}}, \mathbf{q})}{\partial \dot{q}_k} - \frac{\partial L(\dot{\mathbf{q}}, \mathbf{q})}{\partial q_k} = 0 \qquad (2.135)$$

Let us assume that for a coordinate q_j Lagrange's function is not relative to this variable, but only depends on speed $\dot{q}_j$, then

$$\frac{d}{dt}\frac{\partial L(\dot{\mathbf{q}}, \mathbf{q})}{\partial \dot{q}_j} = 0 \quad \text{or} \quad \frac{\partial L(\dot{\mathbf{q}}, \mathbf{q})}{\partial \dot{q}_j} = const \qquad (2.136)$$

and this kind of coordinate is denoted with the term cyclic coordinate. This constant value, which is a derivative of Lagrange's function according to generalized velocity is called generalized momentum

$$\mathbf{p}_j = \frac{\partial L(\dot{\mathbf{q}}, \mathbf{q})}{\partial \dot{q}_j} \qquad (2.137)$$

From equations (2.136) it stems that the generalized momentum for a cyclic coordinate is constant. The occurrence of cyclic coordinates in Lagrange's function depends on the selection of generalized coordinates (coordinate system); consequently, the constancy of certain components of generalized momentum is relative to the adopted reference system. It has been demonstrated [8,12,13] that the total momentum for an isolated system is constant. This also means that for cyclic co-ordinates the components of momentum are constant, while for the remaining components of momentum the exchange of momentum occurs in such a manner that the total momentum of a system remains constant. Let us refer again to La-grange's function for an isolated system (2.134) and calculate the total differential for this function with respect to time

$$\frac{dL}{dt} = \sum_k \frac{\partial L}{\partial q_k} \dot{q}_k + \sum_k \frac{\partial L}{\partial \dot{q}_k} \ddot{q}_k$$

The substitution of $\dfrac{\partial L}{\partial q_k}$ according to Lagrange's equation (2.135) gives

$$\frac{dL}{dt} = \sum_k \frac{d}{dt}\left(\frac{\partial L}{\partial \dot{q}_k}\right) \dot{q}_k + \sum_k \frac{\partial L}{\partial \dot{q}_k} \ddot{q}_k = \sum_k \frac{d}{dt}\left(\frac{\partial L}{\partial \dot{q}_k} \dot{q}_k\right)$$

which results in the relation:

$$\frac{d}{dt}\left(\sum_k \frac{\partial L}{\partial \dot{q}_k} \dot{q}_k - L\right) = 0$$

This can be restated using the definition of momentum (2.137):

$$\sum_k p_k \dot{q}_k - L = const \qquad (2.138)$$

Hence, in an isolated system the following value is constant

$$E = \sum_k p_k \dot{q}_k - L \qquad (2.139)$$

and is called the energy of a system. We can transform the expression used in (2.139)

$$\sum_k p_k \dot{q}_k = \sum_k \dot{q}_k \frac{\partial L}{\partial \dot{q}_k} = \sum_k \dot{q}_k \frac{\partial T}{\partial \dot{q}_k} = 2T \qquad (2.140)$$

This is so due to the fact that the function of kinetic energy $T(\dot{q}, q)$ is a homogenous function of the second order (and, consequently, quadratic form) of the velocity for the case when it is not explicitly relative to time. In accordance with the Euler's for-mula for homogenous functions of n order the following relation is satisfied

$$\sum_i \xi_i \frac{\partial f(\xi)}{\partial \xi_i} = n f(\xi) \qquad (2.141)$$

and for a function of the order $n = 2$ the relation in (2.140) is satisfied. By application of (2.140) in the formula stating energy (2.139) we obtain

$$E = 2T - L = T + U \tag{2.142}$$

This result could have been anticipated as it means that the total energy of an isolated system $(L = L(\dot{\mathbf{q}}, \mathbf{q}), \mathbf{r} = \mathbf{r}(\mathbf{q}))$ consists of the sum of kinetic and potential energy. In an isolated system the law of conservation of generalized momentum is fulfilled [8,13], whereas this property can easily be demonstrated in a Cartesian coordinate system both for particles and rigid bodies in revolute motion. However, if a certain generalized coordinate is a system is expressed in terms of revolute angle: $q_j = \varphi$, then equation (2.137) denotes generalized angular momentum. For a cyclic angular coordinate the appropriate component of angular momentum in an isolated system is constant. As a consequence, Lagrange's equations (2.51), (2.135) denote either the equations of forces for the case when the generalized coordinate q_k has a linear measure (i.e. it is translational) or equations for moments of forces (torques) for generalized coordinates in the form of angular variables (characteristic for revolute motion).

2.1.12.2 Characteristics of Lagrange's Functions and Equations

The equations for both mechanical and electromechanical systems stated in the form of Lagrange's equations possess a number of specific properties useful for the purposes of stating and controlling them as well as in various issues associated with the dynamics applying Lagrange's equations, and in particular for a control of dynamic objects.

1° Lagrange's function does not have a unambiguous expression. For function L and for function

$$L' = L + \frac{d}{dt} F(\mathbf{q}, t) \tag{2.143}$$

the equations of motion expressed using Lagrange's method are in the same form. It is possible to verify by direct substitution of L and L' in (2.144).

2° Lagrange's equation preserves an invariable general form. For a holonomic system it is following

$$\frac{d}{dt} \frac{\partial L}{\partial \dot{q}_k} - \frac{\partial L}{\partial q_k} = P_k \qquad k = 1 \ldots s \tag{2.144}$$

regardless of the selection of the vector of generalized coordinates $\mathbf{q} = (q_1, q_2, \ldots, q_s)$. If unambiguous transformation were to be used

$$u_j = u_j(q_1, \ldots, q_s) \qquad j = 1 \ldots s$$

having a non-zero determinant of Jacobean matrix for this transformation in a specific area of variance, there is a reverse transformation such that $q_k = q_k(u_1, u_2, \ldots, u_s)$.

Then Lagrange's function $L = L(\dot{u}, u)$ takes another specific form, however, the general form of Lagrange's equations

$$\frac{d}{dt}\frac{\partial L}{\partial \dot{u}_j} - \frac{\partial L}{\partial u_j} = V_j, \qquad j = 1 \ldots s$$

remains the same. The specific form of the particular equations for the successive variables of the two vectors of generalized coordinates $\mathbf{q} = (q_1, q_2, \ldots, q_s)$ and $\mathbf{u} = (u_1, u_2, \ldots, u_s)$ is evidently different.

3° The resulting form of Lagrange's equation for a holonomic system with s degrees of freedom, while Lagrange's function is not explicitly relative to time, i.e. $L = L(\dot{q}, q)$, can be stated in the form of a matrix equation, where the respective matrices have the dimension of $s \times s$:

$$\mathbf{B}(\mathbf{q})\ddot{\mathbf{q}} + \mathbf{C}(\dot{\mathbf{q}}, \mathbf{q})\dot{\mathbf{q}} + \mathbf{G}(\mathbf{q}) = \tau \qquad (2.145)$$

The matrices display the following properties [20]:

a) the matrix of inertia $\mathbf{B}(\mathbf{q})$ is a symmetrical and positively determined one and is relative only to the vector of generalized coordinates $\mathbf{q}$. From that it results that matrix of inertia is always reversible. Moreover, there are such scalar quantities $\eta_1(\mathbf{q})$ and $\eta_2(\mathbf{q})$ that the following limitation is satisfied

$$\eta_1(\mathbf{q})\mathbf{I} < \mathbf{B}(\mathbf{q}) < \eta_2(\mathbf{q})\mathbf{I} \qquad (2.146)$$

b) the matrix $\qquad \mathbf{W} = \mathbf{B}(\mathbf{q}) - 2\mathbf{C}(\dot{\mathbf{q}}, \mathbf{q}) \qquad (2.147)$

- is skew-symmetric. This plays a role in the control of various systems and can serve in order to control the correctness of the developed equations of motion.

c) the equations of motion (2.145) resulting from Lagrange's equations are linear due to their structural parameters. This also means that that there is a constant vector Θ of dimension p and a matrix function $\mathbf{Y}(\ddot{\mathbf{q}}, \dot{\mathbf{q}}, \mathbf{q})$ of dimensions $n \times p$ such that

$$\mathbf{B}(\mathbf{q})\ddot{\mathbf{q}} + \mathbf{C}(\dot{\mathbf{q}}, \mathbf{q})\dot{\mathbf{q}} + \mathbf{G}(\mathbf{q}) = \mathbf{Y}(\ddot{\mathbf{q}}, \dot{\mathbf{q}}, \mathbf{q})\Theta = \tau \qquad (2.148)$$

The function $\mathbf{Y}(\ddot{\mathbf{q}}, \dot{\mathbf{q}}, \mathbf{q})$ is denoted with the term regressor, while the vector Θ consists of appropriate combinations of structural parameters such as dimensions, masses and moments of inertia. The dimension p of the vector Θ is not uniformly defined and, hence, the identification of the adequate parameter set for this system in a way that dimension p is minimized plays a practical role.

d) The equations of motion (2.145) possess the property of passivity. This means that the following mapping satisfies the relation:

$$\int_0^T \mathbf{q}(u)\tau(u)du \geq -\beta, \qquad (2.149)$$

for the case of constant β. The evidence of that [3,17] is given by differentiation with respect to time of the total energy of a system (2.142) $E = \frac{1}{2}\dot{\mathbf{q}}^T\mathbf{B}(\mathbf{q})\dot{\mathbf{q}} + U(\mathbf{q})$ and demonstration that due to the property (2.147)

$$\dot{E} = \dot{\mathbf{q}}^T\tau \tag{2.150}$$

which after integration of (2.149) gives $H(T) - H(0) \geq -H(0) = -\beta$ since the total energy $H(T)$ is always greater than zero. The property of passivity (2.149) is relevant in the evidence of the stability of dynamic systems during their control [17].

Example 2.8. Let us undertake the matrix notation of the model consisting *Example 2.5* of a pendulum of mass m_2 attached to mass m_1 sliding along a horizontal bar. The dynamic equations for this object (2.85) in the matrix form are the following:

$$\underbrace{\begin{bmatrix} m_1 + m_2 & m_2 l\cos\varphi \\ m_2 l\cos\varphi & m_2 l^2 \end{bmatrix}}_{\mathbf{B}(\mathbf{q})}\begin{bmatrix} \ddot{y}_1 \\ \ddot{\varphi} \end{bmatrix} + \underbrace{\begin{bmatrix} 0 & -m_2 l\dot{\varphi}\sin\varphi \\ 0 & 0 \end{bmatrix}}_{\mathbf{C}(\dot{\mathbf{q}},\mathbf{q})}\begin{bmatrix} \dot{y}_1 \\ \dot{\varphi} \end{bmatrix} +$$

$$+ \underbrace{\begin{bmatrix} k(y_1 - d_0) \\ m_2 g l\sin\varphi \end{bmatrix}}_{\mathbf{G}(\mathbf{q})} = \underbrace{\begin{bmatrix} Q - D_1\dot{y}_1 \\ Q l\cos\varphi - D_2\dot{\varphi} \end{bmatrix}}_{\tau} \tag{2.151}$$

Since
$$\dot{\mathbf{B}} = \begin{bmatrix} 0 & -m_2 l\dot{\varphi}\sin\varphi \\ -m_2 l\dot{\varphi}\sin\varphi & 0 \end{bmatrix},$$

$$\dot{\mathbf{B}} - 2\mathbf{C} = m_2 l\dot{\varphi}\sin\varphi\begin{bmatrix} 0 & +1 \\ -1 & 0 \end{bmatrix}$$

and hence the last matrix is skew-symmetric. The matrix of inertia $\mathbf{B}(\mathbf{q})$ is positively determined, which results from the value of its determinant

$$\det(\mathbf{B}(\mathbf{q})) = (m_1 + m_2)m_2 l^2 - (m_2 l\cos\varphi)^2 > 0$$

As a result, it is always reversible regardless of the value of angle φ. The equations of motion from the perspective of the linear combination of the parameters (2.148) can be represented for the following vector of parameters

$$\mathbf{\Theta}_1^T = \begin{bmatrix} \Theta_1 & \Theta_2 & \Theta_3 & \Theta_4 \end{bmatrix} = \begin{bmatrix} m_1 + m_2 & m_2 l & k & m_2 l^2 \end{bmatrix} \tag{2.152}$$

In this case the regressor takes the form:

$$\mathbf{Y}_1(\ddot{\mathbf{q}},\dot{\mathbf{q}},\mathbf{q}) = \begin{bmatrix} \ddot{y}_1 & \ddot{\varphi}\cos\varphi - \dot{\varphi}^2\sin\varphi & y_1 - d_0 & 0 \\ 0 & \ddot{y}_1\cos\varphi + g\sin\varphi & 0 & \ddot{\varphi} \end{bmatrix} \tag{2.153}$$

If the parameter $\Theta_4 = m_2 l^2$, which is similar to $\Theta_2 = m_2 l$, were to be omitted from (2.152), which should be associated with the requirement of good familiarity with dimension l, the vector of parameters would take a shorter form

$$\Theta_2^T = [\Theta_1 \quad \Theta_2 \quad \Theta_3] = [m_1 + m_2 \quad m_2 l \quad k] \tag{2.154}$$

Consequently, the matrix of the regressor is following

$$\mathbf{Y}_1(\ddot{\mathbf{q}}, \dot{\mathbf{q}}, \mathbf{q}) = \begin{bmatrix} \ddot{y}_1 & \ddot{\varphi}\cos\varphi - \dot{\varphi}^2 \sin\varphi & y_1 - d_0 \\ 0 & \ddot{y}_1 \cos\varphi + \ddot{\varphi} l + g \sin\varphi & 0 \end{bmatrix} \tag{2.155}$$

However, during the practical process of estimating parameters of the mathematical model a considerable problem may be associated with the determination of the damping of transients. Hence damping coefficients may be incorporated into matrix $\mathbf{C}(\dot{\mathbf{q}}, \mathbf{q})$. As a result $\mathbf{C}'(\dot{\mathbf{q}}, \mathbf{q}) = \begin{bmatrix} D_1 & -m_2 l\dot{\varphi}\sin\varphi \\ 0 & D_2 \end{bmatrix}$. In this case the property (2.147) does not hold. For the new vector of parameters

$$\Theta_3^T = [\Theta_1 \quad \Theta_2 \quad \Theta_3 \quad \Theta_4 \quad \Theta_5] = [m_1 + m_2 \quad m_2 l \quad k \quad D_1 \quad D_2]$$

the regressor amounts to

$$\mathbf{Y}_1(\ddot{\mathbf{q}}, \dot{\mathbf{q}}, \mathbf{q}) = \begin{bmatrix} \ddot{y}_1 & \ddot{\varphi}\cos\varphi - \dot{\varphi}^2 \sin\varphi & y_1 - d_0 & \dot{y}_1 & 0 \\ 0 & \ddot{y}_1 \cos\varphi + \ddot{\varphi} l + g \sin\varphi & 0 & 0 & \dot{\varphi} \end{bmatrix} \tag{2.156}$$

In conclusion, for the mathematical model (2.151) of a pendulum sliding along a horizontal bar (Fig. 2.8) three options for linearity of Lagrange's equations (2.148) were demonstrated on the basis of various structural parameters.

Example 2.9. In a similar manner we shall analyze the equations of manipulator motion from *Example 2.7.* The corresponding matrices (2.145) formed on the basis of equations (2.130) are presented below. The matrix of inertia

$$\mathbf{B}(\mathbf{q}) = \begin{bmatrix} J_1 + (\cos\varphi_2)^2 J_z(q_3) & & \\ & J_y(q_3) & \\ & & m_3 + M \end{bmatrix} \tag{2.157}$$

is diagonal, positively determined and, hence, it is possible to state that it satisfies the requirements (2.146)

$$\begin{bmatrix} J_1 & & \\ & J_{O2y} + J_{S3y} & \\ & & m_3 + M \end{bmatrix} \leq \mathbf{B}(\mathbf{q}) \leq$$

$$\leq \begin{bmatrix} J_1 + J_z(q_{3\max}) & & \\ & J_y(q_{3\max}) & \\ & & m_3 + M \end{bmatrix}$$

The matrix

$$\mathbf{C}(\dot{\mathbf{q}},\mathbf{q}) =$$

$$\begin{bmatrix} -\frac{1}{2}\dot{\varphi}_2 s(2\varphi_2)J_z(q_3)+\dot{q}_3(c2)^2M_3(q_3) & -\frac{1}{2}\dot{\varphi}_1 s(2\varphi_2)J_z(q_3) & \dot{\varphi}_1(c2)^2M_3(q_3) \\ \frac{1}{2}\dot{\varphi}_1 s(2\varphi_2)J_z(q_3) & \dot{q}_3 M_3(q_3) & \dot{\varphi}_2 M_3(q_3) \\ -\dot{\varphi}_1(c2)^2M_3(q_3) & -\dot{\varphi}_2 M_3(q_3) & 0 \end{bmatrix} \quad (2.158)$$

where:
$$\begin{aligned} J_z(q_3) &= J_{O2z}+J_{S3z}+m_3(q_3-a_3)^2+Mq_3^2 \\ J_y(q_3) &= J_{O2y}+J_{S3y}+m_3(q_3-a_3)^2+Mq_3^2 \\ M_3(q_3) &= m_3(q_3-a_3)+Mq_3 \end{aligned}$$

In accordance with (2.147) the matrix $\mathbf{W}=\dot{\mathbf{B}}-2\mathbf{C}$ is a skew-symmetric one:

$$\mathbf{W} = \begin{bmatrix} 0 & \dot{\varphi}_1 s(2\varphi_2)J_z(q_3) & -2\dot{\varphi}_1(c2)^2M_3(q_3) \\ -\dot{\varphi}_1 s(2\varphi_2)J_z(q_3) & 0 & -2\dot{\varphi}_2 M_3(q_3) \\ 2\dot{\varphi}_1(c2)^2M_3(q_3) & 2\dot{\varphi}_2 M_3(q_3) & 0 \end{bmatrix} \quad (2.159)$$

Moreover,

$$\mathbf{G}(\mathbf{q}) = g\begin{bmatrix} 0 \\ c2\big(m_2 l_2+M_3(q_3)\big) \\ s2(m_3+M) \end{bmatrix}$$

$$\tau = \begin{bmatrix} M_1+F_x q_3 s1c2-D_1\dot{\varphi}_1 \\ M_2+F_x q_3 c1s2-D_2\dot{\varphi}_2 \\ F_3-F_x c1c2-D_3\dot{q}_3 \end{bmatrix} \quad (2.160)$$

The equations of motion expressed in terms of the linear regression, in respect to parameters (2.148), may be presented for various compositions of the vector of parameters Θ in the context of the intended program for parameter estimation or for the purposes of design of control. For a six-dimensional vector with parameters

$$\Theta^T = \begin{bmatrix} J_{z1} & J_{O2z}+J_{S3z} & J_{O2y}+J_{S3y} & m_2 l_2 & m_3 & M \end{bmatrix} \quad (2.161)$$

the function of the regressor takes the form:

$$\mathbf{Y}_1(\ddot{\mathbf{q}},\dot{\mathbf{q}},\mathbf{q}) = \quad (2.162)$$

$$\begin{bmatrix} \ddot{\varphi}_1 & (c2)^2\ddot{\varphi}_1-s(2\varphi_2)\dot{\varphi}_1\dot{\varphi}_2 & 0 & 0 & (q_3-a_3)^2\big((c2)^2\ddot{\varphi}_1-s(2\varphi_2)\dot{\varphi}_1\dot{\varphi}_2\big)+2(c2)^2(q_3-a_3)\dot{\varphi}_1\dot{q}_3 \\ 0 & \frac{1}{2}s(2\varphi_2)\dot{\varphi}_1^2 & \ddot{\varphi}_2 & gc2 & (q_3-a_3)^2\big(\ddot{\varphi}_2+\frac{1}{2}\dot{\varphi}_2^2 s(2\varphi_2)\big)+(q_3-a_3)(2\dot{\varphi}_2\dot{q}_3+gc2) \\ 0 & 0 & 0 & 0 & \ddot{q}_3-(q_3-a_3)((c2)^2\dot{\varphi}_1^2+\dot{\varphi}_2^2)+gs2 \end{bmatrix}$$

$$\begin{bmatrix} q_3^2\big((c2)^2\ddot{\varphi}_1-s(2\varphi_2)\dot{\varphi}_1\dot{\varphi}_2\big)+2\dot{\varphi}_1\dot{q}_3 q_3(c2)^2 \\ q_3^2\big(\ddot{\varphi}_2+\frac{1}{2}\dot{\varphi}_2^2 s(2\varphi_2)\big)+q_3(2\dot{\varphi}_2\dot{q}_3+gc2) \\ \ddot{q}_3-q_3((c2)^2\dot{\varphi}_1^2+\dot{\varphi}_2^2)+gs2 \end{bmatrix}$$

Other options for selecting the content of the Θ vector are also available. The selection of them is relative to whether some of the parameters in a model are known with high precision or which should be the subject of a formal estimation and contained in Θ.

2.2 Electromechanical Systems

2.2.1 Principle of Least Action: Nonlinear Systems

The purpose of this chapter is to provide grounds for the statement about the relevance of Lagrange's equations with regard to electromechanical systems. Another objective is to indicate the variability of parameters in the Lagrange's function for systems with non-linear parameters, i.e. the ones in which the parameters are relative to the vector of the system's variables. The principle of least action, also known as principle of stationary action or Hamilton's principle is found in the majority of handbooks in the field of theoretical mechanics [8,12,16] to be central to the synthesis of the laws of motion. In classical mechanics it is derived from general properties of space [13]. Every mechanical system is characterized by function $L(\dot{q},q,t)$ relative to the vector of velocity and generalized coordinates and perhaps time present in the explicit form, known as Lagrange's function. This is the same function that was obtained from the derivation of Lagrange's equations (2.45-2.52) but it can also be derived from Galileo's theory of relativity on the basis of the general properties of homogenous and isotropic space. For any system in motion a functional is defined, which represents a set of functions mapped onto a set of real numbers (Fig. 2.22) called an action.

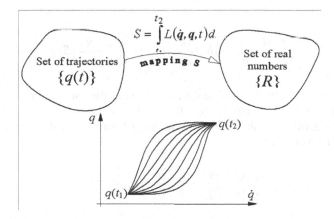

Fig. 2.22 Graphical representation of a functional: mapping S and a set of trajectories $\left\{ \tilde{\dot{q}}, \tilde{q} \right\}$

$$S = \int_{t_1}^{t_2} L(\dot{\mathbf{q}}, \mathbf{q}, t)\, dt \tag{2.163}$$

This functional is a mapping of a set of trajectories $\{\tilde{\dot{\mathbf{q}}}, \tilde{\mathbf{q}}\}$ into a set of real numbers $\{\Re\}$ which result from the integration of Lagrange's function along the particular trajectories. All trajectories belonging to the set have the same class of continuity C^1, which means that they are continuous and possess a continuous derivative; in addition to which, at the instants determined as t_1, t_2 they converge at the same points in space $\mathbf{q}_1, \mathbf{q}_2$; see Fig. 2.22. The principle of least (stationary) action states that for any actual path of motion $\mathbf{q}(t)$ an action (2.163) assumes the smallest value of all possible trajectories belonging to the set $\{\tilde{\dot{\mathbf{q}}}, \tilde{\mathbf{q}}\}$. This also means formally, that in the statement of prerequisites of any motion realized in nature is that the first variation of the functional of action disappears.

$$\delta S = \delta \int_{t_1}^{t_2} L(\dot{\mathbf{q}}, \mathbf{q}, t)\, dt = 0 \tag{2.164}$$

For the variation of integral (2.164) one can further state that:

$$\int_{t_1}^{t_2} \sum_i \left(\frac{\partial L}{\partial q_i} \delta q_i + \frac{\partial L}{\partial \dot{q}_i} \delta \dot{q}_i \right) dt = 0 \tag{2.165}$$

Under the assumption that $\delta \dot{q} = \dfrac{d}{dt}(\delta q)$, the second term of this integral can be integrated by parts

$$\int_{t_1}^{t_2} \left(\frac{\partial L}{\partial \dot{q}_i} \frac{d}{dt} \delta q_i \right) dt = \frac{\partial L}{\partial \dot{q}_i} \delta q_i \Big|_{t_1}^{t_2} - \int_{t_1}^{t_2} \left(\frac{d}{dt} \frac{\partial L}{\partial \dot{q}_i} \right) \delta q_i\, dt \tag{2.166}$$

Since at instants t_1, t_2 all paths of motion converge at the subsequent initial and final point, then: $\delta q_i(t_1) = 0$ and $\delta q_i(t_2) = 0$, $i = 1, \ldots, s$.

After consideration of this fact the first term (2.166) is equal to zero, and integral (2.165) takes the form

$$\int_{t_1}^{t_2} \left(\sum_{k=1}^{s} \left(\frac{\partial L}{\partial q_k} - \frac{d}{dt} \frac{\partial L}{\partial \dot{q}_k} \right) \delta q_k \right) dt = 0$$

The independence of variation of variables (virtual displacements), δq_k leads to the relation

$$\frac{d}{dt} \frac{\partial L}{\partial \dot{q}_k} - \frac{\partial L}{\partial q_k} = 0 \qquad k = 1 \ldots s \tag{2.167}$$

which is a familiar form of Lagrange's equation for a stationary system (without an exchange of energy with the environment). This result means that the actual realization of motion occurs along a trajectory for which the expenditure of action (2.163) of the system is the least possible one. Thus, in the sense of differential description of motion for this case the differential equation of the second order in the form of Lagrange's equation (2.167) has to be fulfilled. This is a familiar result; however, it is more limited in comparison to the equation in (2.51) which accounts for external forces operating in the system. Nevertheless, it was derived on the basis of a general principle, which is valid in classical as well as relativistic mechanics [12,13,16] and in electromagnetism as well. This gives a solid foundation to extend the method of Lagrange's equation onto calculation of electromechanical systems, which is the fundamental object of the study of electrical drives.

2.2.1.1 Electrically Uncharged Particle in Relativistic Mechanics

Deriving from the general postulates associated with relativistic mechanics and in particular from the postulate that action S of a mechanical system may not be relative to the selection of any inertial reference system and, hence, it has to be an invariant of Lorentz transformation [13,16], Lagrange's function for a particle can be stated in the following form

$$L = -mc^2 \sqrt{1 - \left(\frac{v}{c}\right)^2} \tag{2.168}$$

Concurrently, the action on the particle is

$$S = \int_{t_1}^{t_2} L\,dt = -mc^2 \int_{t_1}^{t_2} \sqrt{1 - (v/c)^2}\,dt \tag{2.169}$$

where:

   m - is the rest mass of a particle
   c - is the velocity of light in vacuum

Since always the velocity of a particle v is considerably lower than the speed of light, it is possible to expand Lagrange's function (2.168) so that it takes the form of Taylor's power series of a small quantity (v/c^2). As a result of such expansion we obtain:

$$L = -mc^2 + \tfrac{1}{2}mv^2 + \tfrac{1}{8}mv^2\left(\frac{v}{c}\right)^2 + \tfrac{1}{16}mv^2\left(\frac{v}{c}\right)^4 + \ldots \tag{2.170}$$

For small velocities $v \ll c$ only the first two terms of the power expansion of Lagrange's function are relevant in comparison to the following ones; however, the first term in them, as a constant value, does not contribute anything due to the indefinite form of Lagrange's function (2.143). Hence the result takes the form of familiar classical result well known in mechanics $L = \tfrac{1}{2}\,mv^2$, which denotes kinetic energy of a particle. The calculation of particle's momentum according to the general definition (2.137) results in:

$$\mathbf{p} = \frac{\partial L}{\partial \dot{\mathbf{q}}} = \frac{\partial L}{\partial \mathbf{v}} = \frac{m\mathbf{v}}{\sqrt{1-(v/c)^2}} \tag{2.171}$$

which when expanded into power series with respect to (v/c^2), yields that

$$\mathbf{p} = m\mathbf{v} + \frac{1}{2}m\mathbf{v}\left(\frac{v}{c}\right)^2 + \frac{3}{8}m\mathbf{v}\left(\frac{v}{c}\right)^4 + \dots \tag{2.172}$$

For small velocities $v \ll c$ this results in the classical shape

$$\mathbf{p} = m\mathbf{v} \tag{2.173}$$

The calculation of the total energy of a particle in accordance with relation (2.139) gives the result, which is relevant to our considerations

$$E = \dot{\mathbf{q}}\mathbf{p} - L = \mathbf{v}\mathbf{p} - L = \frac{mc^2}{\sqrt{1-(v/c)^2}} \tag{2.174}$$

By expansion into power series (1.174), this gives

$$E = mc^2 + \frac{1}{2}mv^2 + \frac{3}{8}mv^2\left(\frac{v}{c}\right)^2 + \dots \tag{2.175}$$

This energy consists of rest mass energy $E_0 = mc^2$ and the energy associated with the velocity of particle's motion. The relevance of this result is associated with the fact that the comparison between expressions in (2.170) and (2.175) offers a conclusion that Lagrange's function for a particle in motion is equal to its kinetic energy only for the case of classical mechanics. In the relativistic mechanics the expressions in (2.168) and (2.174) are completely different. The formal reason for this is related to nonlinearity of parameters, or more precisely, nonlinearity of mass, which increases along with speed as illustrated by the formula for the particle's momentum. An illustration of this is found in Fig. 2.23.

From Fig. 2.23 one can conclude that surface areas that illustrate kinetic energy and the supplementary term denoted as co-energy overlap only for small velocities of a particle in motion. This is the case when the function of kinetic energy is a homogenous function of velocity (2.140). Accordingly, from formula (2.175) one can conclude that this is so only when the expansion into power series may omit the terms containing $(v/c)^{2k}$. In practice this means that equality between kinetic energy and kinetic term in Lagrange's function takes place only in case when the mass of the particle is constant, which is represented by a linear system in the sense of involvement of constant parameters. In systems with non-linear parameters relative to velocity $\dot{\mathbf{q}}$, Lagrange's function does not account for kinetic

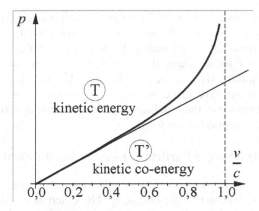

Fig. 2.23 Kinetic energy and co-energy of a particle in motion

energy; however, in accordance with (2.171) the integral of system's momentum takes the form

$$T' = L = \int \mathbf{p}\,d\mathbf{v} = \int \mathbf{p}\,d\dot{\mathbf{q}} \tag{2.176}$$

This integral denotes the kinetic term of Lagrange's function written as T' and is called kinetic complementary energy: co-energy. Its graphical representation (Fig. 2.23) takes the form of the surface area under the curve denoting the relation between the momentum of a system and velocity. From both the formal point of view and its graphical illustration it is clear that kinetic energy T is equal to kinetic co-energy T' only for a system with constant parameters.

$$T' = T \qquad m = const \tag{2.177}$$

Concurrently, the total energy of particle is equal to

$$T = \int \mathbf{v}\,d\mathbf{p} = \int \dot{\mathbf{q}}\,d\mathbf{p} \tag{2.178}$$

which can be integrated and the result takes the form (2.174), and the one in (2.175) after power expansion. Both functions of co-energy (2.176) and kinetic energy (2.178) add up to form a rectangle, which results from formula (2.174) and has also a graphical representation in Fig. 2.23. This also justifies the definition attributed to co-energy, which states that it completes the function of energy to the bound of a rectangle.

$$T' + T = \mathbf{p}\dot{\mathbf{q}} \tag{2.179}$$

The result of these considerations conducted for motion of a particle in relativistic mechanics shows that kinetic co-energy T' (2.176) is involved in the Lagrange's function for a system whereas kinetic energy T is not. The distinction between kinetic co-energy and kinetic energy is only necessary in non-linear systems whose parameters are relative to generalized velocities. In the examined case of a particle in motion mass m relative to the velocity forms the only variable parameter. In engineering practice associated with issues of electric drives we have to do with such low velocities that the relativistic variation in mass is insignificant and, as a result, we assume that $T' = T$ for mechanical variables. The studies conducted here have an even more comprehensive application besides theoretical considerations, as in electromagnetism we have to do with systems with non-linear parameters, in particular with inductance of the windings containing ferromagnetic core, which is relative to the current applied to the windings.

2.2.1.2 Electrically Charged Particle in Electromagnetic Field

In contrast to the previously considered example we shall assume here that a particle with rest mass m has an electric charge Q. The external electromagnetic field is so strong and its source is so remote that the charge Q, which is carried along with the particle, does not affect a change of the field. Scalar potential $\phi(\mathbf{r}, t)$ describes the interaction between the field and immobile charge, while vector potential $\mathbf{A}(A_x, A_y, A_z)$ describes the interaction with electric charge in motion (electric current). It is possible to define the four quantities describing electromagnetic field in the form of a four-vector that is transformed in accordance with to the rules of Lorentz transformations [13]. For an electromagnetic field is called a four-potential, where the scalar potential ϕ is considered as the zero term of this four-potential

$$A^\mu = (\phi, \mathbf{A}) \qquad \mathbf{A} = \mathbf{A}(\mathbf{r}, t) \tag{2.180}$$

Such formal notation referred to in theoretical physics proves very useful in transformation of fields and studies devoted to general properties of electromagnetic interactions. For the purposes of our study it is relevant to note that Lagrange's function for a charged particle in electromagnetic field is expressed with the aid of four-potential of field in the following form:

$$L = -mc^2 \sqrt{1 - \left(\frac{v}{c}\right)^2} + Q\mathbf{v}\mathbf{A} - Q\phi \tag{2.181}$$

The first two terms of Lagrange's function include kinetic co-energy of a particle with an electric charge of Q. This is the familiar mechanical kinetic co-energy (2.168) while the second term is the magnetic kinetic co-energy of the charge in motion. The term kinetic co-energy is used for the reason of its relation to the velocity $\mathbf{v}$. From the two kinetic terms we subtract the potential electric energy of the charged particle, as in the expression of Lagrange's function. On the basis of Lagrange's function we can calculate particle's momentum in the following form

$$\tilde{\mathbf{p}} = \frac{\partial L}{\partial \mathbf{v}} = \frac{m\mathbf{v}}{\sqrt{1 - (v/c)^2}} + Q\mathbf{A} \tag{2.182}$$

The momentum of a charged particle, hence, involves two terms, i.e. the familiar mechanical momentum (2.171) and the magnetic momentum of a particle. The term associated with electric field is absent from the formula, which comes at no surprise, since the interaction with the scalar potential is static. Concurrently, the total energy (2.174) of a charged particle is

$$E = \mathbf{v}\frac{dL}{d\mathbf{v}} - L = \frac{mc^2}{\sqrt{1-(v/c)^2}} + Q\phi \qquad (2.183)$$

In conclusion, the total energy includes the potential energy associated with the scalar potential of the field ϕ beside the energy of the particle's mass. We will proceed to see what will result from Lagrange's equation for a charged particle in the field, whose Lagrange's function is expressed in the form in (2.181). The examined system is stationary (does not account for exchange of energy with environment) and the equation of motion can take the form

$$\underbrace{\frac{d}{dt}\frac{\partial L}{\partial \mathbf{v}}}_{\tilde{\mathbf{p}}} = \frac{\partial L}{\partial \mathbf{r}} \qquad (2.184)$$

The right-hand side of the equation is equal to

$$\frac{\partial L}{\partial \mathbf{r}} = \nabla L = Q\,grad(\mathbf{vA}) - Q\,grad\phi$$

The first term in this equation, in accordance with vector identity [18], may be expressed by the relation

$$\nabla(\mathbf{AB}) = (\mathbf{A}\nabla\mathbf{B}) + (\mathbf{B}\nabla\mathbf{A}) + \mathbf{A}\times(\nabla\times\mathbf{B}) + \mathbf{B}\times(\nabla\times\mathbf{A})$$

The expression which is considered here is $\dfrac{\partial L}{\partial \mathbf{r}}$ and hence $\mathbf{v} = const$. As a consequence

$$grad(\mathbf{vA}) = (\mathbf{v}\nabla\mathbf{A}) + \mathbf{v}\times rot\,\mathbf{A}$$

At the same time, the left hand side of the equation (2.184) takes the form

$$\frac{d}{dt}(\mathbf{p} + Q\mathbf{A}) = \frac{d\mathbf{p}}{dt} + Q\frac{\partial \mathbf{A}}{\partial t} + Q(\mathbf{v}\nabla)\mathbf{A}$$

After the combination and simplification of the two sides of the equation we obtain

$$\frac{d\mathbf{p}}{dt} = Q\left(-\frac{\partial \mathbf{A}}{\partial t} - grad\,\phi + \mathbf{v}\times rot\,\mathbf{A}\right)$$

By defining: $\mathbf{E} = -\dfrac{\partial \mathbf{A}}{\partial t} - grad\,\phi$ $\mathbf{H} = rot\,\mathbf{A}$ \qquad (2.185)

to be the subsequent vectors of electrical and magnetic fields, we finally obtain

$$\frac{d\mathbf{p}}{dt} = Q(\mathbf{E} + \mathbf{v}\times\mathbf{H}) \qquad (2.186)$$

which gives an equation of motion for a charged particle in the electromagnetic field. The right hand side of the equation represents Lorentz force with which a field interacts with an electric charge. The content in this chapter indicates how notions of dynamics in the form of Lagrange's equations can be applied to problems in electrodynamics. The issues dealt with here will be transferred to macroscopic scale for the application in electromagnetic lumped parameter systems.

2.2.2 Lagrange's Equations for Electromechanical Systems in the Notion of Variance

Chapter 2.2.1 presents the principles of variance and their application in respect to a single particle with an aim of indicating that principle of least action involves both mechanics and electromechanical systems. This statement can be further extended from a singular particle to cover electromechanical systems, including macroscopic technical devices serving often for power generation and conversion. Beside machines such as electric power generators and motors used in electric drives, electromechanical devices include electromechanical measurement systems, servo-drives, industrial manipulators and robots as well. In addition, modern engineering tools include Micro/Nano-Electro-Mechanical Systems (MEMs, NEMs) [2], which attract a steadily growing interest. Such devices are different from the standard electromechanical drive in several ways despite the fact that the principle of operation is the same. First of all, the major task attributed to them is not associated with transformation of electric power. Instead, they tend to be used to perform precise action in a coordinated and controlled manner, including manipulation and measurement on a macro- or micro- scale. In addition, they have a greater number of degrees of mechanical freedom than the case is for standard electric drives, where the number of degrees of freedom is normally equal to one. In contrast, in standard manipulators the number of degrees of freedom amounts to 5-6 [1,11,20].

Industrial manipulators, and in particular mobile robots, have a very complex control and information processing technology embedded in them, which is based on a number of internal and external sensors. In addition, they have systems, which apply artificial intelligence algorithms to promote autonomic decision regarding control parameters. Moreover, the driving systems are designed in a way that promotes energy saving in order to enable permanent operation using an internal source of energy in some cases. Such drives are sometimes referred to as actuators, which emphasizes the role of the drive based on articulated joints of the devices. More and more attention and research is dedicated to the application of actuators on a micro- and nano-scale in computer technology, medicine and even biotechnology. Such electromechanical systems apply a great number of electronic and power electronics devices as a result of the need to meet the requirements associated with fast changing motion and necessity of saving power. In fact, this branch of engineering which concurrently involves mechanics, electric drive, electronics and computer engineering for the purposes of control is nowadays referred to as mechatronics [9,14,23].

The transfer from the principles of the motion of particles in electromagnetic field to electromechanical devices used in technology is associated with the need of stating assumptions regarding quasi-stationary character of electromechanical processes and deriving volume integrals from field quantities. Thus, a transfer can be made to the discreet model of the system. This problem is dealt with in the literature in this field [16,24]. The principle of stationary (least) action expenditure for non-stationary electromechanical systems states that

$$\int_{t_1}^{t_2} (\delta L + \delta A)dt = 0 \tag{2.187}$$

where: δL - is the variance of Lagrange's function for an electromechanical system, δA- is not a variance of any function but represents work expenditure performed on virtual displacements, which realize an exchange of energy with the environment, called virtual work of an electromechanical system.

Under the assumption that processes are quasi-stationary in nature, Lagrange's function for an electromechanical system includes only two structural components: a mechanical one associated with the mass and elastic strain of the elements of the system and an electromechanical one related to the electric charge (electric field) and electric current (magnetic field). However, it does not involve a term that accounts for the electromagnetic field. In summary, it is simply a total of Lagrange's functions for the mechanical and the electrical part of the system

$$L = L_m + L_e \tag{2.188}$$

The mechanical term L_m in Lagrange's function is represented by the difference between kinetic mechanical co-energy T'_m and the potential energy U_m. In a similar manner, electrical term of Lagrange's function is expressed as the difference between kinetic co-energy T'_e associated with magnetic field (electric currents) and potential electrical energy U_e associated with electric charges. Hence, the specific form of equation (2.188) takes the form

$$L = (T'_m + T'_e) - (U_m + U_e) \tag{2.189}$$

where: T'_m, T'_e - are kinetic co-energies for the mechanical and electric variables

U_m, U_e - are potential energies: mechanical and electrical.

Mechanical kinetic co-energy and potential energy have already been covered in detail. In the current section we will discuss magnetic kinetic co-energy and electrical potential energy which commonly represent the electrical part of Lagrange's function of the entire system. The number of degrees of freedom is the algebraic sum, which can be obtained by adding the number of degrees of freedom for the mechanical and electrical parts:

$$s = s_m + s_e \tag{2.190}$$

This is so since there aren't any specific variables of electromechanical nature; the variables are discreet and separate. The mutual interaction between the mechanical and electrical parts of the system as well as processes of energy conversion are realized by the so called electromechanical couplings, which result from the fact that electromagnetic energy is relative to the mechanical variable and, as a consequence, the equations are linked. This issue is covered in more detail later in this chapter.

2.2.2.1 Electric Variables

The selection of electric variables for characterizing an electromechanical system is implied by the form of Lagrange's function for a single particle in the field (2.181). The description of potential energy involves charge Q, while the expression of kinetic energy involves a charge $\mathbf{v}Q$ in motion. Hence, the proposed variables in macroscopic description of electromagnetic phenomena are

$$
\begin{aligned}
Q &\quad [C] \\
\dot{Q} = i &\quad [A]
\end{aligned}
\tag{2.191}
$$

which denote electric charge and electric current, respectively. The function of the charge in the electrical term of Lagrange's function is similar to the role of the position in a mechanical system, and $\dot{Q}$ - electric current is an equivalent to mechanical velocity, which in fact is the velocity of the charge. Such a selection of variables is natural and harmonic both in terms of formal similarity and physical role in a system accounting for coordinates in the description of mechanical motion. The electrical variables, including electric charges Q and their time derivatives, i.e. electric currents $\dot{Q} = i$ are present in electric circuits which are also known as electric networks. The role of the variables of the primary description in electric networks is played by charges on the branches Q_b, which occur on the particular branches of the network

$$
\mathbf{Q}_b = (Q_{b1}, Q_{b2}, ..., Q_{bg})
$$

where: g - is a number of branches.

At the same time, the generalized coordinates include the selected charges along the branches, which form the vector of generalized coordinates with the length s_e

$$
\mathbf{q}_e = (Q_1, Q_2, ..., Q_{se})
\tag{2.192}
$$

The role of constraints in an electrical network is played by equations formed in accordance with Kirchhoff's first law, which states that the algebraic sum of currents flowing through a node of an electric network is equal to zero (Fig. 2.24)

$$
\sum_j \dot{Q}_{bj} = 0
\tag{2.193}
$$

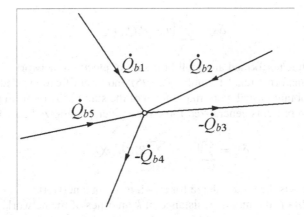

Fig. 2.24 Representation of constraints in electrical network formed in accordance with Kirchhoff's first law

The kinematic equations of constraints (2.193) originating from Kirchhoff's first law can be integrated. Since the nodes in the network do not accumulate electric charge the result is

$$\sum_j Q_{bj} = 0 \qquad \sum_j \delta Q_{bj} = 0 \qquad (2.194)$$

The number of the degrees of freedom s_e in an electrical network is related to the number of branches g and the number of nodes w and is equal to

$$s_e = g - w + 1 \qquad (2.195)$$

The selection of generalized coordinates is subjected to the standard conditions determined by the relations in (2.34-2.36). The transformational formulae type (2.37) are linear, which results from linearity of the equations of constraints (2.193 – 2.194). As a result, they take the same form for electric charges Q, electric currents $\dot{Q}$ and virtual displacements δQ:

$$\mathbf{Q}_b = f(\mathbf{q}_e)$$
$$\dot{\mathbf{Q}}_b = f(\dot{\mathbf{q}}_e) \qquad (2.196)$$
$$\delta \mathbf{Q}_b = f(\delta \mathbf{q}_e)$$

2.2.2.2 Virtual Work of an Electrical Network

Virtual work represents the exchange of energy with the environment. The role of external forces is played by electric voltages u or electromotive forces e. A dissipation of energy takes place under the effect of current flow through resistances R, which for the flowing current $\dot{Q}$ play the similar role as the friction coefficients in resisting mechanical motion.

$$\delta A_e = \sum_{i=1}^{g} \left(u_i - R_i \dot{Q}_{bi} \right) \delta Q_{bi} \qquad (2.197)$$

The summation is performed along all branches of electrical network. After the application of transformational formulae (2.196) and simplification of algebraic expressions, we obtain virtual work in the form of the sum of the terms in parenthesis, each of which represents generalized force for a given generalized coordinate

$$\delta A_e = \sum_{k=1}^{s_e} \left(u_k - \sum_{l=1}^{s_e} R_{kl} \dot{Q}_l \right) \delta Q_k \qquad (2.198)$$

where: u_k - is the total voltage for a k-th mesh in a network
                R_{kl} - is the mutual resistance of k,l meshes of the network.

Example 2.10. Let us apply the presented method for the determination of the number of degrees of freedom, selection of generalized coordinates and virtual work for an electrical network, which corresponds to the topology of a 3-phase bridge rectifier presented in Fig. 2.25.

For the examined network the number of nodes is $w = 6$ and the number of branches is equal to $g = 10$. Hence, the number of degrees of freedom amounts to $s_e = g - w + 1 = 5$. The branches are marked with arrows to show the direction of current flow indicated with a plus sign. The vector of currents in the branches, which play the role of velocity in the primary coordinate system, is the following:

$$\dot{\mathbf{Q}}_b = (\dot{Q}_1, \dot{Q}_2, \ldots, \dot{Q}_{10})$$

Five linearly independent charges are indicated as the generalized coordinates. For example such coordinates include

$$\mathbf{q}_e = (q_1, q_2, q_3, q_4, q_5) = (Q_1, Q_3, Q_4, Q_6, Q_{10})$$

and the corresponding vectors of currents and virtual displacements are:

$$\dot{\mathbf{q}}_e = (\dot{Q}_1, \dot{Q}_3, \dot{Q}_4, \dot{Q}_6, \dot{Q}_{10}) \qquad \delta\mathbf{q}_e = (\delta Q_1, \delta Q_3, \delta Q_4, \delta Q_6, \delta Q_{10}) \qquad (2.199)$$

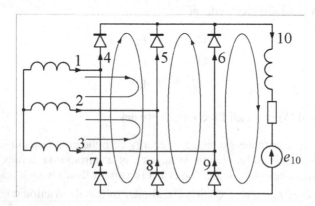

Fig. 2.25 Diagram of electric network representing a 3-phase bridge rectifier

The vector of generalized currents (2.199) plays the role of a set of mesh currents in the method of analysis of electrical networks familiar from electrical engineering. The mesh currents corresponding to vector $\dot{\mathbf{q}}_e$ of generalized currents are indicated in Fig. 2.25. This system contains 6 nodes yielding the relations (2.193) resulting from Kirchhoff's first law:

$$\dot{Q}_1 + \dot{Q}_7 = \dot{Q}_4 \quad \dot{Q}_1 + \dot{Q}_2 + \dot{Q}_3 = 0 \quad \dot{Q}_2 + \dot{Q}_8 = \dot{Q}_5$$

$$\dot{Q}_3 + \dot{Q}_9 = \dot{Q}_6 \quad \dot{Q}_4 + \dot{Q}_5 + \dot{Q}_6 = \dot{Q}_{10} \quad \dot{Q}_7 + \dot{Q}_8 + \dot{Q}_9 = \dot{Q}_{10} \tag{2.200}$$

One of the above equations is linearly dependent on the remaining ones and can serve to play the role of a control node. The resulting transformational formulae (2.196) for the currents take the form

$$\dot{Q}_2 = -\dot{Q}_{1,} - \dot{Q}_3 \quad \dot{Q}_5 = \dot{Q}_{10} - \dot{Q}_4 - \dot{Q}_6 \quad \dot{Q}_7 = \dot{Q}_4 - \dot{Q}_1$$

$$\dot{Q}_8 = \dot{Q}_1 + \dot{Q}_3 + \dot{Q}_{10} - \dot{Q}_4 - \dot{Q}_6 \quad \dot{Q}_9 = \dot{Q}_6 - \dot{Q}_3 \tag{2.201}$$

In accordance with (2.196) the same functional relations are fulfilled for virtual displacements

$$\delta Q_2 = -\delta Q_{1,} - \delta Q_3 \quad \delta Q_5 = \delta Q_{10} - \delta Q_4 - \delta Q_6 \quad \delta Q_7 = \delta Q_4 - \delta Q_1$$

$$\delta Q_8 = \delta Q_1 + \delta Q_3 + \delta Q_{10} - \delta Q_4 - \delta Q_6 \quad \delta Q_9 = \delta Q_6 - \delta Q_3 \tag{2.202}$$

Let us assume that every branch has a certain resistance R_l, where l denotes the number of this branch and supply voltages in branches 1,2,3 are presented along with their sense in Fig. 2.25. The virtual work in the coordinates of primary description, i.e. with the use of branch currents and voltages is equal to

$$\delta A = \sum_{l=1}^{3} (e_l - R_l \dot{Q}_l) \delta Q_l + \sum_{l=4}^{9} (-R_l \dot{Q}_l) \delta Q_l + (-e_{10} - R_{10} \dot{Q}_{10}) \delta Q_{10}$$

Since transformational formulae (2.201-2.202) are linear, the virtual work of this system can be easily obtained by direct substitution. After substitution we have:

$$\delta A = (r_1) \delta Q_1 + (r_3) \delta Q_3 + (r_4) \delta Q_4 + (r_6) \delta Q_6 + (r_{10}) \delta Q_{10}$$

Since virtual displacements (2.199) are independent, each of the terms in parentheses corresponds to the right-hand side of Lagrange's equation for a k-th generalized coordinate. Assuming at the current stage of the study that the system does not account for passive elements but sources and resistances, we obtain $\delta A = 0$. From the independence of δQ_k, $k = 1,...,s_e$ it results that all terms in r_k have to be equal to zero. The result forms the statement of the equations of motion for the system of an electrical network given in the diagram in Fig. 2.25. The equations are found below:

$$
\begin{aligned}
r_1: \quad & e_1 - e_2 - i_1(R_1 + R_2 + R_7 + R_8) - i_3(R_2 + R_8) + \\
& + i_4(R_7 + R_8) + R_8(i_6 - i_{10}) = 0 \\
r_3: \quad & e_3 - e_2 - i_3(R_2 + R_3 + R_8 + R_9) - i_1(R_2 + R_8) + \\
& + i_6(R_8 + R_9) + R_8(i_4 - i_{10}) = 0 \\
r_4: \quad & -i_4(R_4 + R_5 + R_7 + R_8) + i_1(R_7 + R_8) + \\
& + (R_5 + R_8)(i_{10} - i_6) + R_8 i_3 = 0 \\
r_6: \quad & -i_6(R_5 + R_6 + R_8 + R_9) + i_3(R_8 + R_9) + \\
& + (R_5 + R_8)(i_{10} - i_4) + R_8 i_1 = 0 \\
r_{10}: \quad & -e_{10} - i_{10}(R_5 + R_8 + R_{10}) + (R_5 + R_8)(i_4 + i_6) - \\
& - R_8(i_1 + i_3) = 0
\end{aligned}
\tag{2.203}
$$

These equations apply the traditional notation $i = \dot{Q}$ for the electric current. The verification of the equations in (2.203) is quite problematic. However, one can easily note the symmetry of the network and the corresponding selection of generalized variables. This symmetry can be perceived in equations (2.203), which confirms the correctness of the equations gained. It is sufficient to substitute the subscripts $1 \rightarrow 3$, $4 \rightarrow 6$, $7 \rightarrow 9$ in equations r_1, r_3 and a change in their positions will be followed by an unchanged result. In a similar manner we can undertake the procedure in r_4, r_6. Equation r_{10} can also be transformed in a similar manner while remaining in the same form.

2.2.3 Co-energy and Kinetic Energy in Magnetic Field Converters

In electromagnetic converters we have to do with low frequencies of electric current and, as a result, with slowly variable electromagnetic fields. This enables one to consider fields as stationary and separately consider each of the field's component, i.e. magnetic field and electric one. The source of the first one is the vector potential **A** relative to electric currents, while the source of the latter one is the scalar potential ϕ depending on electric charges.

2.2.3.1 Case of Single Nonlinear Inductor

Let us first consider an individual inductor with the current $\dot{Q} = i$, which forms the source of a magnetic linkage Ψ with its coils (Fig. 2.26). Departing from the instantaneous power supplied to the coil

$$
P = e\dot{Q}
$$

where $e = \dfrac{d\Psi}{dt}$ - is electromotive force induced by magnetic field on the coil, we obtain the delivered power as the integral of the instantaneous power

$$T_e = \int\limits_0^t P\, d\tilde{t} = \int\limits_0^t e\dot{Q}\, d\tilde{t} = \int\limits_0^t \frac{d\Psi}{d\tilde{t}}\dot{Q}\, d\tilde{t}$$

The above expression indicates that energy of field T_e is not relative to time; however, it is relative to the current values of variables Ψ and $\dot{Q}$. Hence, it is a function of the state. In the examined integral it is possible to undertake a change of the variables, which gives

$$T_e = \int\limits_0^\Psi \dot{Q}\, d\tilde{\Psi} \tag{2.204}$$

where $\tilde{\Psi}$ is an independent variable and $\dot{Q} = \dot{Q}(\tilde{\Psi})$.

Subsequently, we have to determine Lagrange's function for the examined inductor. This function is represented by the integral of magnetic field momentum with respect to time. The momentum of the field for a single charged particle in accordance with (2.182) is equal to $Q\mathbf{A}$. For the charges massively moving in a conductor it is possible to derive the notion of the density of electric current $\mathbf{j}[A/m^2]$ and the Lagrange's function is expressed as

$$L_e = \int\limits_V \left(\int\limits_0^j \mathbf{a}(\tilde{\mathbf{j}})\,d\tilde{\mathbf{j}} \right) dV = \int\limits_V \mathbf{A}(\mathbf{j})\mathbf{j}\, dV \tag{2.205}$$

The formula in (2.205) is also relevant with regard to non-linear environments with electromagnetic field since the expression for vector potential $\mathbf{A}(\mathbf{j})$ does not assume the superposition from the currents in space V. This potential involves the magnetic properties of a material in space, so that $rot(\mathbf{A}(\mathbf{j})) = \mathbf{B}$, in contrast to vacuum (2.185), where vectors $\mathbf{B}$, $\mathbf{H}$ are linearly dependent. It was indicated in more detail in [16] that this vector based approach can lead to the discreet model of Lagrange's function as a result of the following course of reasoning

$$\int\limits_V \mathbf{A}(\mathbf{j})\mathbf{j}\, dV = \int\limits_S \left(\oint\limits_l \mathbf{A}(\mathbf{j})\,dl \right) \mathbf{j}\, ds = \int\limits_S \Psi(\mathbf{j})(\mathbf{j}\, ds)$$

where S denotes the internal cross-section of the conductor which carries the electric current. As a result, we obtain

$$L_e = T_e' = \int\limits_0^{\dot{Q}} \Psi(\tilde{Q})\, d\tilde{Q} \tag{2.206}$$

This Lagrange's function is equal to the co-energy of the magnetic field T_e'. From the relation (2.206) it results that in the discreet system the magnetic linkage Ψ plays the role of magnetic field momentum and, hence, forms an equivalent of the term for magnetic momentum of a charged particle (2.182). The energy and co-energy of the magnetic field sum up to form a rectangle just as the case of the motion of a particle (2.179). An illustration of this is found in Fig. 2.26.

$$T_e' + T_e = \Psi \dot{Q} \qquad (2.207)$$

The momentum of the magnetic field (magnetic linkage) is the function of the current $\dot{Q}$ and can be determined with the use of a parameter called inductance coefficient

$$\Psi = M(\dot{Q})\dot{Q} \qquad (2.208)$$

where: $M(\dot{Q}) = \dfrac{\partial \Psi}{\partial \dot{Q}}$ - is an incremental inductance coefficient.

The case is similar for that of the mechanical momentum of a particle, which can be determined using a parameter – particle's mass. For the case of the linear magnetization characteristics the inductance coefficient is constant. As a result, $\Psi = M\dot{Q}$ and the energy of the magnetic field is equal to:

$$T_e = \int \dot{Q}\, d(M\dot{Q}) = \tfrac{1}{2}M\dot{Q}^2$$

Thus, co-energy is: $\qquad\qquad\qquad T_e' = \int \Psi\, d\dot{Q} = \tfrac{1}{2}M\dot{Q}^2 \qquad (2.209)$

and the relation $T_e' = T_e$ is fulfilled.

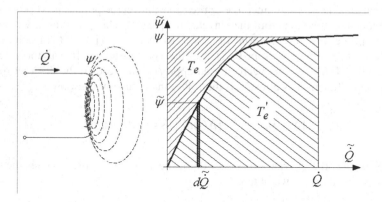

Fig. 2.26 Illustration of the relation between the energy T_e and co-energy T_e• of the magnetic field for an inductor with non-linear magnetization curve

One can note at this point that there is a complete analogy to momentum, energy and kinetic co-energy in mechanics for the case of a particle in motion. The non-linear case in mechanics takes place for sufficiently high velocities (2.171), (2.176-2.179), where the mass m is considerably relative to the velocity v.

2.2.3.2 The Case of a System of Inductors with Magnetic Field Linkage

A very large number of electromechanical devices such as electrical machines and transformers have a number of magnetically linked windings. This is associated with the much more effective transformation of energy for the case of application of multi-phase systems in electric network and machines. Such linkages do not play a small, or even a marginal role; however, it is their function to decide on the operating principle of such devices. Hence, it is essential to take them sufficiently into consideration in the mathematical models and during the design of such systems. In the generalized case magnetic linkage is associated with all windings, which can be presented in an abstract form in Fig. 2.27. The calculation of energy or co-energy for such a linked system usually poses a difficult task and is often conducted using programs for electromagnetic calculations in non-linear environments, which are based on finite elements method or edge variables method. Efficient dedicated 2D and 3D programs have been developed to handle both methods beside accessible free and trial software. Such programs make it possible to determine the field in the form of a spatial distribution of vectors of magnetic potential **A**, field intensity **H**, or the vector of magnetic induction **B**. It is also possible to obtain the integrated parameters such as co-energy and energy of a field, inductance of the windings and ponderomotoric forces encountered in a system.

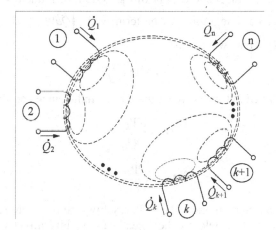

Fig. 2.27 System of n magnetically linked inductors

In the examined case we have to do with the task concerning the statement of generalized forms which enable one to determine the functions of co-energy and energy of a magnetic field in the function of generalized coordinates and their time derivatives – electric currents. The basis for the derivation of such formulae lies in the fact that energy and co-energy of a field are the functions of the state (2.204) and as such are not relative to the means used to obtain a certain state (temporal functions) but depend on the instantaneous state as determined by variables $\Psi, \dot{Q}$. Hence for simplification of the mathematical statement, in a system of linked inductors the particular currents $\dot{Q}_k$ should be successively raised from zero to the final value, while the remaining variables are being kept at a steady level. This is conducted in succession from inductor 1 to the last one, which is denoted as n, while the current variable is denoted with a sign placed above its name $\tilde{\dot{Q}}$ in order to distinguish it from the remaining variables.

$$T_e' = \int_0^{\dot{Q}_1} \Psi_1(\tilde{\dot{Q}}_1, 0, 0, \ldots, 0) d\tilde{\dot{Q}}_1 + \int_0^{\dot{Q}_2} \Psi_2(\dot{Q}_1, \tilde{\dot{Q}}_2, 0, \ldots, 0) d\tilde{\dot{Q}}_2 + \ldots$$

$$+ \int_0^{\dot{Q}_k} \Psi_k(\dot{Q}_1, \dot{Q}_2, \ldots, \tilde{\dot{Q}}_k, 0, \ldots, 0) d\tilde{\dot{Q}}_k + \ldots + \int_0^{\dot{Q}_n} \Psi_n(\dot{Q}_1, \dot{Q}_2, \ldots, \tilde{\dot{Q}}_n) d\tilde{\dot{Q}}_n$$

This can be restated in a more abbreviated form:

$$T_e' = \sum_{k=1}^n \int_0^{\dot{Q}_k} \Psi_k(\dot{Q}_1, \dot{Q}_2, \ldots, \tilde{\dot{Q}}_k, 0, \ldots, 0) d\tilde{\dot{Q}}_k \qquad (2.210)$$

The calculation of the magnetic energy of the system of inductors may involve a similar procedure, i.e. integration of the term type $\int \dot{Q} d\Psi$ or simply the use of formula in (2.207). Thus, we obtain

$$T_e = \Psi \dot{Q} - T_e' \qquad (2.211)$$

For a system of linked inductors we define mutual inductance coefficients

$$M_{kl} = \frac{\partial \Psi_k}{\partial \dot{Q}_l} \qquad (2.212)$$

while in the linear case : $$M_{kl} = \frac{\Psi_{kl}}{\dot{Q}_l},$$

where: Ψ_{kl} is a component of the linkage Ψ_k resulting from current i_l.

The co-energy of the field in the linear case results directly from relation (2.210) and the definition of inductance coefficients (2.212) and is clearly equal to the magnetic energy of the field

$$T_e' = T_e = \int_0^{\dot{Q}_1} M_{11} \tilde{\dot{Q}}_1 d\tilde{\dot{Q}}_1 + \int_0^{\dot{Q}_2} (M_{21} \dot{Q}_1 + M_{22} \tilde{\dot{Q}}_2) d\tilde{\dot{Q}}_2 + \ldots$$

$$+ \int_0^{\dot{Q}_k} (M_{k1} \dot{Q}_1 + M_{k2} \dot{Q}_2 + \ldots + M_{kk} \tilde{\dot{Q}}_k) d\tilde{\dot{Q}}_k + \ldots \qquad (2.213)$$

By integration we obtain:

$$T_e' = T_e = \tfrac{1}{2}M_{11}\dot{Q}_1^2 + \tfrac{1}{2}M_{22}\dot{Q}_2^2 + \ldots + \tfrac{1}{2}M_{kk}\dot{Q}_k^2 + \ldots + \tfrac{1}{2}M_{nn}\dot{Q}_n^2 +$$
$$+ M_{12}\dot{Q}_1\dot{Q}_2 + M_{13}\dot{Q}\dot{Q}_3 + \ldots \qquad \ldots + M_{1n}\dot{Q}_1\dot{Q}_n +$$
$$+ M_{23}\dot{Q}_2\dot{Q}_3 + \ldots \qquad \ldots + M_{2n}\dot{Q}_2\dot{Q}_n + \qquad (2.214)$$
$$\vdots$$
$$+ \ldots + M_{n-1,n}\dot{Q}_{n-1}\dot{Q}_n$$

which can be restated more briefly as:

$$T_e' = T_e = \tfrac{1}{2}\sum_{k=1}^{n}\sum_{l=1}^{n}M_{kl}\dot{Q}_k\dot{Q}_l \qquad (2.215)$$

In the expression (2.215) for energy and co-energy of a system of linked inductors one can apply the relation $M_{kl} = M_{lk}$. As a result of expansion we obtain: for $k = l \rightarrow \tfrac{1}{2}M_{kk}\dot{Q}_k^2$, while for $k \neq l \rightarrow M_{kl}\dot{Q}_k\dot{Q}_l$, which leads to the correct results such as in (2.214).

2.2.4 Potential Energy in Electric Field Converters

Electric potential energy is associated with quasi-static displacement of an electric charge Q in an electric field in the direction of the rising potential. In accordance with (2.183) this energy for a single particle is equal to $Q\,\phi$, where $\phi(\mathbf{r}, t)$ is the scalar potential of a field. This charge is considered as a small (testing) one as it itself does not affect the scalar potential. The discreet element related to the accumulation of electric energy (electric field) is the capacitor, whose ability to accumulate a charge is characterized by parameter C called the electric capacity. The capacity of a capacitor is expressed by the relation of the accumulated charge to the voltage between the electrodes in a capacitor

$$C = \frac{\partial Q}{\partial U} \qquad (2.216)$$

while for a linear capacitor

$$C = \frac{Q}{U} \qquad (2.217)$$

the capacity is a constant value. The potential energy associated with the charged capacitor can be derived from electric power $P = u\dot{Q}$ supplied during charging:

$$U_e = \int_0^t u\dot{Q}d\tilde{t} = \int_0^t u\frac{dQ}{d\tilde{t}}d\tilde{t} = \int_0^Q u\,d\tilde{Q} \qquad (2.218)$$

From relation in (2.218) it results that the potential energy related to an electric field in a capacitor is the function of the state and depends only to the voltage and charge in that capacitor. For this macroscopic case, in contrast to the elementary charge in an external field, charge Q affects the potential in the charged capacitor as a result of charging. Hence, in (2.218) the relation $u = Q/C(Q)$ is fulfilled.

Thereby, using (2.218) we obtain $U_e = \int\limits_0^Q \dfrac{\tilde{Q}}{C(\tilde{Q})}\, d\tilde{Q}$.

For a linear capacitor this gives a form that is popular and often referred to

$$U_e = \int\limits_0^Q \frac{\tilde{Q}}{C(\tilde{Q})}\, d\tilde{Q} = \tfrac{1}{2}\frac{Q^2}{C} \tag{2.219}$$

2.2.5 Magnetic and Electric Terms of Lagrange's Function: Electromechanical Coupling

It is now possible to determine the electromechanical term L_e of Lagrange's function in an electromechanical system. It consists of the kinetic co-energy T_e' related to magnetic field (2.206) and a term for potential energy U_e of the electric field (2.219) deduced from this term

$$L_e = T_e' - U_e \tag{2.220}$$

The application of Lagrange's function in its electromechanical part is associated with a need of making two remarks. For linear magnetic circuits (without magnetic saturation) the co-energy of the magnetic field T_e' is equal to the energy of the field T_e; hence, the distinction is not necessary. The second remark is that in connection with low frequencies of voltage and current alternation in electromechanical systems we have to do with quasi-stationary fields, which means that the fields are virtually not coupled. As a result, we have to do with the magnetic term of the Lagrange's function, the electric term, while the electromagnetic term associated with the coupled fields is absent.

Electromechanical coupling. The ability of electromechanical converters to convert energy occurs as a result of electromechanical couplings. This means that in equations for electrical circuits there are voltages resulting from mechanical motion, while in equations of motion for mechanical variables (displacement, rotation) there are torques or forces resulting from currents or electric charges. The occurrence of couplings is associated with the fact that co-energies or energies of electromagnetic nature are relative to mechanical variables. Let us consider a device with n coupled windings, whose inductance coefficients are relative to a mechanical variable, for instance angle of rotation φ. This is the case in electrical multi-phase machines

$$T_e = \tfrac{1}{2}\sum_{k=1}^{n}\sum_{l=1}^{n} M_{kl}(\varphi)\dot{Q}_k \dot{Q}_l \tag{2.221}$$

In equations involving electrical variables this results in the addition of another term in the k-th equation

$$\frac{d}{dt}\left(\frac{\partial T_e}{\partial \dot{Q}_k}\right) = \frac{d}{dt}\left(\sum_{l=1}^{n} M_{kl}(\varphi)\dot{Q}_l\right) =$$

$$= \sum_{l=1}^{n} M_{kl}(\varphi)\ddot{Q}_l + \dot{\varphi}\sum_{l=1}^{n} \frac{\partial M_{kl}(\varphi)}{\partial \varphi}\dot{Q}_l \tag{2.222}$$

in which we have to do with two terms for voltage induced in the winding: the first term comes from variation of the current in time di_l / dt and is denoted as the voltage of transformation, while the other one associated with angular velocity $\dot{\varphi}$ of converter's motion is denoted with the term rotation induced voltage. In the equation for the mechanical variable φ we have to do with the term

$$\frac{\partial T_e}{\partial \varphi} = \frac{1}{2}\sum_{k=1}^{n}\sum_{l=1}^{n}\left(\frac{\partial}{\partial \varphi}M_{kl}(\varphi)\right)\dot{Q}_k\dot{Q}_l \tag{2.223}$$

It means the torque resulting from the interaction of electric currents or, according to a different interpretation, from the effect of the interaction of magnetic field and electric currents. The latter interpretation is quite self-evident when we take into consideration a converter with non-linear characteristics of magnetization. In this case it is necessary to apply Lagrange's function accounting for co-energy T_e' of the magnetic field and in accordance with (2.210) we can calculate:

$$\frac{\partial L}{\partial \varphi} = \frac{\partial T_e'}{\partial \varphi} = \frac{\partial}{\partial \varphi}\left(\sum_{k=1}^{n}\int_0^{\dot{Q}_1}\Psi_k(\varphi,\dot{Q}_1,\dot{Q}_2,...,\tilde{\dot{Q}}_k,0,...,0)d\tilde{\dot{Q}}_k\right) \tag{2.224}$$

In a similar manner, if the capacity of a capacitor in a converter with electric field is relative to a mechanical variable, for example the distance between the electrodes, we have also to do with electromechanical coupling. The electrical energy of a capacitor with a mobile electrode is expressed by formula

$$U_e = \frac{1}{2}\frac{Q^2}{C(x)} \tag{2.225}$$

where x is a mechanical variable.

In this case in the equation for the electric variable we have an electric term:

$$\frac{\partial U_e}{\partial Q} = \frac{Q}{C(x)}$$

as a result of which the voltage is relative to the position x of an electrode. In the equation for a mechanical variable we obtain the following term:

$$\frac{\partial U_e}{\partial x} = \frac{Q^2}{2}\frac{\partial}{\partial x}\left(\frac{1}{C(x)}\right) = \frac{Q^2}{2}\frac{1}{C^2(x)}\frac{\partial C(x)}{\partial x} = \frac{1}{2}U^2\frac{\partial C(x)}{\partial x} \qquad (2.226)$$

which denotes the mechanical force of attraction of the mobile electrode of a capacitor expressed in [N]. This is confirmed by the dimensional analysis of formula (2.226):

$$\left[V^2\frac{F}{m}\right] = \left[V^2\frac{As}{Vm}\right] = \left[\frac{VAs}{m}\right] = \left[\frac{J}{m}\right] = [N]$$

The similar principle applies to the electrostatic voltmeter for example. Electromechanical couplings are encountered in all examined converters since they constitute the idea governing their operation.

2.2.6 Examples of Application of Lagrange's Equations with Regard to Electromechanical Systems

Example 2.11. For an electrical system forming a network presented in Fig. 2.28 we will state differential equations of motion using Lagrange's equation method. The lumped parameters R,L,C in this network are constant.

The elements present in the branches of the network in Fig. 2.28 are numbered in accordance with the numbers of branches in this network. The number of branches amounts to $g = 5$, while the number of nodes $w = 3$. Hence, the number of the degrees of freedom is equal to

$$s = g - w + 1 = 3$$

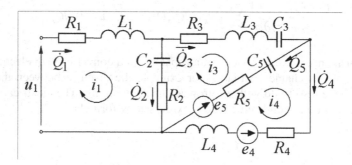

Fig. 2.28 Electrical network with lumped parameters

The vector of the coordinates of the primary description (2.35), i.e. charges Q_i in the particular branches takes the form

$$X = [Q_1, Q_2, ..., Q_5]^T \qquad (2.227)$$

and, by parallel, the vectors of electric currents $\dot{Q}_i$ and virtual charges δQ_i are:

$$\dot{X} = [\dot{Q}_1, \dot{Q}_2, \ldots, \dot{Q}_5]^T \quad \delta X = [\delta Q_1, \delta Q_2, \ldots, \delta Q_5]^T \tag{2.228}$$

The selection of the generalized coordinates can be undertaken in various ways. The selection of currents in the branches with inductances L as generalized velocities leads to the decoupling of the equations of motion with respect to the derivatives of the currents and, as a consequence, we obtain a diagonal matrix of inductance. Given such selection, we have

$$q = [Q_1, Q_3, Q_4]^T$$
$$\dot{q} = [\dot{Q}_1, \dot{Q}_3, \dot{Q}_4]^T \tag{2.229}$$
$$\delta q = [\delta Q_1, \delta Q_3, \delta Q_4]^T$$

The constraints in this system result from the equations for currents established for the Kirchhoff's first law (2.193). Two of them serve for the calculation of transformational formulae (2.196), which in the examined case leads to the relation

$$\dot{Q}_2 = \dot{Q}_1 - \dot{Q}_3 \quad \dot{Q}_5 = \dot{Q}_3 - \dot{Q}_4 \tag{2.230}$$

For the third node, which was not applied for the derivation of transformational formulae for currents (2.230), it is possible to undertake the verification of the formulae by checking whether the balance will be zero as a result of restatement of this formula in the form of currents being generalized velocities. The equations in (2.230) can be integrated and directly lead to the derivation of transformational formulae for electric charges and virtual displacements of the charges:

$$Q_2 = Q_1 - Q_3 \quad \delta Q_2 = \delta Q_1 - \delta Q_3$$
$$Q_5 = Q_3 - Q_4 \quad \delta Q_5 = \delta Q_3 - \delta Q_4 \tag{2.231}$$

Lagrange's function for the examined system takes the form:

$$L = T_e - U_e = \tfrac{1}{2}L_1\dot{Q}_1^2 + \tfrac{1}{2}L_3\dot{Q}_3^2 + \tfrac{1}{2}L_4\dot{Q}_4^2 - $$
$$-\tfrac{1}{2}\frac{(Q_1 - Q_3)^2}{C_2} - \tfrac{1}{2}\frac{Q_3^2}{C_3} - \tfrac{1}{2}\frac{(Q_3 - Q_4)^2}{C_5} \tag{2.232}$$

This Lagrange's function has already been expressed in generalized coordinates using (2.231). The virtual work in the coordinate system of the primary description takes the form:

$$\delta A = (u_1 - R_1\dot{Q}_1)\delta Q_1 + (-R_2\dot{Q}_2)\delta Q_2 + (-R_3\dot{Q}_3)\delta Q_3 + $$
$$+ (-e_4 - R_4\dot{Q}_4)\delta Q_4 + (-e_5 - R_5\dot{Q}_5)\delta Q_5$$

After the application of transformational formulae (2.230-2.231), we obtain:

$$\delta A = \left(u_1 - (R_1 + R_2)\dot{Q}_1 + R_2\dot{Q}_3\right)\delta Q_1 +$$
$$+\left(-e_5 - (R_2 + R_3 + R_5)\dot{Q}_3 + R_2\dot{Q}_1 + R_5\dot{Q}_4\right)\delta Q_3 + \qquad (2.233)$$
$$+\left(e_5 - e_4 - (R_4 + R_5)\dot{Q}_4 + R_5\dot{Q}_3\right)\delta Q_4 = \sum_{k=1,3,4} \tilde{P}_k \delta Q_k$$

The correctness of thus calculated virtual work can be verified by using closed loop currents method. The equations of motion for this electrical network can be formulated in the following manner:

1° $\qquad\qquad$ for $q_1 = Q_1$: $\qquad \dfrac{d}{dt}\dfrac{\partial L}{\partial \dot{Q}_1} - \dfrac{\partial L}{\partial Q_1} = \tilde{P}_1$

$$\frac{d}{dt}\left(L_1\dot{Q}_1\right) + \frac{Q_1 - Q_3}{C_2} = u_1 - (R_1 + R_2)\dot{Q}_1 + R_2\dot{Q}_3$$

or: $\qquad\qquad L_1\dfrac{di_1}{dt} + \dfrac{Q_1 - Q_3}{C_2} = u_1 - (R_1 + R_2)i_1 + R_2 i_3$

2° Similarly, for: $\qquad q_2 = Q_3$: $\qquad \dfrac{d}{dt}\dfrac{\partial L}{\partial \dot{Q}_3} - \dfrac{\partial L}{\partial Q_3} = \tilde{P}_3 \qquad (2.234)$

$$L_3\frac{di_3}{dt} - \frac{Q_1 - Q_3}{C_2} + \frac{Q_3}{C_3} + \frac{Q_3 - Q_4}{C_5} =$$
$$= -e_5 - (R_2 + R_3 + R_5)i_3 + R_2 i_1 + R_5 i_4$$

3° and for: $\qquad\qquad q_3 = Q_4$: $\qquad \dfrac{d}{dt}\dfrac{\partial L}{\partial \dot{Q}_4} - \dfrac{\partial L}{\partial Q_4} = \tilde{P}_4$

$$L_4\frac{di_4}{dt} - \frac{Q_3 - Q_4}{C_5} = e_5 - e_4 - (R_4 + R_5)i_4 + R_5 i_3$$

These equations should be complemented by the following defining equations:

4° - 6° $\qquad\qquad\qquad \dot{Q}_1 = i_1 \quad ; \quad \dot{Q}_3 = i_3 \quad ; \quad \dot{Q}_4 = i_4 \qquad\qquad (2.235)$

Let us further assume that between the inductors L_1 and L_3 there is a magnetic linkage in the form drafted in Fig. 2.29.

This is a negative linkage described by mutual inductance $-M_{13}$. This means that the current in the first and third branch in the direction of current flow marked with an arrow leads to the decrease of the energy of the magnetic field. Accordingly, the new form of Lagrange's function $\tilde{L}$ is the following

$$\tilde{L} = L - M_{13}\dot{Q}_1\dot{Q}_3$$

where L is the Lagrange's function (2.232). As a consequence, the equations 1° and 2° in the system in (2.234) have changed:

$$1° \quad L_1\frac{di_1}{dt} - M_{13}\frac{di_3}{dt} + \frac{Q_1-Q_3}{C_2} = u_1 - (R_1+R_2)\dot{Q}_1 + R_2\dot{Q}_3$$

$$2° \quad L_3\frac{di_3}{dt} - M_{13}\frac{di_1}{dt} - \frac{Q_1-Q_3}{C_2} + \frac{Q_3}{C_3} + \frac{Q_3-Q_4}{C_5} = \quad\quad (2.236)$$

$$= -e_5 - (R_2+R_3+R_5)i_3 + R_2i_1 + R_5i_4$$

The resulting equations are linked by their derivatives whereas the remaining equations in the system (2.234-2.235) remain unchanged. Figs. 2.30 and 2.31 present examples of waveforms for this network supplied with constant voltage of $u_1 = 80\,[V]$, $e_4 = 40\,[V]$ and $e_5 = 20\,[V]$. The values of the parameters are as follows:

$$R_1 = 10\,[\Omega] \quad R_2 = 20\,[\Omega] \quad R_3 = 5\,[\Omega] \quad R_4 = 20\,[\Omega] \quad R_5 = 40\,[\Omega]$$
$$L_1 = 10\,[mH] \quad\quad L_3 = 20\,[mH] \quad\quad L_4 = 80\,[mH]$$
$$C_2 = 20\,[\mu F] \quad\quad C_3 = 50\,[\mu F] \quad\quad C_5 = 30\,[\mu F]$$

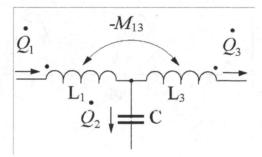

Fig. 2.29 Magnetic linkage of windings L_1 and L_3 in the network in Fig. 2.28

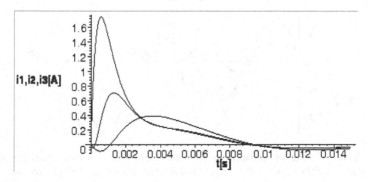

Fig. 2.30 Curves of electric currents for the network in Fig. 2.28 under constant supply voltage, switched under zero initial conditions

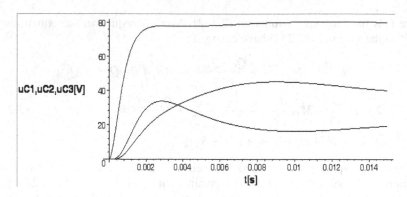

Fig. 2.31 Curves of voltages on capacitors for network in Fig 2.28 under constant supply voltages

Following Figs. 2.32 and 2.33 present the curves for switching under alternating voltage u_1 with the amplitude of $U_1 = 80[V]$, $f = 50$ [Hz], $\varphi_0 = 0.36$, while the remaining voltages are constant $e_4 = 40$ [V], and $e_5 = 20$ [V] and the circuits' parameters remain unchanged.

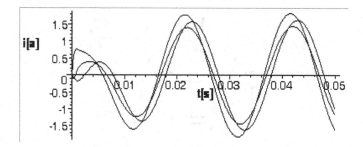

Fig. 2.32 Curves of electric currents for network in Fig 2.28 after switching alternating voltage u_1, while the remaining voltages are constant and circuits' parameters unchanged

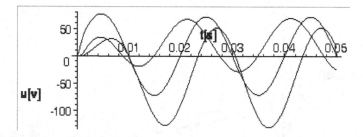

Fig. 2.33 Curves of voltages on capacitors for network in Fig. 2.28 after switching alternating voltage u_1 while the remaining voltages are constant and circuits' parameters unchanged

Example 2.12. The examined electromechanical system is a symmetrical cylinder of mass m with a coil wound around it. This cylinder is rolling down along an inclined plane with an angle of α. The coil consists of z turns and is fused in contact into the side surface of the cylinder in a way that does not interfere with potential motion of the cylinder. The cylinder is made of non-magnetic material whose magnetic permeability is equal to that of the air. The motion occurs in a constant magnetic field, whose induction vector **B** is perpendicular to the base of the plane. First, we will establish the equations of motion for the system, whose kinematic diagram is presented in Fig. 2.34.

Let us assume that the motion is slipless and the axis of the cylinder for the duration of the motion is parallel to y axis which goes along the horizontal edge of the plane. This system is characterized by the following parameters:

m - mass of cylinder including coil
J - cylinder's moment of inertia in relation to the central axis
r - cylinder's radius
l - length of cylinder along the axis
R - resistance of the winding
L_w- inductance coefficient of the winding
z - number of the coil's turns
α - angle of plane's inclination

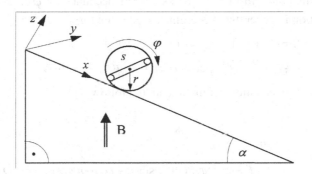

Fig. 2.34 Cylinder with coil wound around it on an inclined plane under constant magnetic field

Under such assumptions the constraints imposed on the system are holonomic and the system has two independent virtual displacements:

$\delta\varphi$ - for the rev olute motion of a cylinder
δQ - for the electric charge in an coil

From the above we can derive generalized coordinates in the form

$$\mathbf{q} = (\varphi, Q) \tag{2.237}$$

The virtual work of the system is

$$\delta A = (-D\dot{\varphi})\delta\varphi + (-R\dot{Q})\delta Q$$

and the kinetic mechanical energy is

$$T_m = \tfrac{1}{2}m\dot{x}^2 + \tfrac{1}{2}J\dot{\varphi}^2 \tag{2.238}$$

Kinetic co-energy of the magnetic field, which is an equivalent to the magnetic energy due to linearity of the environment, has two terms and just as the field is a superposition of two fields: one from the current $\dot{Q}$ in the coil and the external field with constant flux density **B**.

$$T_e' = T_e = \tfrac{1}{2}L_w\dot{Q}^2 + \dot{Q}\Psi_z(\varphi) = \tfrac{1}{2}L_w\dot{Q}^2 + \dot{Q}\Psi_{zm}\cos(\varphi - \varphi_0) \tag{2.239}$$

where: $\Psi_z(\varphi)$ - denotes the flux associated with the coil originating from the external field. The potential energy U_m is associated with the mass of the cylinder in the gravitational field

$$U_m = mgh(\varphi) = mg(h_0 - \varphi r \sin\alpha) \tag{2.240}$$

The above relation contains the following terms:

φ_0 - initial value of the angle of revolution

h_0 - initial elevation of the cylinder's center of mass S above the mount

$\Psi_{zm} = 2Blrz$ - maximum value of the external field linked with the coil.

After transformational formulae are taken into account $x = \varphi r$, $\dot{x} = \dot{\varphi} r$, Lagrange's function in generalized coordinates takes the form:

$$L = (T_e + T_m) - U_m = \tfrac{1}{2}(J + mr^2)\dot{\varphi}^2 + \tag{2.241}$$
$$+ \tfrac{1}{2}L_w\dot{Q}^2 + \dot{Q}\Psi_{zm}\cos(\varphi - \varphi_0) - mg(h_0 - \varphi r \sin\alpha)$$

Hence, the resulting equations of motion are the following

1° $q_1 = \varphi$: $\dfrac{d}{dt}\dfrac{\partial L}{\partial \dot{\varphi}} - \dfrac{\partial L}{\partial \varphi} = D\dot{\varphi}$

$$(J + mr^2)\ddot{\varphi} + \dot{Q}\Psi_{zm}\sin(\varphi - \varphi_0) - mgr\sin\alpha = -D\dot{\varphi}$$

hence: $\ddot{\varphi} = \dfrac{-\dot{Q}\Psi_{zm}\sin(\varphi - \varphi_0) + mgr\sin\alpha - D\dot{\varphi}}{J + mr^2}$

2° $q_2 = Q$: $\dfrac{d}{dt}\dfrac{\partial L}{\partial \dot{Q}} - \dfrac{\partial L}{\partial Q} = -R\dot{Q}$ (2.242)

$$\dfrac{d}{dt}\left(L_w\dot{Q} + \Psi_{zm}\cos(\varphi - \varphi_0)\right) = -R\dot{Q}$$

$$\dfrac{d}{dt}i_w = \dfrac{\Psi_{zm}}{L_w}\dot{\varphi}\sin(\varphi - \varphi_0) - T_w i_w$$

where: $i_w = \dot{Q}$ $T_w = R/L_w$

The expression: $$M_e = -i_w \Psi_{zm} \sin(\varphi - \varphi_0) \qquad (2.243)$$

denotes electromagnetic torque which brakes the cylinder during its rolling motion down the plane. The mean value of this torque is different from zero since the current i_w in the coil is the alternating current. It is possible to select Ψ_{zm} in such a way that the mean velocity Ω_{av} is constant and in that case the time function of the current i_w is periodical, as resulting from (2.242). Given an approximated curve of the current i_w in the steady state in the form

$$i_w = \sum_k i_{mk} \sin(\Omega_{av} kt + \theta_k)$$

and for the rotation angle $\varphi = \Omega_{av} t + \varphi_0$, it is possible to determine electromagnetic torque acting upon the cylinder from the formula (2.243)

$$M_e = -\tfrac{1}{2}\Psi_{zm} *$$
$$* \left\{ \sum_k i_{mk} \left(\cos(\Omega_{av}(k-1)t + \theta_k + \varphi_0) - \cos(\Omega_{av}(k+1)t + \theta_k - \varphi_0) \right) \right\} \qquad (2.244)$$

The basic components of this torque for $k = 1$ are the constant torque and the variable one corresponding to the double of the mean angular velocity:

$$M_{e|k=1} = -\tfrac{1}{2}\Psi_{zm} i_{m1} \left(\cos(\theta_1 + \varphi_0) - \cos(2\Omega_{av} t + \theta_1 - \varphi_0) \right)$$

For the mathematical model of the system presented in (2.242) the parameters have been calculated for a hollow cylinder with the density of $\rho = 1.2$ [kg/dcm^3]. The basic dimensions of the structure are presented in Fig. 2.35. The parameters of the system are following:

$S_{Cu} = 1.5$ [cm^2] - cross-section of the winding
$m = 7.0$ [kg]   - mass of the cylinder
$z = 150$ - number of turns in the coil
$R = 4.5$ [Ω]   - resistance of the winding
$L_w = 0.04$ [H]   - inductance of the winding
$J = 0.032$ [Nms^2] - central moment of inertia of the cylinder
$J_z = J + mr^2$ [Nms^2] - moment of inertia of the cylinder in respect to tangency line of rolling motion
$r = 0.075$ [m]   - radius of the cylinder
$cz = 12.42$ [Wb/T] - constant value which relates flux with flux density
$\alpha = 0.3$ - angle of plane's inclination

The illustrations of the results of computer simulations for this system are presented in Figs. 2.36-2.43. They involve two types of motion. For the first set (Figs. 2.36-2.39) the motion takes place under a strong magnetic field, i.e. for $\mathbf{B} = 0.25$ [T] and is stationary in the sense of possessing periodic characteristics.

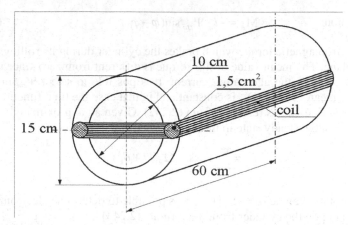

Fig. 2.35 Schematic diagram of cylinder containing coil

However, it is clearly non-uniform in the function of the angle of revolution and the current in the winding differs much from the sinusoidal shape due to the considerable effect of the angular velocity, which is periodic but far from steady. Under a weaker field (**B** = 0.12 [T]) the motion is much faster but also stationary, velocity has a variable term with a lower value and the current presents a curve more reminding a sinusoid.

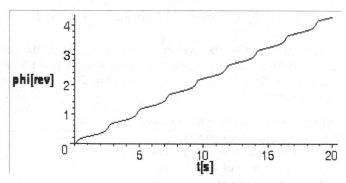

Fig. 2.36 Angle of cylinder revolution for flux density **B** = 0.25[T]

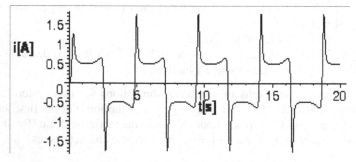

Fig. 2.37 Current curve in coil for flux density **B** = 0.25[T]

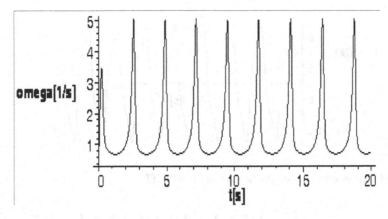

Fig. 2.38 Angular velocity of cylinder's motion for flux density **B** = 0.25[T]

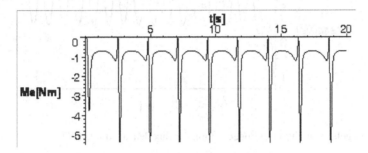

Fig. 2.39 Electromagnetic torque decelerating cylinder's motion for flux density **B** = 0.25[T]

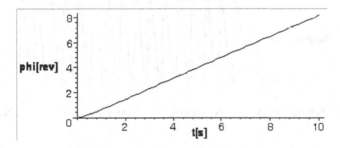

Fig. 2.40 Angle of cylinder revolution for flux density **B** = 0.12[T]

Example 2.13. Dynamics of a contactor with electromechanical drive. Fig. 2.44 presents a model of a contactor with non-linear magnetization characteristics. It consists of an electromagnet, movable jumper of mass m, a system of constant springs with stiffness k_1, coefficient of viscous damping D_1, and springs and dampers acting within the range of small width of the gap x for decelerating and

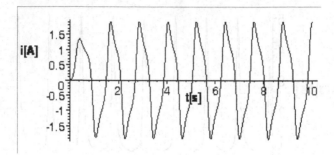

Fig. 2.41 Current curve in coil for flux density **B** = 0.12[T]

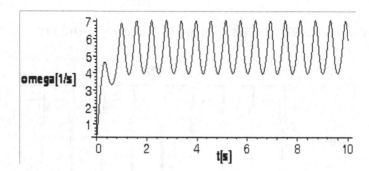

Fig. 2.42 Angular velocity of cylinder motion for flux density **B** = 0.12[T]

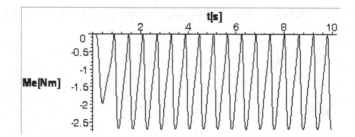

Fig. 2.43 Electromagnetic torque decelerating the motion of cylinder for flux density **B** = 0.12[T]

damping the elastic collision of the jumper against the core. Their stiffness coefficients are defined as k_2 and k_3 and damping ones as D_2 and D_3. An adequate selection of the springs and dampers plays a key role in the securing the correct operation of the electromechanical system of the contactor beside the characteristics of magnetization in the function of the width of the gap x as presented in Fig. 2.45.

This system has two degrees of freedom: one associated with the mechanical motion and the other one for the electric charge in the coil of the winding. Hence, the generalized coordinates follow in the form

$$\mathbf{q} = (x, Q)$$

where: x - is the width of the air gap

Q- is the electric charge in the winding

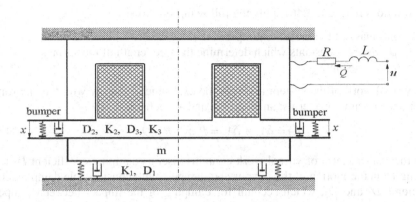

Fig. 2.44 Electromechanical system consisting of a contactor with a movable jumper

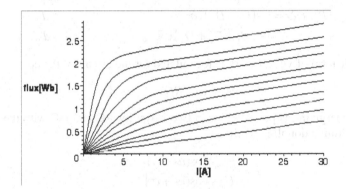

Fig. 2.45 Characteristics of magnetization of contactor's circuit for various values of air gap from the opening to closing of the gap

Lagrange's function for the system involves the kinetic co-energy of the magnetic field, kinetic energy of the moveable jumper and potential energy of the jumper in the gravitational field as well as potential energy of the system of springs given by the function $U_s(x)$.

$$L = \int_{0}^{\dot{Q}} \Psi(\tilde{\dot{Q}}, x) d\tilde{\dot{Q}} + \tfrac{1}{2} m \dot{x}^2 - mg(c - x) - U_s(x) \qquad (2.245)$$

$$U_s(x) = \begin{cases} \frac{1}{2}k_1(x-d_1)^2 & x \le d_1 \\ \frac{1}{2}k_1(x-d_1)^2 + \frac{1}{2}k_2(x-d_2)^2 & x \le d_2 \\ \frac{1}{2}k_1(x-d_1)^2 + \frac{1}{2}k_2(x-d_2)^2 + \frac{1}{2}k_3(x-d_3)^2 & x \le d_3 \end{cases} \qquad (2.246)$$

The relations in (2.245-246) apply the following symbols:

m - mass of the movable jumper
d_1, d_2, d_3 - constants which determine the free length of the springs
c - a constant.

The virtual work of the system involves the exchange of energy with the environment and contains two terms: an electrical and a mechanical one

$$\delta A = \delta A_m + \delta A_e = P_x \delta x + P_Q \delta Q \qquad (2.247)$$

The mechanical term of virtual work contains viscous damping coefficient D_1 resulting from the motion of the jumper and dampers with considerable damping coefficients D_2 and D_3, which account for damping of the impact between jumper and electromagnetic core for small gap width just as for the case of the springs

$$\delta A_m = P_x(x)\delta x = \begin{cases} -D_1 \dot{x}\delta x & x \le d_1 \\ -(D_1 + D_2)\dot{x}\delta x & x \le d_2 \\ -(D_1 + D_2 + D_3)\dot{x}\delta x & x \le d_3 \end{cases} \qquad (2.248)$$

The electric term of virtual work is formulated for the circuit of the coil

$$\delta A_e = (u - R\dot{Q})\delta Q \qquad (2.249)$$

while the energizing is provided for in two ways: by AC and DC with the voltage converted from a doubling rectifier

$$u = \begin{cases} U_{m1}\cos(\omega t + \alpha) \\ U_{m2}|\cos(\omega t + \alpha)| \end{cases} \qquad (2.250)$$

The equation of motion for the mechanical variable takes the form:

$1°$ $\qquad\qquad q_1 = x: \qquad \dfrac{d}{dt}\left(\dfrac{\partial L}{\partial \dot{x}}\right) - \dfrac{\partial L}{\partial x} = P_x(x)$

and after the introduction of Lagrange's function (2.245):

$$\frac{d}{dt}(m\dot{x}) - \int_0^{\dot{Q}} \frac{\partial}{\partial x}\tilde{\Psi}(x,\tilde{Q})d\tilde{Q} - mg + \frac{\partial U_s(x)}{\partial x} = P_x(x)$$

which can be transformed to take the final form:

$$m\ddot{x} = F_e + mg +$$

$$+ \begin{cases} -k_1(x-d_1)-D_1\dot{x} & x \leq d_1 \\ -k_1(x-d_1)-k_2(x-d_2)-(D_1+D_2)\dot{x} & x \leq d_2 \\ -k_1(x-d_1)-k_2(x-d_2)-k_3(x-d_3)-(D_1+D_2+D_3)\dot{x} & x \leq d_3 \end{cases} \qquad (2.251)$$

where:
$$F_e = \int_0^{\dot{Q}} \frac{\partial}{\partial x} \Psi(x,\tilde{\dot{Q}}) d\tilde{\dot{Q}} \qquad (2.252)$$

- is the pull force of the electromagnet.

The equation for variable Q - electric charge results from the general form:

2° $\quad q_2 = Q: \qquad \dfrac{d}{dt}\left(\dfrac{\partial L}{\partial \dot{Q}}\right) - \dfrac{\partial L}{\partial Q} = P_Q$

or: $\qquad \dfrac{d}{dt}\big(\Psi(x,\dot{Q})\big) = u - R\dot{Q}$

After differentiation, we obtain

$$\ddot{Q}\frac{\partial \Psi(x,\dot{Q})}{\partial \dot{Q}} + \dot{x}\frac{\partial \Psi(x,\dot{Q})}{\partial x} = u - R\dot{Q} \qquad (2.253)$$

In the equations of motion we have to do with derivatives of the characteristics of magnetization with respect to the current of the coil $\dfrac{\partial \Psi(x,\dot{Q})}{\partial \dot{Q}}$ and with respect to the width of the air gap $\dfrac{\partial \Psi(x,\dot{Q})}{\partial x}$, which after integration (2.252) corresponds to the pull force of the actuator. In the examined case of the characteristics of magnetization (Fig. 2.45) approximation was carried out analytically using spline functions of the third degree, thanks to which the derivatives according to the coil current are continuous. However, the derivatives of the magnetization characteristics with respect to the width of the gap x are calculated as the differential values $\Delta\Psi / \Delta x$ by linear approximation of the quotients between the particular characteristics. Fig. 2.46 presents the force of jumper pull calculated in accordance with relation (2.252) in the function of the relative width of the gap x, expressed as the per cent of the gap width in the state of complete opening of the contactor, for several values of the current $\dot{Q} = 5,10,\ldots,30\,[A]$. On the basis of the mathematical model with the equations of motion (2.250-2.253) and characteristics of magnetization from Fig. 2.45 a number of simulations was conducted for the examined system of the contactor with electromechanical drive involving shutting of the movable jumper after the voltage is applied. The calculations were performed for the supply of both alternating current as well as the voltage converted from a doubling rectifier, which leads to the flow of direct current in the winding with only a small alternating component. The appropriate operation of a contactor is associated with the necessity of an adequate selection of supply voltages and parameters regarding both springs and dampers. For the supply of the contactor with alternating voltage the amplitude during switching it on is $U_{m1} = 230$ [V], while after the jumper has shut, the value increases to $U_{m1} = 400$ [V] in order to maintain the closure state.

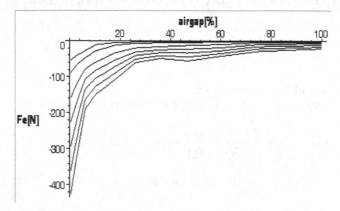

Fig. 2.46 Characteristics of pull force of contactor in the function of relative width of air gap for constant values of coil current $\dot{Q} = 5,10,\ldots,30$ [A]

For the case of supply with rectified voltage the situation is reversed: for the switching on of contactor the amplitude of alternating voltage undergoing rectification is equal to $U_{m2} = 32$ [V] while after the shutting of the jumper is reduced by a half. The various values of the supply voltages are necessary to preserve the appropriate operation of the contactor during the shutting of the jumper and subsequently to keep it in the closed state. They result from the required flux density in the air gap and an indispensable value of the current necessary to initiate the fast motion of the jumper and subsequently to its stable maintenance in the closed state, overcoming strong springs' push, with only small power losses.

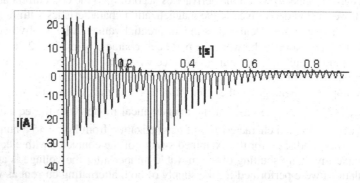

Fig. 2.47 Current waveform for powering on contactor with alternating voltage 230 [V] during shutting of the jumper and switching voltage to 400 [V] after shutting

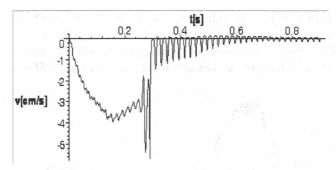

Fig. 2.48 Velocity curve of the jumper for powering on contactor with alternating voltage 400 [V] during shutting of jumper and subsequent switching of voltage to 400 [V] after shutting

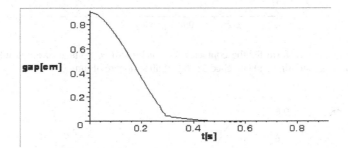

Fig. 2.49 Position of the jumper (width of air gap) for powering on contactor with alternating voltage for conditions presented in Figs. 2.47 and 2.48

The presented illustrations adequately characterize the operation of a contactor under the supply of alternating voltage and indicate a need to increase the voltage after shutting of the jumper in order to ensure the sufficient force for the maintenance of the closed position under small oscillations of the position.

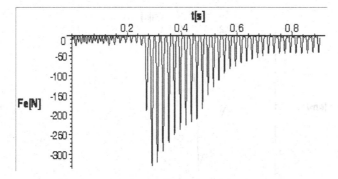

Fig. 2.50 Electromagnet's pull-in force for powering on contactor with alternating voltage 230 [V] during jumper shutting and switching voltage to 400 [V] after shutting

The instant of switching the voltage to a higher one is indicated by all presented curves; however, it is most clearly discernible for the case of the curves of current, velocity of the jumper and the force of electromagnetic pull. The case of the application of rectified voltage to the contactor is illustrated in Figs. 2.51 – 2.54.

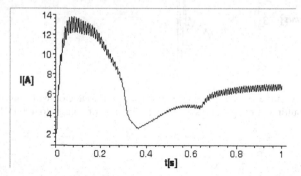

Fig. 2.51 Current waveform for the contactor to which rectified voltage of amplitude 32 [V] was applied during shutting and switched to 16 [V] after closure of jumper

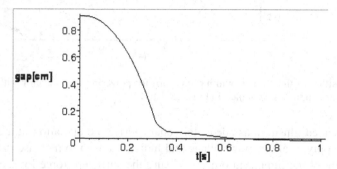

Fig. 2.52 Position of the jumper for powering on contactor with rectified voltage of amplitude 32 [V] during shutting and switched to 16 [V] afterwards

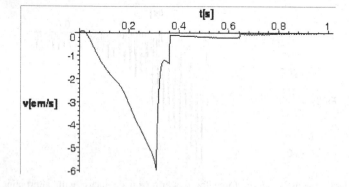

Fig. 2.53 Jumper's velocity for energizing contactor with rectified voltage of amplitude 32 [V] during jumper shutting and switched to 16 [V] after shutting

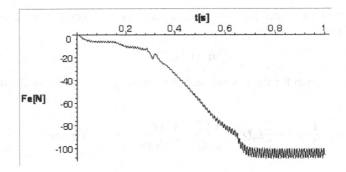

Fig. 2.54 Pull-in force of the electromagnet for energizing contactor with rectified voltage for conditions described as in previous figures

Example 2.14. Parallel plate motion electrostatic actuator. Fig. 2.55 presents the diagram of an electromechanical actuator with a parallel plate motion in a capacitor. In the discussed system the moving electrode of mass m and surface S moves in the vertical direction and is subjected to the force of gravity. The dielectric found between the plates of the capacitor has a dielectric constant ε, which in the calculations regarding the motion of this actuator is equal to the dielectric constant of vacuum ε_0. The circuit of the actuator supply includes: resistance R, additional capacitor with the capacity of C and a small inductance of the circuit L_s. The role of the capacity C consists in the stabilization of the motion of the electrode and, more precisely, an increase of the range of the stationary operation of the actuator. Inductance L_s only assumes a small value since it is formed by the inductance of the energizing wires. In addition, studies have shown that the application of appropriate method of integration for stiff differential equations makes it possible to disregard this inductance.

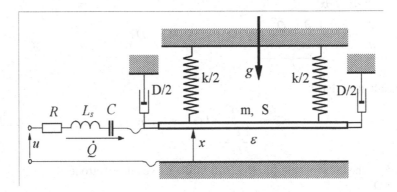

Fig.2.55 Diagram of electrostatic actuator with parallel plate motion in vertical direction

Under the assumption of the full symmetry of the system and resulting parallel motion of the movable electrode, the system has only two degrees of freedom: one for the mechanical motion described by generalized coordinate x, which denotes the distance between the electrodes and the other one for electric charge defined by the generalized coordinate Q. Hence, the vector of generalized coordinates takes the form

$$\mathbf{q} = (x, Q) \qquad (2.254)$$

Lagrange's function for the system described in the generalized coordinates is defined as:

$$L = \frac{1}{2}m\dot{x}^2 + \frac{1}{2}L_s\dot{Q}^2 - \frac{1}{2}\frac{Q_1^2}{C} - \frac{1}{2}\frac{Q_2^2}{C(x)} - \frac{1}{2}k(x - x_s) - mgx \qquad (2.255)$$

where:

x_s - is the free length of the springs

$Q_1 = Q_2 = Q$ - electric charge of each of the capacitors; electrical generalized coordinate

$C(x) = \dfrac{\varepsilon S}{x}$ - capacity of a capacitor with moveable electrode (actuator)

$C_0 = C(x_0) = \dfrac{\varepsilon S}{x_0}$ - nominal capacity of the actuator for $x = x_0$.

The virtual work of the system, which realizes the exchange of energy with the environment for the examined system, is equal to

$$\delta A = (u - R\dot{Q})\delta Q + (-D\dot{x})\delta x \qquad (2.256)$$

The differential equations of motion of the system are following:

for $q_1 = x$

$$\frac{d}{dt}\frac{\partial L}{\partial \dot{x}} - \frac{\partial L}{\partial x} = P_x$$

1°

$$\frac{d}{dt}(m\dot{x}) + \underbrace{\frac{1}{2}\frac{\partial}{\partial x}\left(\frac{Q^2}{C(x)}\right)}_{F_c} + k(x - x_s) + mg = -D\dot{x}$$

As a result, this gives:

$$m\ddot{x} = -\underbrace{\frac{1}{2}\frac{Q^2}{\varepsilon S}}_{F_c} - k(x - x_s) - mg - D\dot{x} \qquad (2.257)$$

where F_c is the electrostatic force of pull of the moveable electrode;

for $q_2 = Q$

$$\frac{d}{dt}\frac{\partial L}{\partial \dot{Q}} - \frac{\partial L}{\partial Q} = P_Q$$

2°

$$\frac{d}{dt}(L_s\dot{Q}) + Q\left(\frac{1}{C} + \frac{1}{C(x)}\right) = u - R\dot{Q}$$

which yields:

$$L_s\ddot{Q} + Q\frac{1}{C_0}\left(\frac{C_0}{C} + \frac{C_0}{C(x)}\right) = u - R\dot{Q}$$

Following the statement that: $C_0/C = \alpha$ and by assuming that

$$\frac{C_0}{C(x)} = \frac{x}{x_0} \qquad (2.258)$$

one obtains

$$L_s\frac{di}{dt} + \frac{Q}{C_0}\left(\alpha + \frac{x}{x_0}\right) = u - Ri \qquad (2.259)$$

For the case when we assume that $L_s = 0$, the equation of motion for the electric circuit takes the following form:

$$\frac{dQ}{dt} = \frac{Q}{T_c}\left(\alpha + \frac{x}{x_0}\right) + \frac{u}{R} \qquad (2.260)$$

where: $\qquad \dfrac{dQ}{dt} = i \qquad T_c = RC_0$

Stationary state, stability of the system. Under the assumption that the stationary state exists for $x = const$, $Q = const$, from equations (2.257, 2.259) we can establish the conditions for the stability of the system:

$$F_c = F_s + F_g \qquad (2.261)$$

where: $\qquad F_c = \dfrac{1}{2}Q^2\dfrac{1}{\varepsilon S} \quad ; \quad F_s = -k(x - x_s) \quad ; \quad F_g = -mg$

Concurrently, $\qquad Q = \dfrac{UC_0}{\alpha + z} \qquad z = \dfrac{x}{x_0} \qquad (2.262)$

The examination of the possible solutions to equation (2.261) in the function of the distance between the electrodes and for actuator parameters, such as the supply voltage U and $\alpha = C_0/C$, leads to the graphical representation of forces F_c and F_s in Figs. 2.56 and 2.57. Their points of intersection represent the possible equilibrium points for the system (for $F_g = 0$, which denotes the actuator in the horizontal motion). These are the three possible cases of the actuator's operation range $0 < z \leq 1$ determined by variable $z = x / x_0$. The first case is relevant for the situation when there are two points of intersection of curves representing forces F_c and F_s. The stationary point of operation is the one for which the relation

$$\frac{\partial F_c}{\partial z} > \frac{\partial F_s}{\partial z} \qquad (2.263)$$

is fulfilled, since in such a case the resultant force restores the system to its initial position after it has been put out of balance. This condition is satisfied for the points of the intersection between curves for F_c and F_s, whose location is closer to the value of $z = 1$. In the second case there is just a single point of the intersection of the curves within the area of actuator operation, which is stationary, as it fulfills the condition in (2.263). The third case is encountered when there are no points of intersection between the two curves and the reason for this is too high a voltage applied to the system followed by a complete short circuit between the plates in the actuator. Fig. 2.56 presents static characteristics for $\alpha = 1.2$ and various values of the voltage $U = 40, 50, 60, 64.5, 70$ [V]. The loss of the stability of the operation for actuator takes place for $U = 66$ [V]. Fig. 2.57 presents the curves for a single value of the supply voltage $U = 50$ [V] and ratio of the capacities equal to $\alpha = 0.8, 0.9, 1.0, 1.2, 1.4, 1.6$. For $\alpha = 1.4$ there are two intersection points: a stationary one for $z = 0.6$ and a non-stationary state for $z = 0.12$. In order to examine more precisely the control of actuators by means of changing the supply voltage, that is, in order to determine the characteristics, the equation (2.261) has been transformed using (2.262) to take the form

$$y^3 k + \beta y^2 + U^2 C_0 / 2x_0^2 = 0 \qquad (2.264)$$

where: $\beta = (mg - k (x_s - \alpha x_0))/x_0$ - is a design constant.

This equation (2.264) involves variable y, which is a linear function of the distance x between the electrodes of an actuator and depends also on the ratio α

$$y = z + \alpha = x / x_0 + \alpha \qquad (2.265)$$

The calculation of characteristics and curves for dynamic states that follow have been conducted for actuator with the following parameters:

$k = 0.0001$ [Nm] - for a horizontal motion actuator
$k = 0.1$ [Nm] - for a vertical motion actuator
$m = 0.4$ [g], $x_0 = 5$ [mm], $C_0 = 1.8$ [pF], $S = 10$ [cm^2],
$\varepsilon = \varepsilon_0 = 8.85 \, E\text{-}12$ [C/Vm], $g = 9.81$ [m/s^2]

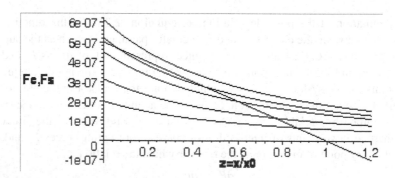

Fig. 2.56 Characteristics of electrostatic force F_c and spring force F_s in the function of the relative distance z between electrodes for $\alpha = 1.2$ and for supply voltages $U = 40,50,60,64.5,70$ [V]. The case of the horizontal motion actuator (without the effect of gravity)

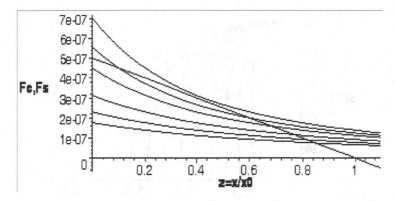

Fig. 2.57 Characteristics of electrostatic force F_c and spring force F_s in the function of the relative distance between the electrodes for supply voltage $U = 50$ [V] and $\alpha = 0.8, 0.9, 1.0, 1.2, 1.4, 1.6$. The case of the horizontal motion actuator

The free length of spring x_s has been selected in such a way that in the initial state under zero voltage the width of the gap is $x = x_0$, hence

$$x_s = x_0 + mg / k \qquad (2.266)$$

The vertical motion of the actuator is a drawback since it is associated with the need to install a relatively stiff spring, whose free length x_s (2.266) should not be ridiculously large in comparison to the travel of the actuator, i.e. in the range $0 < x \leq x_0$, which corresponds to 5 [mm] in the examined case. Hence, the large stiffness of the spring for a vertical motion system results in the need to supply high voltage to control the actuator. The horizontal motion system needs a relatively less tense spring and this results in the considerably lower voltages for the control of the moving electrode.

Fig. 2.58 presents the characteristics of control in the function of voltage U for a system in vertical motion (accounting for gravity, $k = 0.1$ [N/m]) for various values of parameter $\alpha = C_0/C$. The line that goes across the curves is the boundary line of stability of the control. If the boundary is exceeded by the moving electrode a short-circuit with the permanent electrode follows. The next Fig. 2.59 presents similar characteristics for a horizontal motion actuator (with no gravity, $k = 0.0001$ [N/m]). The characteristics are similar, however, for a system in the horizontal motion of the electrode the control voltages tend to be considerably lower.

The two characteristics indicate that the use of a capacitor C put in a series with an actuator results in a beneficial extension of the range of the balanced control and is improved in terms of the precision; however, it is associated with a need of applying higher voltage. The dynamic curves have been modeled on the basis of the equations of motion (2.257-2.259) and solved using Rosenbrock procedure for stiff differential equation systems. Such a need results from the large span of the time constant values encountered in the system, which is associated with the small value of capacity C_0.

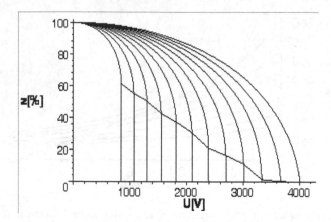

Fig. 2.58 Control characteristics $z = f(U)$ for a capacity vertical motion actuator (subjected to gravity force), for various values of parameter $\alpha = 0.2, 0.4, \ldots, 2.4$. The line illustrating the boundary of the stability is presented

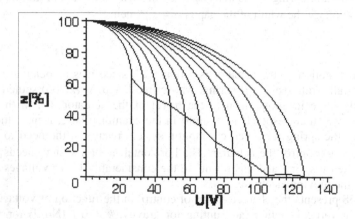

Fig. 2.59 Control characteristics $z = f(U)$ for a capacity actuator as in Fig. 2.58, but for a horizontal motion system which applies a less tense spring

The application of this procedure makes it possible to disregard parameter L_s, which denotes residual inductance in the circuit and, consequently, calculate charge Q of the capacitors from equation (2.260). Thus, one gains very similar curves of the actuator's motion in both cases. The set of curves in Figs. 2.60-2.68 presents the actuator motion operating in a horizontal motion system for a slowly increasing supply voltage $u = U_0 + \Delta u\, t = 50 + 0.025t$ [V], for $\alpha = 1.2$. The loss of stability is recorded for voltage around 70 [V], i.e. for a value similar to the one gained for static characteristics (Fig.2.59), which amounts to $U = 66$ [V].

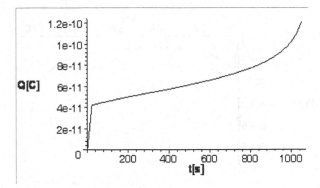

Fig. 2.60 Waveform for charge Q in the function of time for slowly increasing voltage for the horizontal motion actuator

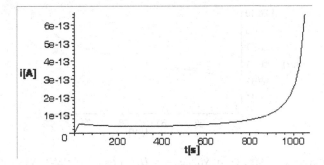

Fig. 2.61 Waveform for current $i = \dot{Q}$ in the function of time for slowly increasing voltage for the horizontal motion actuator

In contrast, one can compare the results for the same actuator for a fast increasing voltage, i.e. $u = U_0 + \Delta u\, t = 50 + 2.5t$ [V], for which case the loss of the stability occurs for the voltage of $U = 160$ [V] after around 50 [s], much above the threshold of static stability.

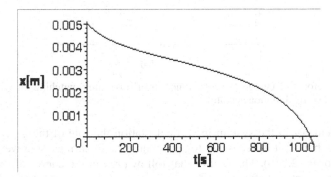

Fig. 2.62 Distance x between electrodes in the actuator in the function of time for slowly increasing voltage for the horizontal motion actuator

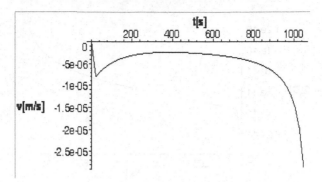

Fig. 2.63 Velocity of motion $\dot{x}$ of moving electrode in the actuator in the function of time for slowly increasing voltage for the horizontal motion actuator

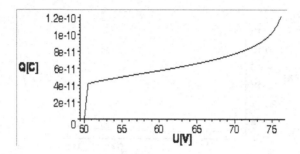

Fig. 2.64 Trajectory of charge Q for voltage $u = U_o + u\ t = 50+0.025\ t$ [V] (slow increase) in the horizontal motion actuator

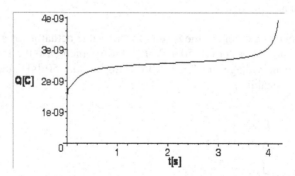

Fig. 2.65 Waveform of charge Q in the function of time after switching on voltage $U = 2100$ [V] in an vertical motion actuator

For the actuator in the system in vertical motion the role of the gravity pull is considerable; hence, it is necessary to apply more tense springs to prevent the use of too long ones (2.266). The figures that follow present the curves of the motion of the moving electrode after an abrupt application of the voltage of $U = 2100$ [V], $\alpha = 1.2$. This value exceeds a little the value of $U = 2091$ [V] determined on the

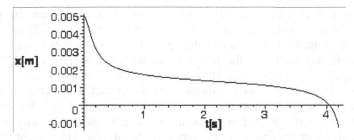

Fig. 2.66 Position x of the moving electrode in the function of time after switching on voltage $U = 2100$ [V] in the vertical motion actuator

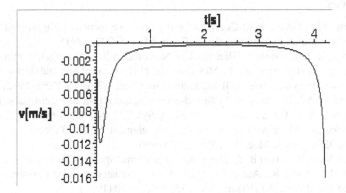

Fig. 2.67 Velocity $\dot{x}$ of the moving electrode in the function of time after switching on voltage $U = 2100$ [V] in the vertical motion actuator

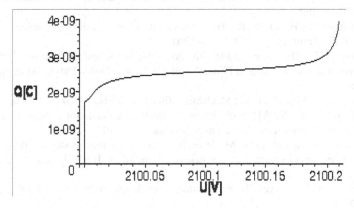

Fig. 2.68 Trajectory of electric charge after switching on voltage $U = 2100$ [V] in the vertical motion actuator and loss of stability of the position of the moving electrode. The boundary of the stability for this case is $U = 2091$ [V]

basis of equation (2.264) as the value of the static stability. In this case we have to do with a fast loss of stability (within 4 [s]) and short-circuit between the electrodes of the actuator.

The presented characteristics and dynamic curves of the electromechanical actuator with variable width of the air gap indicate the extent of the difficulty associated with the control of such an actuator. It tends to operate better in the system with the horizontal motion of the electrode and for stable operation it requires the use of a capacitor in series which affects the range of the control in a very beneficial manner. The cost to cover as a result of the use of such a capacitor with a adequate capacity $\alpha = (1\div1.2\div1.4)$ is associated with the need to supply higher values of the voltage for the control of the actuator's displacement. Capacity actuators play nowadays increasing role in various MicroElectroMechanical Systems (MEMS) like head positioning systems or micro switches etc.[2,4,7,10,14,15].

References

[1] Baharin, I.B., Green, R.J.: Computationally-effective recursive Lagrangian formulation of manipulation dynamics. Int. J. Contr. 54, 195–202 (1991)

[2] Bhushan, B. (ed.): Springer Handbook of Nano-technology. Springer, Berlin (2004)

[3] Bloch, A.M., Reyhanoglu, M., McClamroch, N.H.: Control and stabilization of non-holonomic dynamic systems. IEEE Trans. Aut. Contr. 37, 1746–1750 (1992)

[4] Castaner, L.M., Senturia, S.D.: Speed-energy optimization of electrostatic actuators based on pull-in. IEEE J. Microelectromech. Sys. 8, 290–298 (1999)

[5] Demenko, A.: Movement simulation in finite element model of electric machine dynamics. IEEE Trans. Mag. 32, 1553–1556 (1996)

[6] Featherstone, R.: Rigid Body Dynamics Algorithms. Springer, Heidelberg (2007)

[7] Fujita, H., Suzuki, K., Ataka, M., et al.: A microactuator for head positioning system of hard disc drives. IEEE Trans. Mag. 35, 1006–1010 (1999)

[8] Greenwood, D.: Principles of Dynamics. Prentice-Hall, Englewood Cliffs (1988)

[9] Hung, E.S., Senturia, S.D.: Generating efficient dynamical models for microelectromechanical systems from a finite-element simulation runs. IEEE J. Microelectromech Sys. 8, 280–289 (1999)

[10] Kobayashi, M., Horowitz, R.: Track seek control for hard disc dual-stage servo systems. IEEE Trans. Mag. 37, 949–954 (2001)

[11] Kozłowski, K.: Modeling and Identification in Robotics. Springer, Berlin (1998)

[12] Lanczos, C.: The Variational Principles of Mechanics. Dover Publications, NewYork (1986)

[13] Landau, L.D., Lifschic, E.M.: Mechanika, Elektrodinamika. Nauka, Moskva (1969)

[14] Li, Y., Horowitz, R.: Mechatronics of electrostatic actuator for computer disc drive dual-stage servo systems. IEEE Trans. Mechatr. 6, 111–121 (2001)

[15] McCarthy, B., Adams, G.G., McGruer, N., et al.: A dynamic model, including contact bounce, of an electrostatically actuated microswitch. IEEE J. Microelectromech. Sys. 11, 276–283 (2002)

[16] Neymark, Y.I., Fufayev, N.A.: Dinamika negolonomnych sistem. Nauka, Moskva (1967)

[17] Ortega, R., Loria, A., Nicklasson, P.J., et al.: Passivity-based Control of Euler-Lagrange Systems. Springer, London (1998)

[18] Paul, C.R., Nasar, S.A.: Introduction to electromagnetic fields. McGraw-Hill, New York (1982)

[19] Pons-nin, J., Rodrigez, A., Castaner, L.M.: Voltage and pull-in time in current drive of electrostatic actuators. IEEE J. Microelectromech. Sys. 11, 196–205 (2002)

[20] Siciliano, B., Khatib, O. (eds.): Springer Handbook of Robotics. Springer, Heidelberg (2008)

[21] Sobczyk, T.J.: An energy based approach to modeling the magnetic non-linearity in AC machines. Archives El. Eng. 48, 219–229 (1999)

[22] Su, C., Stepanienko, J.: Robust motion force control of mechanical systems with classical nonholonomic constraints. IEEE Trans. Aut. Cont. 39, 609–612 (1994)

[23] Uhl, T., Bojko, T., Mrozek, Z., et al.: Rapid prototyping of mechatronic systems. J. Theor. Appl. Mech. 38, 645–651 (2000)

[24] White, D.C., Woodson, H.H.: Electromechanical Energy Conversion. John Wiley & Sons, New York (1959)

[25] You, L., Chen, B.: Tracking control designs for both holonomic and nonholonomic constrained mechanical systems. Int. J. Contr. 58, 587–590 (1993)

Chapter 3
Induction Machine in Electric Drives

Abstract. Chapter deals with electrical drives with induction machines. After short introduction concerning basic construction variants and general problems of modeling induction motor drives, mathematical models are developed on a basis of previously introduced Lagrange's method. Models transformed in a classical way are presented by use of orthogonal transformation along with the options governing free parameters of these transformations. Subsequently, other models are developed, in which only one side electrical variables are transformed – stator's or rotor's, while the other side remains untransformed, with natural variables. These models are applied in simulation and presenting various problems of drive systems with electronic power converters and for other external asymmetry cases. Some of them are classical drive problems like DC breaking, operation of Scherbius drive or soft-start systems. Other concern modern systems with electronic power converters, more precisely and successfully modeled and computed in this way. They are drives with two- and three-level voltage source inverters used in PWM control (SVM, DPWM) as well as current source inverters – CSI. These kinds of models are also applied to present vector control (VC) or field oriented control (FOC) and direct torque control (DTC) of induction drives. Abovementioned problems are widely illustrated by examples computed for various dynamic states, with and without automatic control, and for that purpose four induction motors of different rated power are presented. Finally a problem of structural linearization of induction motor drive is covered, beside a number of useful state observer systems that are discussed.

3.1 Mathematical Models of Induction Machines

3.1.1 Introduction

The history of construction and application of induction motors in electric drive dates back as far as over 100 years and the induction machine constitutes the basic unit energized from alternating current in a symmetrical three-phase power distribution system. Beside the three-phase layout some small size machines can be supplied from a single phase for household applications and two-phase machines can be used in the drives of servomechanisms. The practical meaning of three-phase induction machines is emphasized by the fact that they consume nearly 70%

of the generated electric power. The name given to induction motor originates from the single sided power supply to such machines most frequently occurring from the direction of the stator. Inside the rotor there is a fixed winding which generates electric power as a result of induced voltages resulting from magnetic flux that changes (rotates) in relation to the rotor's windings. The mutual interaction of magnetic field in the air gap with the currents in rotor's windings leads to the origin of electromagnetic torque T_e, which sets the rotor in motion. In the induction motor the air gap between the stator and rotor is as narrow as it is technically possible for the requirements of the mechanical structure, since the energy powering the rotor by the magnetic field has a considerably high value. This field should have a high value of flux density in the gap which requires an adequate magnetizing current, that is approximately proportional to the width of the air gap δ. Three-phase induction motors find an application in all branches of industry and in municipal utility management as well as in farming and service workshops. This group of machines involves devices produced within a wide range of power ratings from under 100 [W] and can reach as much as 20 [MW]. The traditional induction machines were applied in drives that do not require the control of rotational speeds. This was due to the cost and problems associated with the use of such control devices while securing the maintenance of high efficiency of transforming electric energy into mechanical one. The course of events has changed considerably over the past 20 years. The increase of accessibility and relative fall in the prices of power electronic switches such as SCRs, GTOs, MOSFETs and IGBTs was followed by continuous development of diverse electronic converters [2,9,10,14,51,69,83,84,86,95]. Their application makes it possible to transform electric power with the parameters of the supply network into variable parameters required at the input of induction motors to meet the needs of the effective control of rotational speed. Such control sometimes known as scalar regulation is discussed in detail in Section 3.3. The following stage in the development of the control systems of induction machines focused on the improvement of the power electronic devices involved in the execution of commands and, in particular, with the development of processors for transferring information, including signal processors adaptable for industrial applications. As a result, it was possible not only to design and implement dynamic control of drives containing induction motors but also develop control that tracks the trajectory of the position and rotor speed. This type of control is encountered mostly in two varieties [13,18,74,76,96,99,100] Field Oriented Control (FOC) and Direct Torque Control (DTC). The methods applied in this respect are discussed in Section 3.4.

3.1.2 Construction and Types of Induction Motors

A typical induction machine is a cylinder shaped machine whose ratio of the diameter to length is in the range of 1.2-0.8. Induction motors are built to meet the requirements of various numbers of phases; however, most commonly they are three-phase machines. The air gap between the stator and the rotor is as small as it is achievable and windings are located in the slots (Fig 3.1) of the stator and rotor. The ferromagnetic circuit is made of a laminated elastic steel magnetic sheets in order to limit energy losses associated with alternating magnetization of the iron during the operation of the machine. Very important role is played by the windings, which are engineered in several basic types. In the stators of high voltage machines they are often made in the form of bars from isolated rectangular shaped conductors formed into coils inserted into the slots of the stator. In this case the slots are open, rectangular and it is possible to insert the ready made rigid coils into them in contrast to the semi-closed slots applied in windings made of coil wire. The material which conducts the current is the high conductivity copper. Another important classification is associated with the single and two layer windings. For the case of single layer winding the side of the coil occupies the entire space available in the slot, while in two layer windings inside the slot there are two sides belonging to two different coils, one above the other, while the sides could belong to the same or different phases of the winding. In machines with higher capacity we usually have to do with windings in two layers. Still another classification of windings in induction motors is associated with integral and fractional slot windings [101,102,103,104]. Integral slot winding is the one in which the number of slots per pole and phase is an integer number. Most induction machines apply integral slot windings since they offer better characteristics of magnetic field in the air gap. Fractional slot windings are used in the cases when the machine is designed in a way that has a large number of poles but it is not justified to apply too large a number of slots in a small cross section. Another reason for the application of fractional slot windings is associated with economic factors when the same ferromagnetic sheets are used for motors with various numbers of pole pairs. In this case for a given number of slots and certain number of pole pairs we have to do with fractional slot windings.

However, the most important role of an engineer in charge of the design of an induction machine is to focus on the development of such a winding whose magnetic field in the air gap resulting from the flow of current through a winding follows as closely as possible a sine curve (Fig 3.2). The windings in the rotors of induction motors are encountered in two various models whose names are adopted by the types of induction motors: slip-ring motor and squirrel-cage motor. The winding in a slip-ring motor is made of coils just as for a stator in the form of a three phase winding with the same number of pole pairs as a winding in a stator and the terminals of phases are connected to slip rings.

With these rings and by adequate butting contact using brushes slipping over the rings it is possible to connect an external element to the windings in a rotor. This possibility is used in order to facilitate the start-up of a motor and in many cases also to control its rotational speed. The squirrel cage forms the other variety of an induction motor rotor's winding that is more common. It is most often made of cast bars made of aluminum or, more rarely of bars made from welded copper alloys placed in the slots. Such bars are clamped using rings on both sides of the rotor. In this way a cage is formed (Fig 3.3); hence, the name squirrel cage was coned. The cage formed in this way does not enable any external elements or supply sources to be connected. It does not have any definite number of phases, or more strictly speaking: each mesh in the network formed by two adjacent bars and connecting ring segments form a separate phase of the winding. Hence, a squirrel

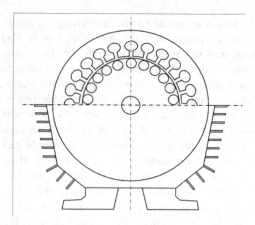

Fig. 3.1 Cross section of an induction machine with semi-closed slots in the stator

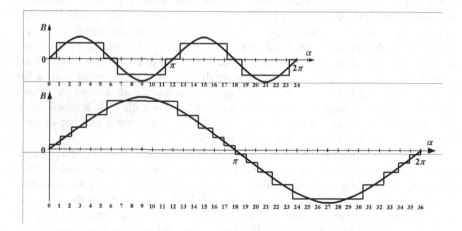

Fig. 3.2 Shape of a magnetic field produced in the air gap of a three-phase induction machine and its fundamental harmonic for a 24 slot stator with the number of pole-pairs $p = 2$ and for a 36 slot stator and $p = 1$

cage winding with m bars in a detailed analysis could be considered as a winding with m phases. Moreover, a squirrel cage winding does not have a defined number of pole pairs. In the most basic analysis of an induction motor one can assume that a squirrel cage winding is a secondary winding that passively adapts in response to the magnetic field as a result of induced voltages and consequently currents. It is possible to further assume that the magnetic field in an air gap with p pole pairs induces in the bars of a cage a system of voltages and currents with p pole pairs as well. Since the number of phases in the rotor is basically arbitrary as the winding is not supplied from an external source this is also a three-phase winding similar to the winding in a stator. Hence, in its basic engineering drawing along the circumference of the stator the magnetic field in the air gap of the induction motor is described by sine curve with p times recurrence during the round of the gap's circumference (Fig 3.2). The difference between the actual shape of the magnetic field in the air gap and the fundamental harmonic of the order $\rho = p$ is approximated by a set of sine curves, forming the higher harmonics of the field, whose spectra and amplitudes can be calculated by accounting for all construction details of the stator and rotor of a machine. The basic reason for the occurrence of higher harmonics of the field in the air gap is associated with the discreetly located conductors in the slots and their accumulation in a small space, the particular span of the coils carrying currents and non-homogenous magnetic permeance in the air gap [80]. This air gap despite having its constant engineering width δ is in the sense of the magnetic permeance relative to the dimensions of the slots in the stator and rotor. The higher harmonics of the magnetic field in the air gap account for a number of undesirable phenomena in induction machines called parasitic phenomena. They involve asynchronous and synchronous parasitic torques that deform the basic characteristics of the electromagnetic torque [80], as well as additional losses resulting from higher harmonics and specific frequencies present in the acoustic signal emitted by the machine.

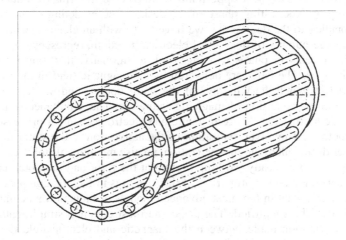

Fig. 3.3 A frequent shape of a squirrel-cage winding of a rotor of induction motor

In the currently manufactured induction motors parasitic phenomena are encountered on a relatively low level and do not disturb the operation of the drive. Hence, in the discussion of the driving characteristics the induction motor is represented by a mathematical model whose magnetic field displays monoharmonic properties. The only harmonic is the fundamental one with the number $\rho = p$, which is equal to the number of pole pairs. The limitation of parasitic phenomena and construction of a machine that is virtually monoharmonic comes as a result of a number of engineering procedures, of which the most basic one involves an appropriate selection of a number of slots in the stator and rotor. The numbers in question are N_s and N_r, respectively and they are never equal to each other and their selection depends largely on the designed number of pole pairs p. Considerable progress has been made in the design and engineering of induction motors over their more than 100 year old history. The measure of this progress not only involves the limitation of parasitic phenomena but also an increase the effectiveness of the structures in terms of the torque rating per kilogram of the machine's mass, long service life, energy efficiency, ecological characteristics, progress in the use of insulation materials, which makes it possible to supply from converters with high frequency and amplitude of voltage harmonics.

3.1.3 Fundamentals of Mathematical Modeling

3.1.3.1 Types of Models of Induction Machines

Mathematical modeling plays a very important role in the design, exploitation and control of electric drives. Modeling and computer simulation, whether with regard to electric drive or in other branches of engineering, that is adequate and effective reduces the time needed and the cost of gaining an optimum design of a drive and its control system. Thus, new opportunities are offered in terms of reducing lead times in the prototype testing phase of the design. The modeling of an induction motor is complex to the degree that we have to do with an electromechanical device with a large number of degrees of electrical freedom, represented by charges and electric currents in phase windings and, additionally, that can account for magnetic linkages. The latter are delivered by the magnetic field in the ferromagnetic material in which the windings of the stator operate and the ferromagnetic core is often in the condition of magnetic saturation. The simultaneous and comprehensive accounting for electromagnetic and electromechanical processes in an induction motor that involves saturation of the active iron in the stator and rotor, energy losses during alternate magnetization, precise mapping of linkages between the windings, the non-steady working regime of the rotor and the potential effect of the heat generated on the properties of the system is in fact too complex and too costly and, hence, even in the most advanced models of induction machines these processes tend to be simplified. The basic and most common simplification consists in the distinction made between the magnetic and electric field due to the small frequencies of the alternation of the field. For that reason, the field is considered to be magnetostatic. Moreover, there is a tendency to simplify the issues associated with energy losses during the alternate magnetization of the iron, and sometimes it is disregarded. Phase windings in a machine are most commonly considered as electric circuits with lumped parameters and their connection with the magnetic field is expressed by flux linkage ψ_k, where subscript k denotes the number of the adequate winding. Overall, the problem is associated with the determination of the flux linkage as the function of electric currents in the particular phase windings of a machine [90]. The issue of the mechanical motion of a rotor is not a complex phenomenon since a typical induction motor has only a single degree of mechanical freedom – angle of rotation of the rotor θ_r. In mathematical modeling of a an induction machine drive we take into consideration two cases: non-homogenous motion of the rotor in the dynamic states – for example during start-up or braking of the motor and motion under a constant angular speed, i.e. in a steady condition of the drive in operation. As a consequence of not accounting for the parasitic torques with synchronic characteristics we do not take into consideration small oscillations of the speed around the balance state; this comes as a consequence of their marginal role in a designed drive. The basic and the common foundation during the development of a mathematical model of an induction machine is the assumption of its geometrical and material symmetry. This allows very largely to simplify the model and it is most often followed in the issues associated with the electric drive. Abandoning of the assumptions of symmetry during the modeling of an induction machine is necessary only in special circumstances, such as modeling of emergency conditions for a drive and for example in the studies devoted to the tolerance of the engineering structure of the machine to its characteristics and potential emergencies. Such an example encountered during the analysis of an induction machine is the study of the effect of the asymmetry of the air gap between the stator and rotor to the resulting forces of magnetic pull and bearing's wear. The assumption of the symmetry also enables one to limit the area of calculation undertaken with an aim of developing field models and determination of boundary conditions for such calculations. Due to the presented impediments and complications the models of induction motors usually account for a number of simplifications which form an adaptation of the examined question and can lead to the statement of an answer. In this respect we can identify three general categories of mathematical modeling of a drive. The categories include: models serving for the *optimization of the construction characteristics* of a motor, secondly, models used for the *determination of electromechanical characteristics* and, thirdly, models whose object is to apply an *induction motor drive control*. The presented three categories of models can be described as follows: a mathematical model of an induction motor aimed at *the optimization of its construction* with regard to the structure of a magnetic circuit is, as a rule, a field based model whose solution is presented in 3D or 2D space, with a particular emphasis on the shape of a ferromagnetic core along with the design of the stator's and rotor's slots as well as spatial distribution of the windings. The ferromagnetic material is considered as non-linear taking into account its characteristics of magnetization. The considerations tend to more frequently involve a magnetic hysteresis loop and less often the occurrence of eddy currents [17,49]. Hence, calculations are performed for fixed positions of the stator in relation to the rotor or a constant speed of the motion, while the current density in the windings is as a rule constant over the entire cross-section of the winding in the slot. For the case of winding bars with large dimensions we have to account for the non-homogenous distribution of the current density in the radial direction. The construction of a typical induction motor due to the plane-parallel field representation enables one to perform field calculations in 2D space without affecting the precision of the results. The calculations apply professional software suites using Finite Elements Method (FEM) or Edge Elements Method (EEM). Such software contains procedures making it possible to gain various data and images regarding field characteristics in a particular subject, to obtain a number of integrated parameters such as the value of energy and co-energy of the magnetic field, electromagnetic torque, forces calculated by means of various methods and inductance of the windings in the area of calculation [24,48]. As one can conclude from this description, field models are applicable not only with an aim of improving the engineering and considering details of material parameters but can also provide valuable data in the form of lumped parameters for the calculation of the problems encountered in the drive. In particular, relevant insight is offered by the data regarding the inductance of the windings and its relation to the magnetic saturation. The mathematical models serving for the determination of *electromechanical characteristics* of a drive, both in static and dynamic states, as a rule are formed as models with lumped parameters. The reason is that in this case the engineering details are related to in an indirect way using a small number of parameters, which subsequently combine a number of physical properties of a machine. During the determination of characteristics, in particular the mechanical ones, the parasitic phenomena are frequently accounted for in the form of additional elements of electromagnetic torque derived from higher harmonics of the magnetic flux and harmonics associated with variable terms expressed by other elements in the permeneance of the air gap. The models which are applicable for stating the characteristics in many cases have to be precise in terms of energy balance since one of their application is in the determination of the losses of energy and efficiency of the drive. The analysis of lumped parameters is performed by a number of specialized calculation methods. This is based on field calculations in the electrical machines for the specific conditions of operation [37,48,91]. The mathematical models applied in the issues associated with *drive control* tend to be the most simplified models. As a principle, they disregard the losses in the iron, the phenomena of magnetic saturation and nuances in the form of multi-harmonic spectrum of the magnetic field in the air gap. Such models take the form of a system of ordinary differential equations. The models are transformed using the properties of the machine's symmetry into systems of equations, in which the form of the equations is relatively simple in the sense of the assumption of constant parameters of a system, the number and structure of expressions. Thus, the models correspond to the requirements of the control system due to its interaction with the transformed measurement signals derived from feedback in the system. The rationale for using the possibly most uncomplicated (in terms of calculations) models in the questions of control is associated with the fact that they are later used for the calculation of the vector of state variables of the drive in real time. The mathematical models of the induction motor find an increasingly wider application in the modern methods of control concerned with linearization through non-linear feedback of the dynamic model of a drive, which, in reality, has a non-linear structure. A type of this kind of control is also named control with inverse dynamics. An arising question is concerned with the practical application of models that do not account for a number of phenomena in induction machines including magnetic saturation. The solution proposed involves the contemporary control methods, also applied in electric drive, which are more resistant to the uncertainty of the parameters of the model and disturbances along the measurement paths. Such models include robust and adaptive control [22,45,53,57,65,75,105], in which case the mathematical model is combined with estimation of the parameters in real time. The currently solved tasks in drive control apply the following procedure: simple and functional control models in terms of calculations are accompanied with the correction of discrepancies resulting from parameter estimation using signals that are easily accessible by way of measurement. From the point of view of the current book the principal interest focuses on the mathematical models designed for determination of the characteristics of the drives and the ones applied for the purposes of control.

3.1.3.2 Number of Degrees of Freedom in an Induction Motor

The question of the number of the degrees of freedom (2.33) is encountered in systems with lumped parameters whose motion (dynamics) is described by a system of ordinary differential equations. For the case of an induction machine (Fig.3.4) this means one degree of freedom of the mechanical motion $s_m = 1$, for variable θ_r denoting the angle of rotor position and the adequate number of the degrees of freedom s_e for electric circuits formed by the phase windings. For the case when both the stator and rotor have three phases and the windings are independent, in accordance with the illustration in Fig. 3.4a, the number of electric degrees of freedom is $s_{el} = 6$. The assumption that electric circuits take the form of phase windings with electric charges Q_i as state variables does not exclude the applicability of a field model for the calculation of magnetic fluxes ψ_i linked with the particular windings of the motor. This possibility results from the decoupling of the magnetic and electric fields in the machine and the consideration of electric currents $i_i = \dot{Q}_i$ in the machine as sources of magnetic vector potential (2.180), (functions that are responsible for field generation).

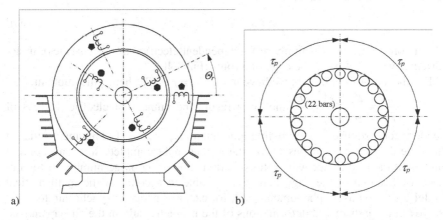

Fig. 3.4 Diagram with cross-section of induction motor: a) 3-phase stator and rotor windings b) rotor squirrel-cage windings

In this case we have to do with field-circuit models [48], in which the model with lumped parameters describing the dynamics of an electromechanical system (in this study the induction motor) is accompanied by an interactively produced model of the electromagnetic field in which the present flux linkages ψ_i are defined. Hence, the model of an induction motor whose diagram is presented in Fig 3.4a has

$$s = s_m + s_e = 7 \qquad (3.1)$$

degrees of freedom. In this place one can start to think about the state encountered in the windings of a squirrel cage motor (Fig. 3.4b), which does not contain a standard three-phase winding, but has a cage with $m = N_r$ number of bars. The squirrel cage winding responds to the MMFs produced by stator winding current. The induced EMFs in squirrel cage windings display the same symmetry properties on condition that the squirrel cage of the rotor is symmetric in the range of angular span corresponding to a single pole of the stator winding or its total multiple. Hence, the resulting number of degrees of freedom s_{sq} for a symmetrical squirrel cage winding [101] is expressed by the quotient

$$\frac{m}{2p} = \frac{u}{v} \qquad s_{sq} = u \qquad (3.2)$$

where: m - the number of bars in the symmetrical cage of a rotor
                u,v - relative prime integers

The number of the degrees of freedom of the electric circuits of a rotor's squirrel-cage winding $s_{sq} = u$ corresponds to the smallest natural number of the bars in a cage contained in a span of a single pole of the stator's winding or its multiple. This is done under the silent assumption that the stator's windings are symmetrical. If the symmetry is not actually the case, the maximum number of the degrees of freedom of a cage is equal to

$$s_{sq} = m + 1 \qquad (3.3)$$

which corresponds to the number of independent electric circuits (meshes), in accordance with (2.195), in the cage of a rotor (Fig 3.4b).

For the motor in Fig. 3.4b, we have $p = 2$, $N_r = m = 22$, hence the quotient:

$$\frac{m}{2p} = \frac{22}{4} = \frac{11}{2} = \frac{u}{v}$$ and, as a result, the number of electric degrees of

freedom for a squirrel cage winding amounts to $s_{sq} = u = 11$. This means that in this case the two pole pitches of the stator contain 11 complete slot scales or slot pitches of the rotor, after which the situation recurs. The large number of the degrees of freedom of the cage makes it possible to account in the mathematical model for the parasitic phenomena [80], for example parasitic synchronic torques. However, if we disregard deformations of the magnetic field in the air gap and assume that it is a plane-parallel and monoharmonic one with the single and basic harmonic equal to $\rho = p$, then in order to describe such a field we either need only two coordinates or two substitutive phase windings, in most simple cases orthogonal ones. For such an assumption of monoharmonic field the number of the degrees of freedom decreases to $s_{sq} = 2$ regardless of the number of bars in the rotor's cage. In the studies of induction motor drives and its control the principle is to assume the planar and monoharmonic field in the air gap. Nevertheless, at the stage when we are starting to develop the mathematical models of induction machines, it is assumed for the slip ring and squirrel cage machines that the rotor's winding is three-phased (as in Fig. 3.4a) for the purposes of preserving a uniform course of reasoning. Hence, as indicated earlier, under the assumptions of a planar and monoharmonic field in the air gap, slip-ring and squirrel-cage motors are equivalent and can be described with a single mathematical model with the only difference that the winding of a squirrel-cage motor is not accessible from outside, in other words, the voltages supplying the phases of the rotor are always equal to zero. According to (2.189) and (2.210), Lagrange's function for a motor with three phase windings in the stator and rotor can take the form:

$$L = \tfrac{1}{2}J\dot{\theta}_r^2 + \sum_{k=1}^{6} \int_0^{\dot{Q}_k} \psi_k(\dot{Q}_1,...\tilde{\tilde{Q}}_k,0...,0,\theta_r)d\tilde{\tilde{Q}}_k \qquad (3.4)$$

and the virtual work (2.198) expressing the exchange of energy is equal to:

$$\delta A = (-T_l - D\dot{\theta}_r)\delta\theta_r + \sum_{k=1}^{6}(u_k - R_k\dot{Q}_k)\delta Q_k \qquad (3.5)$$

where:
$\mathbf{q} = (Q_1, Q_2,...,Q_6, \theta_r)$ - vector of generalized coordinates
J - moment of inertia related to the motor's shaft,
T_l - load torque on the motor's shaft,
D - coefficient of viscous damping of the revolute motion,
R_k - resistance of k-th phase winding,
$\dot{Q}_k = i_k, u_k$ - electric current and supply voltage of k-th phase winding,
ψ_k - magnetic flux linked with k-th winding.

The model in this form already contains two simplifications, i.e. it disregards iron losses associated with magnetization of the core and changes of the windings' resistances following a change in their temperature.

From the above the equations of motion for electric variables follow in the form:

$$\frac{d}{dt}\left(\frac{\partial L}{\partial \dot{Q}_k}\right) - \frac{\partial L}{\partial Q_k} = u_k - R_k\dot{Q}_k \quad k = 1,...,6 \qquad (3.6)$$

with the capacitors missing from the system $\dfrac{\partial L}{\partial Q_k} = 0$.

Whereas, according to (3.4) and using the designation for currents $i_m = \dot{Q}_m$, we obtain

$$\frac{\partial L}{\partial \dot{Q}_k} = \psi_k(i_1,\ldots,i_k,0,\ldots,0,\theta_r) +$$

$$+ \sum_{l=k+1}^{6} \int_0^{i_l} \frac{\partial}{\partial i_k} \psi_l(i_1,\ldots,i_{l-1},\tilde{i}_l,0,\ldots,0,\theta_r) d\tilde{i}_l \tag{3.7}$$

If k-th winding were the final one, as for $k = n = 6$, then

$$\frac{\partial L}{\partial \dot{Q}_k} = \psi_k(i_1,\ldots,i_n,\theta_r) \tag{3.8}$$

The resulting equation takes the form

$$\psi_k(i_1,\ldots,i_n,\theta_r) = \frac{\partial L}{\partial \dot{Q}_k} = \psi_k(i_1,\ldots,i_k,0,\ldots,0,\theta_r) +$$

$$+ \sum_{l=k+1}^{6} \int_0^{i_l} \frac{\partial}{\partial i_k} \psi_l(i_1,\ldots,i_{l-1},\tilde{i}_l,0,\ldots,0,\theta_r) d\tilde{i}_l \tag{3.9}$$

From the comparison of (3.8) –(3.9) it results that for the simplicity of notation we should treat the equation in (3.7), which is currently considered, as the final one. In this case the equation for the circuits of an induction motor takes the form

$$\frac{d}{dt}\big(\psi_k(i_1,\ldots,i_n,\theta_r)\big) = u_k - R_k i_k \tag{3.10}$$

After the differentiation of the left-hand side we obtain

$$\sum_{m=1}^{6} \frac{\partial \psi_k}{\partial i_m} \frac{di_m}{dt} + \frac{\partial \psi_k}{\partial \theta_r} \dot{\theta}_r = u_k - R_k i_k$$

$$k = 1,\ldots,6 \tag{3.11}$$

The left-hand side expressions (3.11) denote electromotive forces induced in k-th phase winding as a result of the variations in time of flux linkage. The first terms are derived from the variations of the currents and are called electromotive forces of transformation, while the final term is related to the angular speed of the rotor $\dot{\theta}_r = \Omega_r$ and is called the electromotive force of rotation. The equation for the torque expressed with variable θ_r takes the following form

$$\frac{d}{dt}\left(\frac{\partial L}{\partial \dot{\theta}_r}\right) - \frac{\partial L}{\partial \theta_r} = -T_l - D\dot{\theta}_r$$

and, consequently,

$$\frac{d}{dt}\left(J\dot{\theta}_r\right) - \frac{\partial}{\partial\theta_r}\sum_{k=1}^{6}\int_{0}^{i_k}\psi_k(i_1,\ldots i_{k-1},\tilde{i}_k,0,\ldots,0,\theta_r)d\tilde{i}_k = -T_l - D\dot{\theta}_r,$$

Which, in case of constant moment of inertia J, can be denoted alternatively as

$$J\ddot{\theta}_r = T_e - T_l - D\dot{\theta}_r \qquad (3.12)$$

where:

$$T_e = \sum_{k=1}^{6}\int_{0}^{i_k}\frac{\partial}{\partial\theta_r}\psi_k(i_1,\ldots i_{k-1},\tilde{i}_k,0,\ldots,0,\theta_r)d\tilde{i}_k \qquad (3.13)$$

is the electromagnetic torque produced by the induction motor. In spite of the fact that the resulting equations of motion are stated for a system with lumped parameters and in this case for 7 variables corresponding to 7 degrees of freedom of the motor, they can find a very broad application. This results from the general form of the flux linkage associated with the particular phase windings $\psi_k = (i_1,\ldots,i_n, \theta_r)$. It could be gained by various methods accounting for the saturation and various engineering parameters of the magnetic circuit. For a squirrel-cage motor, for the case if one needed to account for the existing parasitic torques, it would be necessary to abandon the starting assumption of the monoharmonic image of the field in the air gap and, hence increase, the number of equations for the phases of the rotor from 3 to s_{sq}, as it results from (3.2).

3.1.4 Mathematical Models of an Induction Motor with Linear Characteristics of Core Magnetization

3.1.4.1 Coefficients of Windings Inductance

In a majority of issues associated with the motion of an induction machine, in particular in the issues associated with the control of drives with induction motors, we can assume a simplification involving an approximation of the characteristics of motor magnetization using linear relation. Hence, the definition of the coefficients of self-inductance and mutual inductance of the machine's windings follows in the form

$$\tilde{M}_{kl} = \frac{\partial\psi_k}{\partial i_l} = \frac{\psi_{kl}}{i_l} \qquad (3.14)$$

For $k = l$ this coefficient is named the self-inductance coefficient and includes two terms:

$$\tilde{M}_{kk} = L_{\sigma k} + M_{kk} \qquad (3.15)$$

where $L_{\sigma k}$ is the leakage inductance, which results from the magnetic flux linked solely to k-th phase winding. In contrast, for $k \neq l$ this coefficient is denoted as mutual inductance coefficient

$$\tilde{M}_{kl} = M_{kl} \tag{3.16}$$

The determination of the coefficient of the inductance of an induction machine could be derived from measurements on an existing machine or could be based on calculations, which is already possible at the stage of motor design. The experimental studies, which can serve in order to determine the coefficients of winding inductance, involve the measurements of the characteristics of the idle running of a motor, short circuits – for the purpose of stating the inductance of the leakage and other tests – for instance of the response to a voltage step function. On this basis it is possible to establish the approximate parameters of a machine, including the inductance of the windings as well as to apply the methods for the estimation of the parameters from selected measurement characteristics [37]. The calculation methods involve the calculation of the field in the machine using field programs [92], which provides information regarding integrated parameters, including inductance. For an induction motor it is sufficient to assume calculations of plano-parallel field (2D) with supplementary data and corrections regarding the boundary section of the field in the machine. In particular this concerns leakage inductance of the end winding section of the windings. Moreover, a number of analytical methods has been developed for the calculation of the field and inductance coefficients in an induction motor, thus providing valuable information for induction motor models. However, these tend to be less precise than the ones that result from field calculations since they account only for the major term of the energy of the magnetic field, i.e. the energy of the field in the machine's air gap. In the fundamental notion (Fig. 3.2), under the assumption of monoharmonic distribution of the field in the gap, the coefficients of mutual inductance take the form:

$$M_{kl} = M \cos(\alpha_k - \alpha_l) \tag{3.17}$$

where: M - value of inductance coefficient for phase coincidence
α_k, α_l - angles which determine the positions of the axes of windings k,l

In accordance with Fig. 3.4 these angles are:

$$
\begin{aligned}
&\alpha_1, \alpha_2, \alpha_3 = 0, \ 2\pi/3p, \ -2\pi/3p \\
&\alpha_4, \alpha_5, \alpha_6 = \theta_r, \ \theta_r + 2\pi/3p, \ \theta_r - 2\pi/3p
\end{aligned} \tag{3.18}
$$

- for the stator's windings and rotor's windings, respectively.

The number of the pole pairs p reflects p-time recurrence of the system of windings and spatial image of the field along the circumference of the air gap. On the basis of relations in (3.15 – 3.18), the matrix of the inductance coefficients of stator's windings takes the form

$$\mathbf{M}_{sph} = L_{\sigma s}\mathbf{1}_3 + M_s \begin{bmatrix} 1 & -\frac{1}{2} & -\frac{1}{2} \\ -\frac{1}{2} & 1 & -\frac{1}{2} \\ -\frac{1}{2} & -\frac{1}{2} & 1 \end{bmatrix} \tag{3.19}$$

and similarly, for the windings in the rotor

$$\mathbf{M}_{rph} = L_{\sigma r}\mathbf{1}_3 + M_r \begin{bmatrix} 1 & -\frac{1}{2} & -\frac{1}{2} \\ -\frac{1}{2} & 1 & -\frac{1}{2} \\ -\frac{1}{2} & -\frac{1}{2} & 1 \end{bmatrix} \tag{3.20}$$

At the same time, mutual inductance matrices between stator and rotor windings are relative to the angle of rotation θ_r

$$\begin{aligned} \mathbf{M}_{srph} &= \mathbf{M}^T_{rsph} = \\ &= M_{sr}\begin{bmatrix} \cos p\theta_r & \cos(p\theta_r + 2\pi/3) & \cos(p\theta_r - 2\pi/3) \\ \cos(p\theta_r - 2\pi/3) & \cos p\theta_r & \cos(p\theta_r + 2\pi/3) \\ \cos(p\theta_r + 2\pi/3) & \cos(p\theta_r - 2\pi/3) & \cos p\theta_r \end{bmatrix} \end{aligned} \tag{3.21}$$

In the above equations:

$L_{\sigma s}$, $L_{\sigma r}$ - are leakage inductance coefficients of stator and rotor windings
M_s, M_r - main field inductance coefficients of stator's and rotor's windings
M_{sr} - mutual inductance coefficient of stator's and rotor's windings for full linkage between the windings (aligned position of windings' axes)
$\mathbf{1}_3$ - unitary matrix with dimension 3.

3.1.4.2 Model with Linear Characteristics of Magnetization in Natural (phase) Coordinates

At the beginning it is necessary summarize the simplifying assumptions for this model of the induction motor, starting with the most important ones:

- complete geometrical and material symmetry of the electromagnetic structure of the motor
- linear characteristics of magnetization of the electromagnetic circuit
- planar and monoharmonic distribution of the field in the air gap, resulting in a single harmonic with the number $\rho = p$
- disregarding of the losses in the iron
- disregarding of the external influence (for example temperature) on the parameters of the motor.

Since, in accordance with the second assumption, we consider a linear case of magnetization, on the basis of (3.14) the following relation is satisfied

$$\psi_k = \sum_l \psi_{kl} = \sum_l M_{kl} i_l \tag{3.22}$$

Using the relation (3.22), the flux linkages of windings of the stator and the rotor can be recorded in the matrix form

$$\Psi_{sph} = [\psi_{s1} \quad \psi_{s2} \quad \psi_{s3}]^T = M_{sph} i_s + M_{srph} i_r$$
$$\Psi_{rph} = [\psi_{r1} \quad \psi_{r2} \quad \psi_{r3}]^T = M_{rsph} i_s + M_{rph} i_r \tag{3.23}$$

$$i_s = [i_{s1} \quad i_{s2} \quad i_{s3}]^T$$
$$i_r = [i_{r1} \quad i_{r2} \quad i_{r3}]^T$$

- with vectors of phase currents of the stator and rotor, respectively.

With the aid of (3.23) the equations of the electric circuits of the induction motor can take the following matrix form

$$U_{sph} = R_s i_s + \frac{d}{dt} \left(M_{sph} i_s + M_{srph}(\theta_r) i_r \right)$$
$$U_{rph} = R_r i_r + \frac{d}{dt} \left(M_{rsph} i_s (\theta_r) + M_{rph} i_r \right) \tag{3.24}$$

or

$$\begin{bmatrix} U_{sph} \\ U_{rph} \end{bmatrix} = \begin{bmatrix} R_s I_3 & 0_3 \\ 0_3 & R_r I_3 \end{bmatrix} \begin{bmatrix} i_s \\ i_r \end{bmatrix} + \frac{d}{dt} \left(\begin{bmatrix} M_{sph} & M_{srph}(\theta_r) \\ M_{rsph}(\theta_r) & M_{rph} \end{bmatrix} \begin{bmatrix} i_s \\ i_r \end{bmatrix} \right) \tag{3.25}$$

where: $0_3, 1_3$ - zero matrix and unitary matrix with dimension 3.

$$U_{sph} = [u_{s1} \quad u_{s2} \quad u_{s3}]^T$$
$$U_{rph} = [u_{r1} \quad u_{r2} \quad u_{r3}]^T$$

On the basis of (3.13) and (2.215), the electromagnetic torque of the motor can take the following form:

$$T_e = \frac{\partial}{\partial \theta_r} \left\{ \frac{1}{2} \sum_{k=1}^{6} \sum_{l=1}^{6} M_{kl}(\theta_r) i_k i_l \right\} = \frac{1}{2} \begin{bmatrix} i_s^T & i_r^T \end{bmatrix} \frac{\partial}{\partial \theta_r} \begin{bmatrix} M_{sph} & M_{srph}(\theta_r) \\ M_{rsph}(\theta_r) & M_{rph} \end{bmatrix} \begin{bmatrix} i_s \\ i_r \end{bmatrix} =$$
$$= \frac{1}{2} \left\{ i_s^T \frac{\partial}{\partial \theta_r} M_{srph}(\theta_r) i_r + i_{rph}^T \frac{\partial}{\partial \theta_r} M_{rsph}(\theta_r) i_s \right\} \tag{3.26}$$

Since both terms of the quadratic form (3.26) are equal, then

$$T_e = i_s^T \left(\frac{\partial}{\partial \theta_r} M_{srph}(\theta_r) \right) i_r \tag{3.27}$$

The equation concerning the revolute motion (3.12) remains unchanged, it is necessary only to apply a more detailed expression (3.27) to determine electromagnetic torque of the motor.

3.1.4.3 Transformation of Co-ordinate Systems

Most of the dynamic curves and solutions with regard to control are not conducted in phase coordinates, such as the case of the mathematical model in (3.25), (3.27), but they are stated in the transformed coordinates. There are several reasons for that: an adequately selected transformation is capable of transforming the system (3.24) that contains periodically variable coefficients (trigonometric functions) into a system with constant parameters. Thus, there is no need to apply large computing power and the cost thereof is reduced, which is particularly relevant in the issues of drive control in real time. In addition, as one can conclude from the form of mutual inductance matrix of the windings (3.17, 3.21), that their order is not 3 but only 2. They have one dimension too many, which can be concluded by adding up all the rows in each of the matrices. The physical reason is self-evident: in order to describe a monoharmonic planar field it is sufficient to use two variables. Hence, the field can be produced by currents in two phase windings that are not situated along a single axis (perpendicular axes are most applicable). The early applications of the transformation of the coordinate systems originate from Park and served in order to analyze the operation of synchronous generators. The general theory of transformation of coordinate systems in multi-phase electric machines is based on the Floquet's theorem. From it results that for linear systems of ordinary differential equations with time periodic coefficients it is possible to identify such a transformation $T(\theta_r, \dot{\theta}_r)$ for which in the new defined coordinates the machine's equations are independent of the angle of rotation of the rotor [53]. In the particular cases (monoharmonic field distribution) it is possible to gain this result by the application of transformation $T(\theta_r)$, i.e. only relative to the position of the rotor. The examples of solutions and applications in this area are multiple and can be found in the bibliography [76,80,82]. From the technical point of view it is only sensible to apply orthogonal transformations in electric drive. This means the ones whose matrices fulfill the condition that

$$\mathbf{T}^{-1} = \mathbf{T}^T \tag{3.28}$$

This is the case when the vectors which form the matrix of transformation $\mathbf{T}$ are orthogonal and have an elementary length, that is:

$$\mathbf{v}_i \mathbf{v}_j = 0 \quad \text{for} \quad i \neq j \quad \text{and} \quad \mathbf{v}_i \mathbf{v}_i = 1 \quad \text{for} \quad i = j \quad i, j = 1,2,3 \tag{3.29}$$

where $\mathbf{v}_i$ is the column (row) of the matrix $\mathbf{T}$. This property is indispensable since orthogonal transformations preserve the scalar product and square form for the transformed vectors. The scalar product for the case of an electric machine corresponds to the instantaneous power delivered to the clamps of the machine

$$p = \begin{bmatrix} u_1 & u_2 & u_3 \end{bmatrix} \begin{bmatrix} i_1 \\ i_2 \\ i_3 \end{bmatrix} = u_1 i_1 + u_2 i_2 + u_3 i_3$$

As a result of transformation of the above expression, using orthogonal transformation **T**, we obtain:

$$p = \mathbf{U}^T \underbrace{\mathbf{T}^T \mathbf{T}}_{\mathbf{I}_3} \mathbf{i} = \mathbf{U}^{*T} \mathbf{i}^* = u_\alpha i_\alpha + u_\beta i_\beta + u_\gamma i_\gamma \tag{3.30}$$

where
$$\mathbf{U}^* = \mathbf{TU} = \begin{bmatrix} u_\alpha & u_\beta & u_\gamma \end{bmatrix}^T$$
$$\mathbf{i}^* = \mathbf{Ti} = \begin{bmatrix} i_\alpha & i_\beta & i_\gamma \end{bmatrix}^T$$

are transformations of the vectors of the voltage and current of a three-phase electric machine. Concurrently, electromagnetic torque in the multi-phase machine is in the algebraic sense expressed by the quadratic form

$$T_e = \mathbf{i}^T \frac{\partial}{\partial \theta_r} \mathbf{Mi} = \mathbf{i}^T \underbrace{\mathbf{T}^T \mathbf{T}}_{\mathbf{I}_3} \frac{\partial}{\partial \theta_r} \mathbf{M} \underbrace{\mathbf{T}^T \mathbf{T}}_{\mathbf{I}_3} \mathbf{i} = \mathbf{i}^{*T} \mathbf{M}^{*T} \mathbf{i}^* \tag{3.31}$$

where

$$\mathbf{M}^* = \mathbf{T}^T \tilde{\mathbf{M}} \mathbf{T} \qquad \tilde{\mathbf{M}} = \frac{\partial}{\partial \theta_r} \mathbf{M} \tag{3.32}$$

is the transformed matrix of the derivatives of mutual inductance between the windings. As one can conclude from (3.32), the variables have been transformed into the form $\mathbf{i}^*$, while due to the orthogonality of the matrix of transformation **T** the electromagnetic torque remains unchanged. Under the assumptions adopted at the beginning of this chapter the machine has a monoharmonic field in the air gap, which results in the fact that the matrix of mutual inductance between the stator and rotor is relative only to argument $p\theta_r$ of the periodic functions. This enables one to easily identify the orthogonal transformation of **T** such that orders the mathematical model in the sense of leading to the constant coefficients of the differential equations. This study applies the following orthogonal matrices of the transformation:

- for the quantities relative to 3-phase windings in the stator

$$\mathbf{T}_s = \sqrt{\frac{2}{3}} \begin{bmatrix} \dfrac{1}{\sqrt{2}} & \dfrac{1}{\sqrt{2}} & \dfrac{1}{\sqrt{2}} \\ \cos(\omega_c t) & \cos(\omega_c t - a) & \cos(\omega_c t + a) \\ -\sin(\omega_c t) & -\sin(\omega_c t - a) & -\sin(\omega_c t + a) \end{bmatrix} \tag{3.33}$$

- for quantities relative to 3-phase windings in the rotor

$$
\mathbf{T}_r = \sqrt{\frac{2}{3}} \begin{bmatrix} \dfrac{1}{\sqrt{2}} & \dfrac{1}{\sqrt{2}} & \dfrac{1}{\sqrt{2}} \\ \cos(\omega_c t - p\theta_r) & \cos(\omega_c t - p\theta_r - a) & \cos(\omega_c t - p\theta_r + a) \\ -\sin(\omega_c t - p\theta_r) & -\sin(\omega_c t - p\theta_r - a) & -\sin(\omega_c t - p\theta_r + a) \end{bmatrix} \quad (3.34)
$$

where: ω_c - is an arbitrary pulsation that constitutes the degree of freedom of the planned transformation

$a = 2\pi/3$ - argument of the symmetric phase shift.

The transformation of the particular phase variables occurs in the following way:

$$
\begin{aligned}
\mathbf{Z}_{s0uv} &= \mathbf{T}_s \mathbf{Z}_{sph} = \begin{bmatrix} z_{s0} & z_{su} & z_{sv} \end{bmatrix}^T \\
\mathbf{Z}_{r0uv} &= \mathbf{T}_r \mathbf{Z}_{rph} = \begin{bmatrix} z_{r0} & z_{ru} & z_{rv} \end{bmatrix}^T
\end{aligned} \quad (3.35)
$$

Next, we will proceed to see how this works for a system of symmetric 3-phase sinusoidal voltages of frequency f_s supplying stator's windings:

$$
\mathbf{U}_{s0uv} = \mathbf{T}_s \mathbf{U}_{sph} = \mathbf{T}_s \begin{bmatrix} u_{s1} & u_{s2} & u_{s3} \end{bmatrix}^T \quad (3.36)
$$

that is:

$$
\begin{bmatrix} u_{s0} \\ u_{su} \\ u_{sv} \end{bmatrix} = \sqrt{\frac{2}{3}} \begin{bmatrix} \dfrac{1}{\sqrt{2}} & \dfrac{1}{\sqrt{2}} & \dfrac{1}{\sqrt{2}} \\ \cos(\omega_c t) & \cos(\omega_c t - a) & \cos(\omega_c t + a) \\ -\sin(\omega_c t) & -\sin(\omega_c t - a) & -\sin(\omega_c t + a) \end{bmatrix} *
$$

$$
* U_{mph} \begin{bmatrix} \cos(\omega_s t + \gamma) \\ \cos(\omega_s t + \gamma - a) \\ \cos(\omega_s t + \gamma + a) \end{bmatrix} = \sqrt{\frac{3}{2}} U_{mph} \begin{bmatrix} 0 \\ \cos[(\omega_s - \omega_c)t + \gamma] \\ \sin[(\omega_s - \omega_c)t + \gamma] \end{bmatrix} \quad (3.37)
$$

where: $\omega_s = 2\pi f_s$.

As a result of the transformation the voltage $u_{s0} = 0$, and voltages u_{su} and u_{sv} form an orthogonal system. In general, transformations of $\mathbf{T}_s, \mathbf{T}_r$ (3.33), (3.34) lead to the restatement of the phase variables in the stator's or rotor's windings for two perpendicular axes 'u,v' which are in revolute motion with the arbitrary angular speed of ω_c - as it was presented in Fig. 3.5. The third of the transformed axes – axis '0' is perpendicular to axes 'u,v' and acts in the axial direction; hence, it does not contribute to the planar field of the machine.

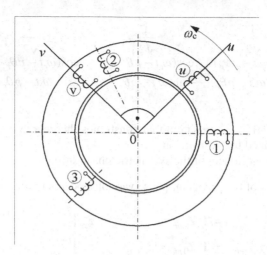

Fig. 3.5 Illustration of the orthogonal transformation of 3-phase windings of AC machine into rotating '*u,v*' axes

3.1.5 Transformed Models of Induction Motor with Linear Characteristics of Core Magnetization

3.1.5.1 Model in Current Coordinates

Firstly, we transform equations (3.24) for electric variables – currents $\mathbf{i}_s$, $\mathbf{i}_r$. For the equations of the stator windings

$$\mathbf{T}_s\mathbf{U}_{sph} = R_s\mathbf{I}_3\mathbf{T}_s\mathbf{i}_s + \frac{d}{dt}\left(\mathbf{T}_s\mathbf{M}_{sph}\mathbf{T}_s^T\overbrace{\mathbf{T}_s\mathbf{i}_s}^{\mathbf{i}_{s0uv}} + \mathbf{T}_s\mathbf{M}_{srph}(\theta_r)\mathbf{T}_r^T\overbrace{\mathbf{T}_r\mathbf{i}_r}^{\mathbf{i}_{r0uv}}\right) -$$

$$-\left(\frac{d}{dt}\mathbf{T}_s\right)\left(\underbrace{\mathbf{T}_s\mathbf{M}_{sph}\mathbf{T}_s^T}_{\mathbf{M}_{s0uv}}\mathbf{T}_s\mathbf{i}_s + \underbrace{\mathbf{T}_s\mathbf{M}_{srph}(\theta_r)\mathbf{T}_r^T}_{\mathbf{M}_{sr0uv}}\mathbf{T}_r\mathbf{i}_r\right) \tag{3.38}$$

The above transformations apply the property of the orthogonal matrix: $\mathbf{T}_s\mathbf{T}_s^T = \mathbf{T}_s^T\mathbf{T}_s = \mathbf{I}_3$ and similarly for the matrix of transformation $\mathbf{T}_r$. From equation (3.38) we obtain:

$$\mathbf{U}_{s0uv} = R_s\mathbf{i}_{s0uv} + \left(\frac{d}{dt} - \omega_c\mathbf{A}_3\right)\left(\mathbf{M}_{s0uv}\mathbf{i}_{s0uv} + \mathbf{M}_{sr0uv}\mathbf{i}_{r0uv}\right) \tag{3.39}$$

where:
$$\mathbf{i}_{s0uv} = \mathbf{T}_s\mathbf{i}_{sph} = \begin{bmatrix} i_{s0} & i_{su} & {}_{sv} \end{bmatrix}^T$$
$$\mathbf{i}_{r0uv} = \mathbf{T}_r\mathbf{i}_{rph} = \begin{bmatrix} i_{r0} & i_{ru} & i_{rv} \end{bmatrix}^T \tag{3.40}$$

$$\mathbf{M}_{s0uv} = \mathbf{T}_s \mathbf{M}_{sph} \mathbf{T}_s^T = \begin{bmatrix} L_{\sigma s} & & \\ & L_s & \\ & & L_s \end{bmatrix}$$

$L_s = L_{\sigma s} + L_m$ - self-inductance of stator's windings

$L_m = \frac{3}{2} M_s$ - inductance of magnetization

$$\mathbf{M}_{sr0uv} = \mathbf{T}_s \mathbf{M}_{srph}(\theta_r) \mathbf{T}_r^T = L_m \begin{bmatrix} 0 & & \\ & 1 & \\ & & 1 \end{bmatrix} \qquad (3.41)$$

$$\frac{d}{dt} \mathbf{T}_s = \sqrt{\frac{2}{3}} \omega_c \begin{bmatrix} 0 & 0 & 0 \\ -\sin(\omega_c t) & -\sin(\omega_c t - a) & -\sin(\omega_c t + a) \\ -\cos(\omega_c t) & -\cos(\omega_c t - a) & -\cos(\omega_c t + a) \end{bmatrix} = \omega_c \mathbf{A}_3 \mathbf{T}_s \qquad (3.42)$$

while: $\quad \mathbf{A}_3 = \begin{bmatrix} 0 & 0 & 0 \\ 0 & 0 & 1 \\ 0 & -1 & 0 \end{bmatrix}$ - is a skew-symmetric matrix

and $\left(\dfrac{d}{dt} - \omega_c \mathbf{A}_3 \right)$ - is an operator acting on the right-hand side expression.

Acting in a similar manner for the phase circuit of the rotor, after transformation, and employing

$$\frac{d}{dt} \mathbf{T}_r = \left(\omega_c - p\dot{\theta}_r \right) \mathbf{A}_3 \mathbf{T}_r$$

we obtain:

$$\mathbf{0} = R_r \mathbf{i}_{r0uv} + \left(\frac{d}{dt} - (\omega_c - p\dot{\theta}_r) \mathbf{A}_3 \right) \left(\mathbf{M}_{rs0uv} \mathbf{i}_{s0uv} + \mathbf{M}_{r0uv} \mathbf{i}_{r0uv} \right) \qquad (3.43)$$

where: $\qquad \mathbf{M}_{r0uv} = \mathbf{T}_r \mathbf{M}_{rph} \mathbf{T}_r^T = \begin{bmatrix} L_{\sigma r} & & \\ & L_r & \\ & & L_r \end{bmatrix} \qquad (3.44)$

$L_r = L_{\sigma r} + L_m$ - self-inductance of rotor's windings

$\left(\dfrac{d}{dt} - (\omega_c - p\dot{\theta}_r) \mathbf{A}_3 \right)$ - operator acting on the right-hand side expression.

The quadratic form of the (3.27) is transformed likewise:

$$T_e = \underbrace{\mathbf{i}_s^T \mathbf{T}_s^T}_{\mathbf{i}_{s0uv}} \mathbf{T}_s \frac{\partial}{\partial \theta_r} \mathbf{M}_{sr}(\theta_r) \mathbf{T}_r^T \underbrace{\mathbf{T}_r \mathbf{i}_r}_{\mathbf{i}_{r0uv}} = p L_m \mathbf{i}_{s0uv}^T (-\mathbf{A}_3) \mathbf{i}_{r0uv}$$

which finally results in a relation

$$T_e = pL_m (i_{sv} i_{ru} - i_{su} i_{rv})$$ (3.45)

From equations (3.39-3.45) after their transformation it results that the zero sequence equations are not involved in the conversion of energy and these equations are autonomic

$$u_{s0} = R_s i_{s0} + \frac{d}{dt} L_{\sigma s} i_{s0}$$

$$0 = R_r i_{0r} + \frac{d}{dt} L_{\sigma r} i_{r0}$$ (3.46)

The axial current i_{s0} occurs only for the case of asymmetry in the power supply (3.37), in addition to which, it is present only in the case when the windings of the stator are not connected in a star. Otherwise, the equations of constraints eliminate the possibility of the occurrence of i_{s0} even in case of the existing voltage u_{s0}. For this reason in further part of mathematical modeling of induction motor we will account only for two terms representing electric quantities in the axes 'u,v':

$$\mathbf{U}_{suv} = R_s \mathbf{i}_{suv} + \left(\frac{d}{dt} - \omega_c \mathbf{A}_2 \right) \left(\mathbf{L}_s \mathbf{i}_{suv} + \mathbf{L}_m \mathbf{i}_{ruv} \right)$$

$$\mathbf{0} = R_r \mathbf{i}_{ruv} + \left(\frac{d}{dt} - (\omega_c - p\dot{\theta}_r) \mathbf{A}_2 \right) \left(\mathbf{L}_m \mathbf{i}_{suv} + \mathbf{L}_r \mathbf{i}_{ruv} \right)$$ (3.47)

and for the equation for the revolute motion

$$J \frac{d\Omega_r}{dt} = \underbrace{pL_m (i_{sv} i_{ru} - i_{su} i_{rv})}_{T_e} - T_l - D\Omega_r$$ (3.48)

where: $\Omega_r = \dot{\theta}_r$ - is the angular speed of rotor's revolution

$$\mathbf{i}_{suv} = \begin{bmatrix} i_{su} & i_{sv} \end{bmatrix}^T \qquad \mathbf{i}_{ruv} = \begin{bmatrix} i_{ru} & i_{rv} \end{bmatrix}^T \qquad \mathbf{U}_{suv} = \begin{bmatrix} u_{su} & u_{sv} \end{bmatrix}^T$$

$$\mathbf{A}_2 = \begin{bmatrix} 0 & 1 \\ -1 & 0 \end{bmatrix} \text{ - is a skew-symmetric matrix of dimension 2.}$$

The dimension of the state vector of the model (3.47–3.48) of the induction motor is $s = 5$, which comes as a consequence of the lack of equations for the zero currents i_{s0}, i_{r0}.

3.1.5.2 Models in Mixed Coordinates

The transformation of flux linkages of phase winding (3.23) in a similar manner to the procedure with equations (3.38) leads to relations which determine axial fluxes in the function of axial currents

$$\begin{bmatrix} \Psi_{suv} \\ \Psi_{ruv} \end{bmatrix} = \begin{bmatrix} L_s & L_m \\ L_m & L_r \end{bmatrix} \begin{bmatrix} \mathbf{i}_{suv} \\ \mathbf{i}_{ruv} \end{bmatrix} = L_m \begin{bmatrix} \dfrac{1}{k_s} & 1 \\ 1 & \dfrac{1}{k_r} \end{bmatrix} \begin{bmatrix} \mathbf{i}_{suv} \\ \mathbf{i}_{ruv} \end{bmatrix} \tag{3.49}$$

where: $k_s = \dfrac{L_m}{L_s}$ $k_r = \dfrac{L_m}{L_r}$ are called coefficients of magnetic coupling for stator and rotor windings.

Equations (3.47) using (3.49) can be restated and take the form

$$\mathbf{U}_{suv} = R_s \mathbf{i}_{suv} + \left(\frac{d}{dt} - \omega_c \mathbf{A}_2 \right) \Psi_{suv}$$

$$0 = R_r \mathbf{i}_{ruv} + \left(\frac{d}{dt} - (\omega_c - p\Omega_r) \mathbf{A}_2 \right) \Psi_{ruv} \tag{3.50}$$

In equations (3.50) we have to do both with axial currents $\mathbf{i}_{suv}$, $\mathbf{i}_{ruv}$ and axial fluxes Ψ_{suv}, Ψ_{ruv} - that is with a double set of variables. In order to eliminate one of them, we need a relation that is the reverse to the one in (3.49):

$$\begin{bmatrix} \mathbf{i}_{suv} \\ \mathbf{i}_{ruv} \end{bmatrix} = \frac{1}{L_m} \frac{1-\sigma}{\sigma} \begin{bmatrix} \dfrac{1}{k_r} & -1 \\ -1 & \dfrac{1}{k_s} \end{bmatrix} \begin{bmatrix} \Psi_{suv} \\ \Psi_{ruv} \end{bmatrix} \tag{3.51}$$

where: $\sigma = 1 - k_s k_r$ is called the coefficient of windings' leakage.

a) *Mixed coordinates* $\Psi_{suv}, \mathbf{i}_{suv}$

For the elimination of variables $\Psi_{ruv}, \mathbf{i}_{ruv}$ from equations (3.50) we apply the following relations:

$$\Psi_{ruv} = \frac{1}{k_r} \left(\Psi_{suv} - L_s \sigma \mathbf{i}_{suv} \right) \qquad \mathbf{i}_{ruv} = \frac{1}{L_m} \Psi_{suv} - \frac{1}{k_s} \mathbf{i}_{suv} \tag{3.52}$$

Using (3.52) and transforming the result into the standard form, we obtain:

$$\dot{\Psi}_{suv} = \mathbf{U}_{suv} - R_s \mathbf{i}_{suv} + \omega_c \mathbf{A}_2 \Psi_{suv}$$

$$\mathbf{i}_{suv} = \frac{1}{L_s \sigma} (\alpha_r + p\Omega_r \mathbf{A}_2) \Psi_{suv} - \frac{1}{\sigma} (\alpha_s + \alpha_r) \mathbf{i}_{suv} + \tag{3.53}$$

$$+ (\omega_c - p\Omega_r) \mathbf{A}_2 \mathbf{i}_{suv} + \frac{\mathbf{U}_{suv}}{L_s \sigma}$$

By stating (3.53) in the form of equations with a single dimension and supplementing with the equation of motion we obtain the model:

$$\dot{\psi}_{su} = u_{su} - R_s i_{su} + \omega_c \psi_{sv}$$

$$\dot{\psi}_{sv} = u_{su} - R_s i_{sv} - \omega_c \psi_{su}$$

$$i_{su} = \frac{1}{L_s \sigma}(\alpha_r \psi_{su} + p\Omega_r \psi_{sv}) - \frac{1}{\sigma}(\alpha_s + \alpha_r)i_{su} + (\omega_c - p\Omega_r)i_{sv} + \frac{u_{su}}{L_s \sigma} \quad (3.54)$$

$$i_{sv} = \frac{1}{L_s \sigma}(\alpha_r \psi_{sv} - p\Omega_r \psi_{su}) - \frac{1}{\sigma}(\alpha_s + \alpha_r)i_{sv} - (\omega_c - p\Omega_r)i_{su} + \frac{u_{sv}}{L_s \sigma}$$

$$\dot{\Omega}_r = \frac{1}{J}[\underbrace{p(\psi_{su}i_{sv} - \psi_{sv}i_{su})}_{T_e} - T_l - D\Omega_r]$$

The above calculation of the torque applies following property of the skew-symmetric matrix $\mathbf{A}_2$:

$$\mathbf{i}^T \mathbf{A}_2 \mathbf{i} = 0$$

b) *Mixed coordinates* $\mathbf{i}_{suv}$, Ψ_{ruv}

This time we should eliminate variables $\mathbf{i}_{ruv}$, Ψ_{suv} from equations (3.50). This is done using the following relations:

$$\mathbf{i}_{ruv} = \frac{1}{L_r}\Psi_{ruv} - k_r \mathbf{i}_{suv} \qquad \Psi_{suv} = k_r \Psi_{ruv} - \sigma L_m \mathbf{i}_{suv}$$

After substitution and transformations we obtain:

$$\dot{\Psi}_{ruv} = -\alpha_r \Psi_{ruv} + L_m \alpha_r \mathbf{i}_{suv} + (\omega_c - p\Omega_r)\mathbf{A}_2 \Psi_{ruv}$$

$$\mathbf{i}_{suv} = \frac{k_r}{L_s \sigma}(\alpha_r + p\Omega_r \mathbf{A}_2)\Psi_{ruv} - \gamma \mathbf{i}_{suv} + \omega_c \mathbf{A}_2 \mathbf{i}_{suv} + \frac{\mathbf{U}_{suv}}{L_s \sigma} \quad (3.55)$$

which results in a single dimensional equations and the overall model containing the equation of motion:

$$\dot{\psi}_{ru} = -\alpha_r \psi_{ru} + L_m \alpha_r i_{su} + (\omega_c - p\Omega_r)\psi_{rv}$$

$$\dot{\psi}_{rv} = -\alpha_r \psi_{rv} + L_m \alpha_r i_{sv} - (\omega_c - p\Omega_r)\psi_{ru}$$

$$i_{su} = \frac{k_r}{L_s \sigma}(\alpha_r \psi_{ru} + p\Omega_r \psi_{rv}) - \gamma i_{su} + \omega_c i_{sv} + \frac{U_{su}}{L_s \sigma} \quad (3.56)$$

$$i_{sv} = \frac{k_r}{L_s \sigma}(\alpha_r \psi_{rv} - p\Omega_r \psi_{ru}) - \gamma i_{sv} - \omega_c i_{su} + \frac{U_{sv}}{L_s \sigma}$$

$$\dot{\Omega}_r = \frac{1}{J}[\underbrace{pk_r(\psi_{ru}i_{sv} - \psi_{rv}i_{su})}_{T_e} - T_l - D\Omega_r]$$

Below is a summary of coefficients used in the above models:

$$k_s = \frac{L_m}{L_s} \qquad k_r = \frac{L_m}{L_r} \qquad \alpha_s = \frac{R_s}{L_s} \qquad \alpha_s = \frac{R_s}{L_s}$$

$$\sigma = 1 - k_s k_r \qquad \gamma = \frac{1}{\sigma}(\alpha_s + \alpha_r(1-\sigma)) \tag{3.57}$$

The models presented using the system of equations (3.54), (3.56) serve not only for the purposes of the calculations of the dynamics of the drive but also are applied in modern methods of induction motor control.

3.1.5.3 Model in Flux Coordinates

In order to obtain a model in flux coordinates it is necessary to eliminate currents i_{suv}, i_{ruv}, from equation (3.50) by using for this purpose relations in (3.51). As a result we obtain:

$$\dot{\Psi}_{suv} = \mathbf{U}_{suv} + \frac{1}{\sigma}\alpha_s(k_r\Psi_{ruv} - \Psi_{suv}) + \omega_c\mathbf{A}_2\Psi_{suv}$$

$$\dot{\Psi}_{ruv} = \frac{1}{\sigma}\alpha_r(k_s\Psi_{suv} - \Psi_{ruv}) + (\omega_c - p\Omega_r)\mathbf{A}_2\Psi_{ruv} \tag{3.58}$$

The complete model accompanied by the equation of the revolute motion is presented below:

$$\dot{\psi}_{su} = u_{su} + \frac{1}{\sigma}\alpha_s(k_r\psi_{ru} - \psi_{su}) + \omega_c\psi_{sv}$$

$$\dot{\psi}_{sv} = u_{sv} + \frac{1}{\sigma}\alpha_s(k_r\psi_{rv} - \psi_{sv}) - \omega_c\psi_{su}$$

$$\dot{\psi}_{ru} = \frac{1}{\sigma}\alpha_r(k_s\psi_{su} - \psi_{ru}) + (\omega_c - p\Omega_r)\psi_{rv} \tag{3.59}$$

$$\dot{\psi}_{rv} = \frac{1}{\sigma}\alpha_r(k_s\psi_{sv} - \psi_{rv}) - (\omega_c - p\Omega_r)\psi_{ru}$$

$$\dot{\Omega}_r = \frac{1}{J}[\underbrace{\frac{pk_r}{L_s\sigma}(\psi_{sv}\psi_{ru} - \psi_{su}\psi_{rv})}_{T_e} - T_l - D\Omega_r]$$

3.1.5.4 Special Cases of Selecting Axial Systems 'u,v'

Angular speed ω_c encountered in transformations $\mathbf{T}_s$, $\mathbf{T}_r$ (3.33), (3.34) as a free parameter makes it possible to comfortably conduct calculations and interpret the results for the statement of the speed of the revolution of orthogonal axes 'u,v' (Fig. 3.5). For this purpose we apply three basic substitutions encountered in the following stages

a) $\omega_c = 0$. *Equations in α,β axes*

One of the commonly applied solution regarding the selection of an axial system is associated with immobilization of the system 'u,v' with respect to the motor's stator. In this case we have to do with the system of axes denoted as α,β combined with the stator of the machine. Before substituting $\omega_c = 0$ the equations of motion for the phase circuits (3.47) are presented in a matrix form:

$$
\begin{bmatrix} \mathbf{U}_{suv} \\ \mathbf{0} \end{bmatrix} = \begin{bmatrix} R_s & 0 \\ 0 & R_r \end{bmatrix} \begin{bmatrix} \mathbf{i}_{suv} \\ \mathbf{i}_{ruv} \end{bmatrix} + \begin{bmatrix} L_s & L_m \\ L_m & L_r \end{bmatrix} \frac{d}{dt} \begin{bmatrix} \mathbf{i}_{suv} \\ \mathbf{i}_{ruv} \end{bmatrix} -
$$
$$
- \begin{bmatrix} \omega_c L_s & \omega_c L_m \\ (\omega_c - p\Omega_r)L_m & (\omega_c - p\Omega_r)L_r \end{bmatrix} \begin{bmatrix} \mathbf{A}_2 \mathbf{i}_{suv} \\ \mathbf{A}_2 \mathbf{i}_{ruv} \end{bmatrix}
$$

(3.60)

After the substitution $\omega_c = 0$ and rearrangement we obtain equations in α,β axes in the standard form:

$$
\frac{d}{dt} \begin{bmatrix} \mathbf{i}_{s\alpha\beta} \\ \mathbf{i}_{r\alpha\beta} \end{bmatrix} = \frac{1}{\sigma L_s} \begin{bmatrix} \mathbf{U}_{s\alpha\beta} \\ -k_r \mathbf{U}_{s\alpha\beta} \end{bmatrix} -
$$
$$
- \frac{1}{\sigma} \left(\begin{bmatrix} \alpha_s & -k_s \alpha_r \\ -k_r \alpha_s & \alpha_r \end{bmatrix} \begin{bmatrix} \mathbf{i}_{s\alpha\beta} \\ \mathbf{i}_{r\alpha\beta} \end{bmatrix} + p\Omega_r \begin{bmatrix} -k_s k_r & -k_s \\ k_r & 1 \end{bmatrix} \begin{bmatrix} \mathbf{A}_2 \mathbf{i}_{s\alpha\beta} \\ \mathbf{A}_2 \mathbf{i}_{r\alpha\beta} \end{bmatrix} \right)
$$

(3.61)

Voltages $\mathbf{U}_{s\alpha\beta}$ result from the system of sine voltages (3.37) after applying transformation $\mathbf{T}_s$ with $\omega_c = 0$. The curves for currents $\mathbf{i}_{s\alpha\beta}$, $\mathbf{i}_{r\alpha\beta}$ have a pulsation ω_s in the steady state, which results from the voltage $\mathbf{U}_{s\alpha\beta}$. The complete mathematical model of the induction motor in the axial coordinates is found below:

$$
\dot{i}_{s\alpha} = \frac{1}{L_s \sigma} u_{s\alpha} - \frac{1}{\sigma} \left(\alpha_s i_{s\alpha} - k_s \alpha_r i_{r\alpha} - p\Omega_r (k_s k_r i_{s\beta} + k_s i_{r\beta}) \right)
$$

$$
\dot{i}_{s\beta} = \frac{1}{L_s \sigma} u_{s\beta} - \frac{1}{\sigma} \left(\alpha_s i_{s\beta} - k_s \alpha_r i_{r\beta} + p\Omega_r (k_s k_r i_{s\alpha} + k_s i_{r\alpha}) \right)
$$

$$
\dot{i}_{r\alpha} = -\frac{k_r}{L_s \sigma} u_{s\alpha} - \frac{1}{\sigma} \left(-k_r \alpha_s i_{s\alpha} + \alpha_r i_{r\alpha} + p\Omega_r (k_r i_{s\beta} + i_{r\beta}) \right)
$$

(3.62)

$$
\dot{i}_{r\beta} = -\frac{k_r}{L_s \sigma} u_{s\beta} - \frac{1}{\sigma} \left(-k_r \alpha_s i_{s\beta} + \alpha_r i_{r\beta} - p\Omega_r (k_r i_{s\alpha} + i_{r\alpha}) \right)
$$

$$
\dot{\Omega}_r = \frac{1}{J} [\underbrace{pL_m (i_{s\beta} i_{r\alpha} - i_{s\alpha} i_{r\beta})}_{T_e} - T_l - D\Omega_r]
$$

In a similar manner on the basis of equations (3.56, 3.59) following substitution $\omega_c = 0$, we may have the equations of motion for the motor in α,β axes for mixed and flux coordinates.

b) $\omega_c = p\Omega_r$. *Equations in d,q axes*

Following the substitution $\omega_c = p\Omega_r$ in the equations of motion, the system of orthogonal equivalent axes 'u,v' is stiffly related to the rotor and, hence, it has been given the name 'd,q' system. This system is most commonly applied in the description of the dynamics of synchronous machines and axis 'd' is then situated consistently with the longitudinal axis of the machine's rotor. The system of axes

'd,q' in this case can be effectively applied with regard to the induction motor as well. In this case for powering stator windings with symmetric sinusoidal system of 3-phase voltages (3.37) in stator the currents' curves we have to do with pulsation ω_r

$$\omega_r = \omega_s - p\Omega_r = \omega_s s$$
$$s = 1 - p\Omega_r / \omega_s = 1 - \Omega_r / \omega_f \tag{3.63}$$

which is called the slip pulsation of an induction motor. Slip s reflects a relative delay in rotation speed Ω_r of the rotor in respect to the magnetic field rotation speed $\omega_f = \omega_s/p$. Slip s during the standard operation of an induction motor ranges from a fraction of a percent to a few per cent of one. Concurrently, ω_r is the physical pulsation of rotor currents.

For instance one can note the equations for a motor expressed in axes 'd,q' in mixed coordinates i_{suv}, Ψ_{ruv}. On the basis of equations (3.56) and following the substitution $\omega_c = p\Omega_r$, we obtain:

$$\dot{\psi}_{rd} = -\alpha_r \psi_{rd} + L_m \alpha_r i_{sd}$$
$$\dot{\psi}_{rq} = -\alpha_r \psi_{rq} + L_m \alpha_r i_{sq}$$
$$i_{sd} = \frac{k_r}{L_s \sigma}(\alpha_r \psi_{rd} + p\Omega_r \psi_{rq}) - \gamma i_{sd} + p\Omega_r i_{sq} + \frac{U_{sd}}{L_s \sigma} \tag{3.64}$$
$$i_{sq} = \frac{k_r}{L_s \sigma}(\alpha_r \psi_{rq} - p\Omega_r \psi_{rd}) - \gamma i_{sq} - p\Omega_r i_{sd} + \frac{U_{sq}}{L_s \sigma}$$
$$\dot{\Omega}_r = \frac{1}{J}[\underbrace{pk_r(\psi_{rd}i_{sq} - \psi_{rq}i_{sd})}_{T_e} - T_l - D\Omega_r]$$

c) $\omega_c = \omega_s$. *Equations of motion in 'x,y' axes*

The axial system 'x,y' is a system which rotates with the speed $\omega_s = p\omega_f$, that is the p-multiple of the magnetic field speed in the air gap. In case of supplying phase windings with a symmetric sinusoidal system of 3-phase voltages, after the application of transformation (3.37), voltage U_{xy} is constant. As a result, the electric variables i_{xy}, Ψ_{xy} in the steady state play the role of constant functions. In the transient state their variability is reflecting envelope curves of alternating quantities in the phase windings. The mathematical model based on the system of coordinates 'xy' is beneficial to conduct numerical calculations of dynamic curves, since the practical computing concern the envelopes of the phase curves and it is possible to conduct them with a much larger integration step. As an example, it is possible to take into consideration the mathematical model of a motor in flux coordinates (3.59) expressed in the 'xy' axial system:

$$\dot{\psi}_{sx} = \frac{\alpha_s}{\sigma}\left(k_r\psi_{rx} - \psi_{sx}\right) + \omega_s\psi_{sy} + u_{sx}$$

$$\dot{\psi}_{sy} = \frac{\alpha_s}{\sigma}\left(k_r\psi_{ry} - \psi_{sy}\right) - \omega_s\psi_{sx} + u_{sy}$$

$$\dot{\psi}_{rx} = \frac{\alpha_r}{\sigma}\left(k_s\psi_{sx} - \psi_{rx}\right) + s\omega_s\psi_{ry} \tag{3.65}$$

$$\dot{\psi}_{ry} = \frac{\alpha_r}{\sigma}\left(k_s\psi_{sy} - \psi_{ry}\right) - s\omega_s\psi_{rx}$$

$$\dot{\Omega}_r = \frac{1}{J}\left[\frac{pk_r}{L_s\sigma}(\psi_{sy}\psi_{rx} - \psi_{sx}\psi_{ry}) - T_l - D\Omega_r\right]$$

The models of the induction motor presented in 'xy' system are particularly applicable in the control, which comes as a consequence of the simplicity of the equations for the motor in this system. This is especially discernible for Field Oriented Control (FOC) techniques.

3.1.6 Mathematical Models of Induction Motor with Untransformed Variables of the Stator/Rotor Windings

There is a number of practical reasons why it is beneficial to preserve untransformed variables on one of the sides of an induction motor. This means that the variables which define the electric state in the stator's or rotor's windings remain in the form of natural variables, while the state of the connections between the windings is maintained by the introduction of adequate equations with constraints resulting from Kirchhoff's laws (2.193). The preservation of the untransformed variables and the resulting equations of motion for a single side enables one to derive the so-called internal asymmetry within windings, which are defined using natural variables. Hence, it is possible either to incorporate arbitrary lumped elements in the particular phase windings or apply asymmetrical supply voltages with arbitrary waveforms. In particular, as a result of this, it is possible to perform the calculations for the braking with direct current for any connections between phase windings, operation under single-phase supply, operating with an auxiliary phase for capacitor starting a single-phase motor, analysis of a series of emergency issues and select safety measures. However, the most important application of the mathematical models of this type is in the modeling of electronic power converters in combination with the supplied machine, in which power transistors or silicon controlled rectifiers (SCRs) are designed for the control of voltages and currents in the particular windings. This type of modeling, which has been the object in numerous research, can be most effectively conducted in the circumstances of preserving a fixed structure of an electric system by the introduction of resistances with variable values corresponding to the state of the examined power electronic switches in the particular branches of electric circuits. In the blocking state they assume high values limiting the flow of current across a certain branch, while in the conducting state the values are small, i.e. ones which correspond to the parameters of conduction calculated on the basis of the data for such components taken from manufacturers' catalogues. This type of modeling is associated with a number of impediments in numerical calculations of the curves of examined variables due to time constants in the particular electric circuits, whose values differ by several orders of magnitude, as a result of application of high value blocking resistance. For this reason these models apply stiff methods of integration for Ordinary Differential Equations (ODFs). Another solution to be applied involves the use of simple single-step procedures for the solution of stiff systems with a small step size. In any case, however, power semiconductor switches encountered in the branches of electric motors are more easily modeled for the case when variables for a given circuit are the phase variables. The reason is associated with the fact that the state of a given power semiconductor switch is relative to the control signals and forward current in this element. In the issues of control of squirrel-cage induction motors semiconductor systems are members of the circuits of the stator in a machine, while the variables concerning squirrel-cage windings of the rotor are transformed to the orthogonal axes u,v. However, if control occurs in the rotor of a slip-ring motor and the control elements are situated there (including the converter), it is beneficial to have untransformed variables (phase currents) in the rotor. But then the electric variables of the stator's windings could as well be transformed into orthogonal axes u,v for the purposes of succinct notation.

3.1.6.1 Model with Untransformed Variables in the Electric Circuit of the Stator

We shall assume that we deal with a 3-phase motor with windings connected in a star. As a result, when the currents of the phase windings 1 and 2 are considered as state variables, the system of stator windings is characterized with constraints (Fig. 3.6)

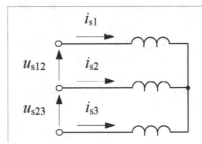

Fig. 3.6 Star-connected 3-phase stator windings

$$\begin{bmatrix} i_{s1} \\ i_{s2} \\ i_{s3} \end{bmatrix} = \underbrace{\begin{bmatrix} 1 & 0 \\ 0 & 1 \\ -1 & -1 \end{bmatrix}}_{\mathbf{W}_i} \begin{bmatrix} i_{s1} \\ i_{s2} \end{bmatrix}$$

or more briefly: $\mathbf{i}_s = \mathbf{W}_i \, \mathbf{i}_{s12}$ (3.66)

$$\begin{bmatrix} u_{s12} \\ u_{s23} \end{bmatrix} = \underbrace{\begin{bmatrix} 1 & -1 & 0 \\ 0 & 1 & -1 \end{bmatrix}}_{\mathbf{W}_u} \begin{bmatrix} u_{s1} \\ u_{s2} \\ u_{s3} \end{bmatrix}$$

or more briefly: $\mathbf{U}_{s12} = \mathbf{W}_u \, \mathbf{U}_{sph}$ (3.67)

As a result of the introduction of the equations of constraints (3.66-3.67) and the transformation of the currents of stator windings, we obtain for the stator's system of equations (3.24):

$$\mathbf{W}_u *\Big| \mathbf{U}_{sph} = R_s \mathbf{i}_s + \frac{d}{dt}\Big(\mathbf{M}_{sph}\mathbf{i}_s + \mathbf{M}_{srph}(\theta_r)\mathbf{i}_r\Big)$$

$$\underbrace{\mathbf{W}_u\mathbf{U}_{sph}}_{\mathbf{U}_{s12}} = \underbrace{R_s\mathbf{W}_u\mathbf{W}_i}_{\mathbf{R}_{s12}}\mathbf{i}_{s12} + \frac{d}{dt}\Big(\underbrace{\mathbf{W}_u\mathbf{M}_{sph}\mathbf{W}_i}_{\mathbf{M}_{s12}}\mathbf{i}_{s12} + \underbrace{\mathbf{W}_u\mathbf{M}_{srph}\mathbf{T}_r^T}_{\mathbf{M}_{sr12uv}}\underbrace{\mathbf{T}_r\mathbf{i}_r}_{\mathbf{i}_{ruv}}\Big)$$

As a consequence:

$$\mathbf{U}_{s12} = \mathbf{R}_{s12}\mathbf{i}_{s12} + \frac{d}{dt}\Big(\mathbf{M}_{s12}\mathbf{i}_{s12} + \mathbf{M}_{sr12uv}\mathbf{i}_{ruv}\Big)$$ (3.68)

In equations (3.68):

$$\mathbf{R}_{s12} = \begin{bmatrix} R_{s1} & -R_{s2} \\ R_{s3} & R_{s2}+R_{s3} \end{bmatrix}$$ (3.69)

makes it possible to account for non-homogenous resistances in the stator's phase windings, and for the case of symmetry $R_{s1} = R_{s2} = R_{s3}$ and, hence:

$$\mathbf{R}_{s12} = R_s\begin{bmatrix} 1 & -1 \\ 1 & 2 \end{bmatrix}$$ (3.70)

$$\mathbf{M}_{s12} = L_s\begin{bmatrix} 1 & -1 \\ 1 & 2 \end{bmatrix} \quad L_s = L_{\sigma s} + L_m$$ (3.71)

$$\mathbf{M}_{sr12uv} = L_m\begin{bmatrix} \sqrt{\dfrac{3}{2}} & -\dfrac{\sqrt{2}}{2} \\ 0 & \sqrt{2} \end{bmatrix}$$ (3.72)

Similar transformations are performed for the system of equations in the rotor's circuits (3.24):

$$\mathbf{T}_r *\Big| 0 = R_r\mathbf{i}_r + \frac{d}{dt}\Big(\mathbf{M}_{rsph}(\theta_r)\mathbf{i}_s + \mathbf{M}_{rph}\mathbf{i}_r\Big)$$

and, consequently:

$$0 = R_r \underbrace{\mathbf{T}_r \mathbf{i}_r}_{\mathbf{i}_{ruv}} + \frac{d}{dt}\left(\underbrace{\mathbf{T}_r \mathbf{M}_{rph} \mathbf{T}_r^T}_{\mathbf{M}_{rsuv}} \mathbf{T}_r \mathbf{i}_r\right) - \left(\frac{d}{dt}\mathbf{T}_r\right)\mathbf{M}_{rph}\mathbf{T}_r^T \underbrace{\mathbf{T}_r \mathbf{i}_r}_{\mathbf{i}_{ruv}} +$$

$$+\frac{d}{dt}\left(\underbrace{\mathbf{T}_r \mathbf{M}_{rsph} \mathbf{W}_i}_{\mathbf{M}_{rsuv12}} \mathbf{i}_{s12}\right) - \left(\frac{d}{dt}\mathbf{T}_r\right)\mathbf{M}_{rsph}\mathbf{W}_i \mathbf{i}_{s12}$$

In the following transformations we apply two relations:

1° Connection of stator's windings in a star makes it possible to eliminate the zero sequence equation.

2° $\quad \dfrac{d}{dt}(\mathbf{T}_r) = (\omega_c - p\dot{\theta}_r)\mathbf{A}_2 \mathbf{T}_r$

which for the system $\quad \alpha, \beta \rightarrow \omega_c = 0 \quad$ gives

$$\frac{d}{dt}(\mathbf{T}_r) = -p\dot{\theta}_r \mathbf{A}_2 \mathbf{T}_r \tag{3.73}$$

As a result, the equations of the rotor's circuits take the form:

$$0 = R_r \mathbf{i}_{ruv} + \frac{d}{dt}\left(\mathbf{M}_{ruv}\mathbf{i}_{ruv} + \mathbf{M}_{rsuv12}\mathbf{i}_{s12}\right)+$$

$$+ p\dot{\theta}_r \mathbf{A}_2\left(\mathbf{M}_{ruv}\mathbf{i}_{ruv} + \mathbf{M}_{rsuv12}\mathbf{i}_{s12}\right) \tag{3.74}$$

where: $\qquad \mathbf{M}_{ruv} = \begin{bmatrix} L_r & \\ & L_r \end{bmatrix} \qquad \mathbf{M}_{rsuv12} = L_m \begin{bmatrix} \sqrt{\dfrac{3}{2}} & 0 \\ \dfrac{\sqrt{2}}{2} & \sqrt{2} \end{bmatrix} \tag{3.75}$

It is important to note that: $\mathbf{M}_{rsuv12} \neq \mathbf{M}_{sr12uv}^T$, since $\mathbf{W}_i \neq \mathbf{W}_u^T$.

As a result, the system of equations for the electric circuits of the induction motor for this case can be noted as

$$\begin{bmatrix} u_{s12} \\ u_{s23} \\ 0 \\ 0 \end{bmatrix} = \begin{bmatrix} \begin{bmatrix} R_{s1} & -R_{s2} \\ R_{s3} & R_{s2}+R_{s3} \end{bmatrix} & \begin{bmatrix} 0 & 0 \\ 0 & 0 \end{bmatrix} \\ p\dot{\theta}_r L_m \begin{bmatrix} \frac{\sqrt{2}}{2} & \sqrt{2} \\ -\sqrt{\frac{3}{2}} & 0 \end{bmatrix} & \begin{bmatrix} R_r & p\dot{\theta}_r L_r \\ -p\dot{\theta}_r L_r & R_r \end{bmatrix} \end{bmatrix} \begin{bmatrix} i_{s1} \\ i_{s2} \\ i_{ru} \\ i_{rv} \end{bmatrix} +$$

$$+ \begin{bmatrix} L_s \begin{bmatrix} 1 & -1 \\ 1 & 2 \end{bmatrix} & L_m \begin{bmatrix} \sqrt{\frac{3}{2}} & -\frac{\sqrt{2}}{2} \\ 0 & \sqrt{2} \end{bmatrix} \\ L_m \begin{bmatrix} \sqrt{\frac{3}{2}} & 0 \\ \frac{\sqrt{2}}{2} & \sqrt{2} \end{bmatrix} & L_r \begin{bmatrix} 1 & 0 \\ 0 & 1 \end{bmatrix} \end{bmatrix} \frac{d}{dt}\begin{bmatrix} i_{s1} \\ i_{s2} \\ i_{ru} \\ i_{rv} \end{bmatrix} \tag{3.76}$$

The electromagnetic torque expressed in these variables takes the form:

$$T_e = pL_m[(\tfrac{\sqrt{2}}{2}i_{s1} + \sqrt{2}\,i_{s2})i_{ru} - \sqrt{\tfrac{3}{2}}i_{s1}i_{rv}] \tag{3.77}$$

The equations for an induction motor in the form (3.76) make it possible to account for semiconductor elements of converters using variable values of resistances R_{s1}, R_{s2}, R_{s3} encountered in phase windings in a fixed structure system.

3.1.6.2 Model with Untransformed Variables of Electric Circuit of the Rotor

This model finds application in the issues regarding internal asymmetries and control of drives with a slip-ring induction machine. For example, it is applied in the calculations of the start-up of a slip-ring motor with asymmetric resistances during the start-up, atypical systems of connections between phase windings of the rotor, analysis of cascade systems with a slip-ring motor, for instance the Scherbius cascaded system. In the discussed example the electric variables of the stator's circuits undergo axial transformation along u,v axes, while the variables of the rotor's circuits remain untransformed. Under the assumption of the star connection of rotor windings, we have to introduce equations of constraints. This time to ensure symmetry of the resulting equations the reference phase is the one denoted with the number 2 and, as a consequence, the current constraints offer the elimination of the current i_{r2}, while voltage constraints refer inter-phase voltages to the terminals of phase winding 2:

$$\begin{bmatrix} u_{r12} \\ u_{r32} \end{bmatrix} = \underbrace{\begin{bmatrix} 1 & -1 & 0 \\ 0 & -1 & 1 \end{bmatrix}}_{\mathbf{W}_{ur}} \begin{bmatrix} u_{r1} \\ u_{r2} \\ u_{r3} \end{bmatrix} \tag{3.78}$$

or

$$\mathbf{U}_{r13} = \mathbf{W}_{ur}\,\mathbf{U}_{rph} \tag{3.79}$$

In this case, in contrast to (3.66, 3.67) the following is fulfilled:

$$\mathbf{W}_{ir} = \mathbf{W}_{ur}^T \tag{3.80}$$

which results in the symmetry of matrices in equations (3.88). The assumption of the similar course of action as in section 3.1.6.1 leads to the transformation of the system of equations in the following way:

$$\mathbf{T}_s * \left| \mathbf{U}_{sph} = R_s \mathbf{i}_s + \frac{d}{dt}\left(\mathbf{M}_{sph}\mathbf{i}_s + \mathbf{M}_{srph}(\theta_r)\mathbf{i}_r\right)\right.$$

$$\underbrace{\mathbf{T}_s\mathbf{U}_{sph}}_{\mathbf{U}_{s0uv}} = R_s\underbrace{\mathbf{T}_s\mathbf{i}_s}_{\mathbf{i}_{s0uv}} + \frac{d}{dt}\left(\underbrace{\mathbf{T}_s\mathbf{M}_{sph}\mathbf{T}_s^T}_{\mathbf{M}_{s0uv}}\underbrace{\mathbf{T}_s\mathbf{i}_s}_{\mathbf{i}_{s0uv}} + \underbrace{\mathbf{T}_s\mathbf{M}_{srph}(\theta_r)\mathbf{W}_{ir}}_{\mathbf{M}_{sr0uv13}}\mathbf{i}_{r13}\right)$$

where:

$$\mathbf{M}_{sr0uv13} = \frac{\sqrt{2}}{2} L_m \begin{bmatrix} 0 & 0 \\ \sin p\theta_r + \sqrt{3}\cos p\theta_r & 2\sin p\theta_r \\ -\cos p\theta_r + \sqrt{3}\sin p\theta_r & -2\cos p\theta_r \end{bmatrix} \tag{3.81}$$

$$\mathbf{i}_{r13} = \begin{bmatrix} \mathbf{i}_{r1} & \mathbf{i}_{r3} \end{bmatrix} \qquad \mathbf{i}_r = \mathbf{W}_{ir}\mathbf{i}_{r13}$$

After the equation for the zero sequence is disregarded, we obtain:

$$\mathbf{U}_{suv} = R_s\mathbf{i}_{suv} + \mathbf{M}_{suv}\frac{d}{dt}\mathbf{i}_{suv} + \mathbf{M}_{sruv13}\frac{d}{dt}\mathbf{i}_{r13} - p\dot{\theta}_r\mathbf{A}_2\mathbf{M}_{sruv13}\mathbf{i}_{r13} \tag{3.82}$$

where:
$$\mathbf{A}_2 = \begin{bmatrix} 0 & 1 \\ -1 & 0 \end{bmatrix}$$

is a skew-symmetric matrix applied with regard to the relation:

$$\frac{d}{dt}\left(\mathbf{M}_{sruv13}\right) = -p\dot{\theta}_r\mathbf{A}_2\mathbf{M}_{sruv13}$$

For rotor's equations we obtain:

$$\mathbf{W}_{ur}*\Big|\mathbf{U}_{rph} = R_r\mathbf{i}_r + \frac{d}{dt}\left(\mathbf{M}_{rph}\mathbf{i}_r + \mathbf{M}_{rs}\mathbf{i}_s\right)$$

$$\underbrace{\mathbf{W}_{ur}\mathbf{U}_{rph}}_{\mathbf{U}_{r13}} = \underbrace{\mathbf{W}_{ur}R_r\mathbf{W}_{ir}}_{\mathbf{R}_{r13}}\mathbf{i}_{r13} +$$

$$+\frac{d}{dt}\left(\underbrace{\mathbf{W}_{ur}\mathbf{M}_{rph}\mathbf{W}_{ir}}_{\mathbf{M}_{r13}}\mathbf{i}_{r13} + \underbrace{\mathbf{W}_{ur}\mathbf{M}_{rs}\mathbf{T}_s^T}_{\mathbf{M}_{rs130uv}}\underbrace{\mathbf{T}_s\mathbf{i}_s}_{\mathbf{i}_{s0uv}}\right) \tag{3.83}$$

For an asymmetric matrix of resistance of rotor's windings $\mathbf{R}_r$ we obtain

$$\mathbf{R}_{r13} = \mathbf{W}_{ur}R_r\mathbf{W}_{ir} = \begin{bmatrix} R_1+R_2 & R_2 \\ R_2 & R_2+R_3 \end{bmatrix} \tag{3.84}$$

which for the case of the symmetry of the resistance can be reduced to the form:

$$\mathbf{R}_{r13} = R_r\begin{bmatrix} 2 & 1 \\ 1 & 2 \end{bmatrix} \tag{3.85}$$

In a similar way matrix $\mathbf{M}_{r13}$ takes the form:

$$\mathbf{M}_{r13} = L_r\begin{bmatrix} 2 & 1 \\ 1 & 2 \end{bmatrix}$$

Concurrently, matrix $\mathbf{M}_{rs130uv}$ due to its symmetry (3.80) fulfills the relation:

$$\mathbf{M}_{rs130uv} = \mathbf{M}_{sr0uv13}^T \tag{3.86}$$

After abandoning the zero sequence equation for the stator currents the equations for the electric circuits of the rotor take the form:

$$\mathbf{U}_{r13} = \mathbf{R}_{r13}\mathbf{i}_{r13} + \mathbf{M}_{r13}\frac{d}{dt}\mathbf{i}_{r13} + \mathbf{M}_{rs13uv}\frac{d}{dt}\mathbf{i}_{suv} + p\dot{\theta}_r\,\mathbf{M}_{rs13uv}\mathbf{A}_2\mathbf{i}_{suv} \tag{3.87}$$

The above equation applies the relation $\mathbf{A}_2^T = -\mathbf{A}_2$.

As a result of the combination of (3.82) with (3.87), we obtain a system of equations for electric circuits of the slip-ring motor with untransformed variables of the electric circuits of the rotor:

$$
\begin{bmatrix} u_{su} \\ u_{sv} \\ u_{r12} \\ u_{r32} \end{bmatrix} =
\begin{bmatrix} R_s\begin{bmatrix}1 & 0 \\ 0 & 1\end{bmatrix} & v\mathbf{K}(\theta_r) \\ v\mathbf{K}^T(\theta_r) & R_r\begin{bmatrix}2 & 1 \\ 1 & 2\end{bmatrix} \end{bmatrix}
\begin{bmatrix} i_{su} \\ i_{sv} \\ i_{r1} \\ i_{r3} \end{bmatrix} +
$$
$$
+ \begin{bmatrix} L_s\begin{bmatrix}1 & 0 \\ 0 & 1\end{bmatrix} & x\mathbf{N}(\theta_r) \\ x\mathbf{N}^T(\theta_r) & L_r\begin{bmatrix}2 & 1 \\ 1 & 2\end{bmatrix} \end{bmatrix}
\frac{d}{dt}\begin{bmatrix} i_{su} \\ i_{sv} \\ i_{r1} \\ i_{r3} \end{bmatrix}
\tag{3.88}
$$

where:
$$v = \frac{\sqrt{2}}{2}p\dot{\theta}_r L_m \qquad x = \frac{\sqrt{2}}{2}L_m$$

$$\mathbf{K}(\theta_r) = \begin{bmatrix} (\cos p\theta_r - \sqrt{3}\sin p\theta_r) & 2\cos p\theta_r \\ (\sin p\theta_r + \sqrt{3}\cos p\theta_r) & 2\sin p\theta_r \end{bmatrix}$$

$$\mathbf{N}(\theta_r) = \begin{bmatrix} (\sin p\theta_r + \sqrt{3}\cos p\theta_r) & 2\sin p\theta_r \\ (-\cos p\theta_r + \sqrt{3}\sin p\theta_r) & -2\cos p\theta_r \end{bmatrix}$$

The expression for the electromechanical torque takes the form:

$$
\begin{aligned}
T_e = \sqrt{2}pL_m[\cos p\theta_r(\tfrac{1}{2}i_{r1}(\sqrt{3}i_{sv} + i_{su}) + i_{r3}i_{su}) + \\
+ \sin p\theta_r(\tfrac{1}{2}i_{r1}(i_{sv} - \sqrt{3}i_{su}) + i_{r3}i_{sv})]
\end{aligned}
\tag{3.89}
$$

The characteristic property of equations (3.88-3.89) for a slip-ring induction motor with untransformed circuits of the rotor is the dependence of these equations on the angle of the rotation of the rotor θ_r.

3.2 Dynamic and Static Characteristics of Induction Machine Drives

The mathematical models of induction machines developed in the preceding sections will be applied here in the simulations and calculations of typical dynamic states and static characteristics of the drive. In the first stage the presentation will involve dynamic characteristics calculated for standardized electromechanical parameters of a drive. In the subsequent section the equations of motion will be reduced to the steady state and on this basis an equivalent circuit diagram of an induction motor will be derived together with static characteristics, typical parameters and graphical images for the characteristics.

3.2.1 Standardized Equations of Motion for Induction Motor Drive

Despite the common operating principle and a unique description in the form of a mathematical model, induction machines form a class that is considerably distinctive. The range of the rated powers varies from a fraction of a [kW] to the machines exceeding 10 [MW]. Concurrently, speed ratings resulting from the number of pole pairs applied in the construction, typically range for a machine from $p=1$ to $p=6$ pole pairs, and particular manufacturers offer machines with a higher number of pole pairs. Rated voltages applied to supply the primary windings also tend to very across the machines in accordance with the standardized series of voltages, while the majority of the motor run off a 230/400 [V] supply voltage or a high voltage of 6 [kV]. In addition, induction machines are differentiated by the structure of the windings of the secondary side (rotor) and in particular by the shape and profile of the cross-section of the bars in the squirrel-cage rotor. This part is responsible for the increase in the value of resistance R_r, which is the basic parameter which characterizes the mathematical model and for the fact that this parameter is considerably relative to the frequency of the currents in the cage's bars. Squirrel-cage machines, which are distinguished by the tall and slender shape or particular profiles that tend to become thinner towards the air gap are characterized by resistance R_r, whose value increases for higher current frequencies in the cage occurring during motor's start-up. For high frequencies of the current in the cage the leakage reactance definitely dominates in the impedance of the bar in the cage rotor and the current is displaced towards the air gap, which brings a reduction of the active cross-section of the bar and increases resistance. Section 3.1 has dealt with the development of the mathematical models of a motor with constant resistance of rotor's windings R_r hence, they are relevant for slip-ring and single-cage motors with a weak effect of current displacement. Their application with regard to squirrel-cage motors, for instance deep slot motors introduces a considerable error in particular in terms of the characteristics at the phase of start-up and small angular velocities of the rotor. In industrial practice, double-cage induction motors are applied in order to improve the start-up properties [36]. In the upper cage such motors have bars with a smaller diameter, which for higher current frequencies have higher resistance levels. As a result, more complex mathematical models are necessary for their modeling.

In order to rationally perform the standardization of the parameters [82] of various induction motors the equations of motion in flux coordinates (3.56) will be inconsiderably transformed and expressed in the system α,β, ($\omega_c = 0$). It is necessary to rescale the variables regarding the axial fluxes by dividing them by the rated voltage U_{sn}:

$$\varphi = \psi / U_{sn} \tag{3.90}$$

As a result, the equations for the axial fluxes take the form:

$$\dot{\varphi}_{su} = \frac{1}{\sigma}\alpha_s\left(k_r\varphi_{ru} - \varphi_{su}\right) + u_{su}/U_{sn}$$

$$\dot{\varphi}_{sv} = \frac{1}{\sigma}\alpha_s\left(k_r\varphi_{rv} - \varphi_{sv}\right) + u_{sv}/U_{sn}$$

$$\dot{\varphi}_{ru} = \frac{1}{\sigma}\alpha_r\left(k_s\varphi_{su} - \varphi_{ru}\right) - p\Omega_r\varphi_{rv} \tag{3.91}$$

$$\dot{\varphi}_{rv} = \frac{1}{\sigma}\alpha_r\left(k_s\varphi_{sv} - \varphi_{rv}\right) + p\Omega_r\varphi_{ru}$$

In a similar manner, the equation of rotor's motion is transformed and it is multiplied by p, hence standardized to the reference of one pole pair machine:

$$p\dot{\Omega}_r = \frac{p}{J}\left[\frac{pk_rU_{sn}^2}{L_s\sigma}(\varphi_{sv}\varphi_{ru} - \varphi_{su}\varphi_{rv}) - T_l - D\Omega_r\right]$$

This equation is transformed to take the form:

$$\dot{\omega}_e = c_{em}(\varphi_{sv}\varphi_{ru} - \varphi_{su}\varphi_{rv}) - \frac{T_l'}{J'} - \frac{D'}{J'}\omega_e \tag{3.92}$$

where:

$$c_{em} = \frac{p}{J}\frac{pk_rU_{sN}^2}{L_s\sigma} = \frac{p}{J}\frac{2T_b\omega_{sN}^2}{k_s} \tag{3.93}$$

is the electromechanical constant for an induction motor, while

$$J' = \frac{J}{p^2} \qquad D' = \frac{D}{p^2} \qquad T_l' = \frac{T_l}{p} \tag{3.94}$$

represent, respectively: moment of inertia, coefficient of viscous damping and load torque derived for the number of pole pairs $p = 1$.

T_b - is the break torque of the motor in a steady state (3.130)

$\omega_e = p\Omega_r$ - is the angular velocity of the rotor expressed in terms of a motor with a single pole pair, called 'electrical angular speed'.

This version of the equations of motion (3.91) represents standardized equations for an induction motor for which the entire class of single-cage motors, not accounting for current displacement in the rotor's cage, are expressed in terms of a single synchronous velocity (for $p = 1$) regardless of the values of the rated voltage and the number of pole pairs. This is achieved as a result of dividing axial fluxes ψ by the rated voltage U_{sn} (3.90). In this model we have to do with the following independent parameters relative to the design of induction motors:

k_s, k_r - coefficients of magnetic coupling of windings (3.57)

α_s, α_r - coefficients of damping (3.57) (inverse of the time constants)

c_{em} electromechanical constant for a drive (3.93)

The above parameters are standardized, synthetic parameters of the mathematical model of an induction motor in its most simple version that does not account for the magnetic saturation of the core and current displacement in the rotor's cage. They cover each individual motor and make it possible to calculate its dynamic and static characteristics. Other parameters used in the modeling of an induction motor, for example coefficient of windings' leakage σ (3.57) are relative to the ones presented in (3.94) or are involved in them already (for instance number of pole pairs p). The presented method of standardization of equations and parameters is based on [82]. On the basis of the calculations of parameters (3.94) conducted on of data gained from industrial catalogues from several meaningful manufacturers for a few dozen of squirrel-cage motors with various power ratings and rated voltages, the table found below has been developed (Table 3.1).

Table 3.1 Standardized parameters of typical induction squirrel-cage motors

Rated power/ standarized parameters	1	10	100	1000	3000
	[kW]	[kW]	[kW]	[kW]	[kW]
$k_s = k_r$ [-]	0.96	0.975	0.978	0.981	0.983
α_s [1/s]	13.0	5.0	2.3	1.4	1.1
α_r [1/s]	15.0	6.0	3.0	1.7	1.3
$c_{em} \times 10^{-8}$ $[s^{-4}]$	40.0	18.0	4.5	1.8	0.9
σ [-]	0.080	0.050	0.042	0.035	0.030
γ [-]	210	190	80	55	42

It contains standardized parameters for a wide range of squirrel-cage induction motors with basic design and a small influence of current displacement in the rotor's cage. Beside the basic parameters (3.94) it contains relative parameters (3.57), which are encountered in several versions of the mathematical model. The data in Table 3.1 give mean values of the parameters for the group of examined motors with the number of pole pairs $p = 1...5$. The research that follows serves for the purpose of setting an example and illustration and is based on four motors from the group of 60 motors that were used for the preparation of Table 3.1. The selected motors are representative of the groups of small, medium and large power ones, respectively. Their parameters are presented below; one can make an effort to compare them with the parameters in Table 3.1.

Small power motor (S1)

$P_n = 5.5 \, [kW] \quad U_n = 400 \, [V] \quad I_n = 11.3 \, [A] \quad T_n = 55.3 \, [Nm]$

$n_n = 950 \, [r \, / \min] \quad p = 3 \quad s_n = 0.05$

$\alpha_s = 5.4 \; [1 \, / \, s] \quad \alpha_r = 7.0 \; [1 \, / \, s] \quad L_s = 0.167 \; [H] \quad L_m = 0.1627 \; [H]$

$\sigma = 0.052 \quad k_s = k_r = 0.974 \quad R_s = 0.90 \; [\Omega] \quad R_r = 1.17 \; [\Omega]$

$c_{em} = 32 \times 10^8 \; [s^{-4}] \quad J = 0.04 \, [Nms^2]$ (3.95)

$T_b / T_n = 3.5 \quad T_{st} / T_n = 2.7 \quad I_{st} / I_n = 5.8$

Medium power motor - high voltage (S2H)

$P_n = 315 \, [kW] \quad U_n = 6000 \, [V] \quad I_n = 36.5 \, [A] \quad T_n = 2034 \, [Nm]$

$n_n = 1480 \, [r \, / \min] \quad p = 2 \quad s_n = 0.014$

$\alpha_s = 1.6 \; [1 \, / \, s] \quad \alpha_r = 3.1 \; [1 \, / \, s] \quad L_s = 1.2393 \; [H] \quad L_m = 1.208 \; [H]$

$\sigma = 0.0492 \quad k_s = k_r = 0.975 \quad R_s = 1.93 \; [\Omega] \quad R_r = 3.85 \; [\Omega]$ (3.96)

$c_{em} = 3.15 \times 10^8 \; [s^{-4}] \quad J = 5.5 \, [Nms^2]$

$T_b / T_n = 2.1 \quad T_{st} / T_n = 1.0 \quad I_{st} / I_n = 4.8$

Medium power motor – low voltage (S2L)

$P_n = 560 \, [kW] \quad U_n = 400 \, [V] \quad I_n = 975 \, [A] \quad T_n = 3590 \, [Nm]$

$n_n = 1450 \, [r \, / \min] \quad p = 2 \quad s_n = 0.034$

$\alpha_s = 1.7 \; [1 \, / \, s] \quad \alpha_r = 2.5 \; [1 \, / \, s] \quad L_s = 3.342 \; [mH] \quad L_m = 3.275 \; [mH]$

$\sigma = 0.040 \quad k_s = k_r = 0.98 \quad R_s = 5.7 \; [m\Omega] \quad R_r = 8.4 \; [m\Omega]$ (3.97)

$c_{em} = 3.1 \times 10^8 \; [s^{-4}] \quad J = 14.0 \, [Nms^2]$

$T_b / T_n = 2.8 \quad T_{st} / T_n = 1.2 \quad I_{st} / I_n = 5.7$

High power motor (S3)

$P_n = 2500 \, [kW] \quad U_n = 3400 \, [V] \quad I_n = 490 \, [A] \quad T_n = 8000 \, [Nm]$

$n_n = 2920 \, [r \, / \min] \quad p = 1 \quad s_n = 0.028$

$\alpha_s = 1.2 \; [1 \, / \, s] \quad \alpha_r = 1.6 \; [1 \, / \, s] \quad L_s = 0.0719 \; [H] \quad L_m = 0.07085 \; [H]$

$\sigma = 0.030 \quad k_s = k_r = 0.985 \quad R_s = 0.086 \; [\Omega] \quad R_r = 0.115 \; [\Omega]$ (3.98)

$c_{em} = 1.1 \times 10^8 \; [s^{-4}] \quad J = 45.0 \, [Nms^2]$

$T_b / T_n = 2.9 \quad T_{st} / T_n = 1.0 \quad I_{st} / I_n = 5.7$

3.2.2 Typical Dynamic States of an Induction Machine Drive – Examples of Trajectories of Motion

Equations of motion for an induction machine drive presented in Section 3.1.5 for various coordinate systems have 5 degrees of freedom: four for electric variables in axes u,v of the transformed system of electric variables of a machine and one degree of freedom for the mechanical variable in the form of the angle of rotation of the rotor. It is possible to have to do with more than one coordinate of the mechanical motion, for instance under the assumption of a flexible drive shaft, where the angles of shaft rotation on the side of the motor and on the side of the drive are different and this difference corresponds to the torsion angle of the shaft. The particular dynamic states formally constitute distinct initial conditions for a system of ordinary differential equations that form the mathematical model of a drive. The number of these states can be infinite for various initial conditions; however, from the practical point of view a few of them are encountered most frequently and hence they deserve a more in-depth analysis here. The typical dynamic states for an induction machine drive include: start-up from standstill, motor start-up under non-zero rotational speed, change of a load, reversal – i.e. change in the direction of rotation and electrical braking. In the first order we will discuss dynamic states, which can be easily and effectively solved using models with transformed variables of the stator and rotor. The presentation will cover, respectively: start-up from a standstill, start-up under angular speed different from zero (repeated start-up and reversal) and drive regime of operation under cyclic variable load. The following stage will include the presentation of dynamic states, which can be conveniently calculated on the basis of models in which single side of a motor is untransformed. Such cases include the issue of a soft-start and DC braking of an induction motor.

3.2.2.1 Start-Up during Direct Connection to Network

A computer simulation of this dynamic state is performed for zero initial conditions. It is possible to conduct calculations by application of various versions of the mathematical model in a transformed coordinate system. It is beneficial to apply the model in current coordinates in α,β axes (3.61) or in flux coordinates (3.59). Both systems are in the standard form and the application of the α,β system enables one to achieve the natural frequency of voltages in the transformed system. The calculations of the current curves require the application of a reverse transformation $\mathbf{T}_s^T$ (3.33) under the assumption that $\omega_c = 0$, $i_{s0} = 0$. For the case of the model in (3.59) with flux coordinates, the transformation of $\mathbf{T}_s^T$, which leads to transformation of currents in the axial α,β system has to be preceded by the transformation of (3.51) to convert flux variables into axial currents. As a result, we obtain:

$$
\mathbf{i}_s = \begin{bmatrix} i_{s1} \\ i_{s2} \\ i_{s3} \end{bmatrix} = \sqrt{\frac{2}{3}} \frac{1}{L_m} \frac{1-\sigma}{\sigma} \begin{bmatrix} 1 & 0 \\ -\dfrac{1}{2} & \dfrac{\sqrt{3}}{2} \\ \dfrac{1}{2} & -\dfrac{\sqrt{3}}{2} \end{bmatrix} \begin{bmatrix} \dfrac{1}{k_r}\psi_{s\alpha} - \psi_{r\alpha} \\ \dfrac{1}{k_r}\psi_{s\beta} - \psi_{r\beta} \end{bmatrix} \tag{3.99}
$$

The examples of trajectories of the motion presented in this section apply a model in the flux coordinates (3.91-3.92), while phase currents are obtained using (3.99). The calculations of the trajectories of start-up from a stall was conducted for three induction motors with respective, small, medium and large power, whose parameters are given in (3.95), (3.97) and (3.98), respectively. Fig. 3.7 presents the curves of the phase current, electromagnetic torque and angular speed for an unloaded small power motor whose moment of inertia on the shaft is $J = 3J_s$, where J_s denotes the moment of inertia of the motor's rotor. The waveforms of the same type are presented in Figs. 3.8 and 3.9 for medium and large power motors. In Fig. 3.8 for the medium power motor the trajectory of electromagnetic torque, i.e. the relation of the torque and angular speed is additionally presented.

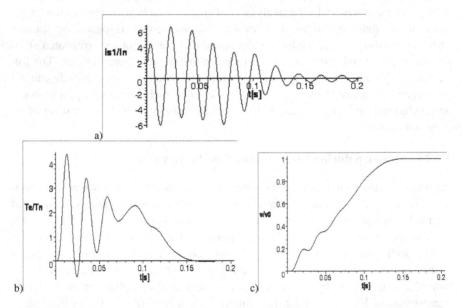

Fig. 3.7 a) phase current b) electromagnetic torque c) angular velocity during free acceleration after direct connecting to the supply network, for the small power motor. Motor is unloaded and $J = 3J_s$

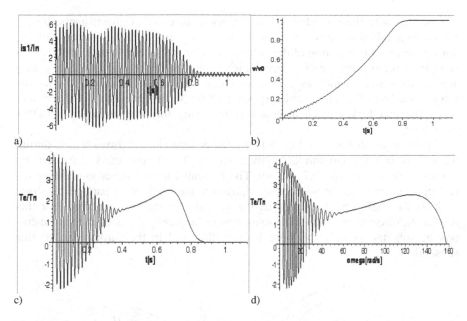

Fig. 3.8 a) phase current b) angular velocity c) electromagnetic torque time history d) torque trajectory, during free acceleration after direct connecting to the supply network, for the medium power motor, while $T_l = 0$ and $J = 2J_s$

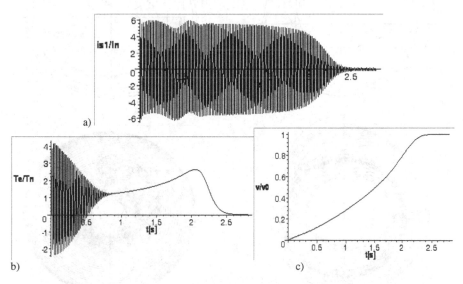

Fig. 3.9 a) phase current b) electromagnetic torque c) angular velocity, during free acceleration after direct connecting to the supply network, for the high power motor, while $T_l = 0$ and $J = 2J_s$

One can note the considerable oscillatory changes of electromagnetic torque with a large initial value $(4...6\ T_n)$ and a frequency similar to the network voltage during the direct connection of the induction motor to the supply network. This results from the occurrence of an aperiodic component of the magnetic flux generated by the stator's windings Ψ_s in association with slowly increasing flux of the rotor's windings Ψ_r. The oscillatory state of the torque occurs until the instant when the two fluxes reach a steady state during the rotation over a circular trajectory.

This is well illustrated in Fig. 3.10 by the presentation of fluxes Ψ_s, Ψ_r during the start-up of the small and large power motors. This figure refers to the start-up curves presented in Figs. 3.7 and 3.9. The presented torque waveforms during direct connection to the network pose a hazard to the mechanical parts of the drive such as the shaft, clutch as well as the very device that is connected. For this reason the direct connection is more and more frequently replaced with the methods of soft-start, which are more widely discussed in the further part of this subsection.

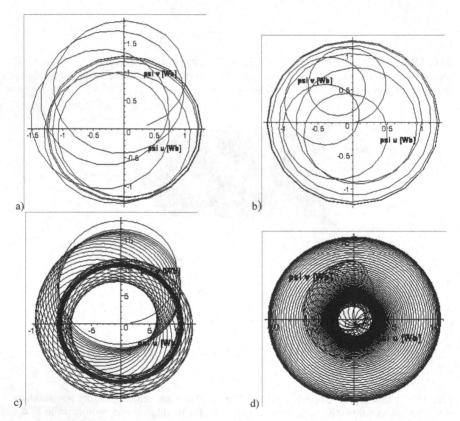

Fig. 3.10 Magnetic flux vector trajectory in the air gap of induction motor during free acceleration after direct connection to the network: a) Ψ_s b) Ψ_r for the small power motor c) Ψ_s d) Ψ_r for the high power motor

3.2.2.2 Reconnection of an Induction Motor

Reconnection is the term which denotes the dynamic state encountered during engaging a motor during coasting i.e. for a non-zero angular speed. One can note the difference between a reconnection: resulting from a breakdown in power supply for 0.3...1 [s], when the magnetic field in the motor from the weakening current of the cage does not decay completely and a reconnection after a breakdown of power supply of over 1 [s], when the magnetic field decays completely. For fast

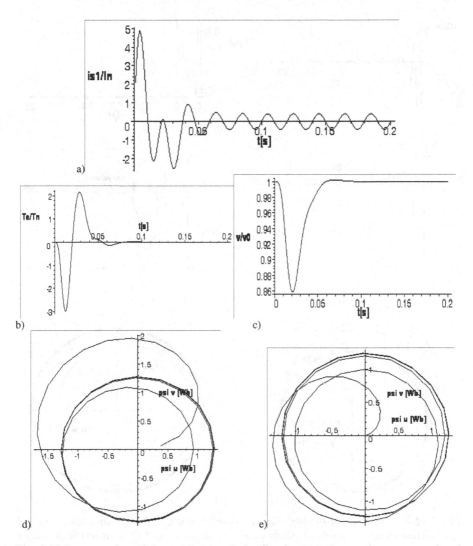

Fig. 3.11 Reconnection of the small power induction motor at the synchronous speed and zero current initial conditions: a) stator current b) electromagnetic torque c) angular speed d) stator flux Ψ_s e) rotor flux Ψ_r trajectory

reconnections one has to take into account non-zero initial conditions for electric variables on the rotor's windings (Ψ_r, $\mathbf{i}_r$) since the formed electromagnetic torque is then considerably dependent on the phase of the voltage connected to the stator's terminals. Just as in the case of synchronizing a synchronous machine with the network it is possible to undertake a reconnection in accordance with the

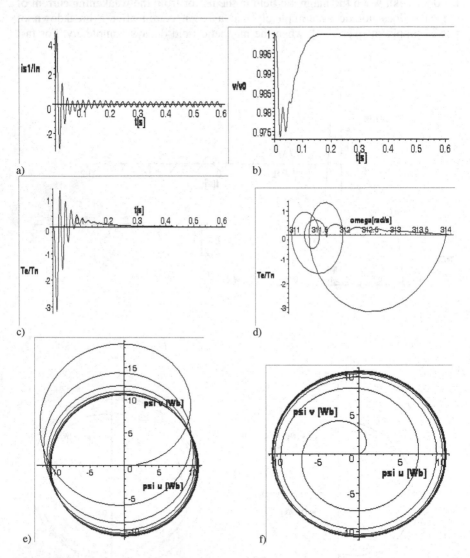

Fig. 3.12 Reconnection of the high power induction motor at the synchronous speed and zero current initial conditions: a) stator current b) angular speed c) electromagnetic torque d) torque-speed trajectory e) stator flux Ψ_s f) rotor flux trajectory Ψ_r

phase, whose characteristics include low values of the connection currents and electromagnetic torque. In contrast, in the most adverse case, such reconnection can occur in the circumstances of the opposition between the phases of network voltages and the one on the motor's terminals. In the latter case we have to do with a large connection current, which is hazardous for the drive due to a torque surge. For this reason it is best to avoid fast and direct reconnections of an induction motor into the network. For large power motors the duration period of the hazardous reconnection lasts for about 0.8-1.0 [s], while for small and medium size ones the breakdown the time is 0.5 [s]. The reconnection of the motor after the period of the voltage breakdown over 1 [s] could be considered as the connection from the zero initial conditions of electric variables. The curve of the current and torque after such a reconnection is relative to the angular speed $\Omega_r(0)$ after which the reconnection has occurred; however, it does not exceed the values that are present during the direct connection of a motor during standstill.

The curves of the currents, electromagnetic torque and angular speed and magnetic fluxes in the stator and rotor after connection to the network for synchronous speed are presented in Figs. 3.11 and 3.12 for small and large power motors, respectively.

3.2.2.3 Drive Reversal

The term reversal denotes turning on a drive under a speed that is reverse to the direction of the rotation resulting from the sequence of phases of the supply network after turning on. A reversal may be associated with the needs of a technology or

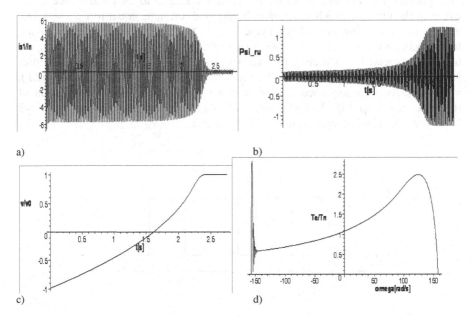

Fig. 3.13 Reversal of the medium power induction motor drive ($J = 2J_s$): a) stator current b) rotor flux c) relative rotor speed d) torque-speed trajectory

may form a type of braking resulting from a counter current. In this case the reversal should be discontinued when the rotational speed of the drive is close to zero and before the drive starts its rotation in the opposite direction. The dynamic curves of the current and electromagnetic torque in the initial phase of the reversal are similar to the values of the curves for these values during start-up under direct turning on. The calculations of the trajectories of the drive motion apply zero initial conditions for currents (fluxes), by assuming an adequately long interval in the supply (1 [s] or more) and an angular speed similar to the synchronous speed but with a negative sign. Examples of curves during the reversal of a middle power induction motor (3.97) are presented in Fig 3.13.

3.2.2.4 Cyclic Load of an Induction Motor

Load on a motor may contain a variable term. In this case steady operation state, understood as fixed point, is not achieved by a drive on its characteristics. In contrast, the drive operates in a closed trajectory when the operating regime becomes steady. For high inertia of the drive and a relatively small variable term of the load torque the trajectory of the motion is close to a fixed point. In the opposite case the trajectory of the drive's motion forms a curve that considerably diverges from static characteristic. The trajectory is relative to the value of the variable term, frequency of the load variation and moment of inertia relative to the motor's shaft. The examples of the drive regime of operation for a large component of variable load are presented in Figs. 3.15, 3.16, 3.17. Fig. 3.14 that precedes them presents the stepwise variable load torque acting on the induction machine's shaft. Fig. 3.15 presents the dynamic curve for the mean load equal to the rated torque in the cycle of the load, in which for $\tau_m = 0.2\tau_l$ the load torque is equal to $T_{lmax} = 4.2T_n$, and in the remaining part of the cycle $T_{lmin} = 0.2T_n$ with the frequency of the torque variation $f_l = 3$ [Hz]. The trajectory of the electromagnetic torque in respect to angular speed forms a closed curve with the shape of an eight. Fig. 3.16 presents the cases of the identical load on a drive but for frequency $f_l = 6$ [Hz] and $f_l = 15$ [Hz]. As a consequence, there is a considerable reduction and limitation of the trajectory loop. Fig. 3.17 presents a variable load with the frequency of $f_l = 3$ [Hz] and $f_l = 2$ [Hz] at the boundary of drive break.

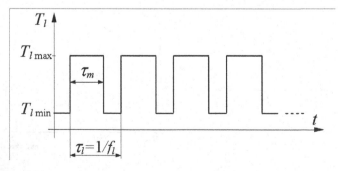

Fig. 3.14 The cyclic load torque of the drive

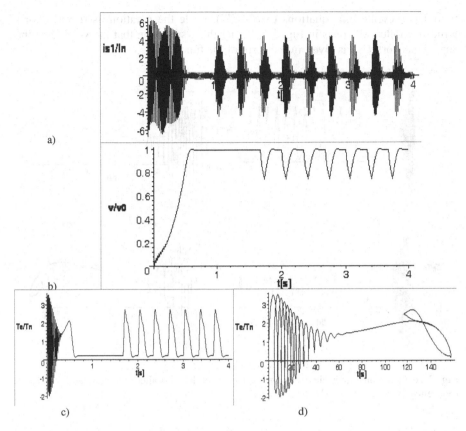

Fig. 3.15 Induction motor drive characteristics under periodic load changes: $f_l = 3[\text{Hz}]$, $T_{lav} = T_n$, $T_{lmax} = 4.2\ T_n$: a) phase current b) rotational speed c) torque time history d) torque-speed trajectory

3.2.2.5 Soft-Start of an Induction Motor for Non-simultaneous Connection of Stator's Windings to the Network

As it was mentioned earlier (3.2.2.1), the direct connection of an induction motor to the supply network results in the high value of an oscillatory component of electromagnetic torque during start-up. Besides, there is considerable value of the start-up current. This is well illustrated in Figs. 3.7-3.9 for small, medium and high power motors. The oscillations of the electromagnetic torque can be considerably limited and, hence, it is virtually possible eliminate their effect as a result of the application of synchronized connection of phase windings in the network. In the first stage, two clamps of the stator's windings are connected with a suitable synchronization with the network and in the second stage the third clamp is connected with an adequate phase delay. A computer simulation of the examples that illustrate this issue can be conveniently conducted by use of a model of an induction motor with untransformed electric variables of the stator's windings. This model is presented in equations (3.76-3.77), while the notation used for stator's windings follow the ones in Fig. 3.6. Under the assumption that the voltage of the supply network U_{s12} is given in the form of the function

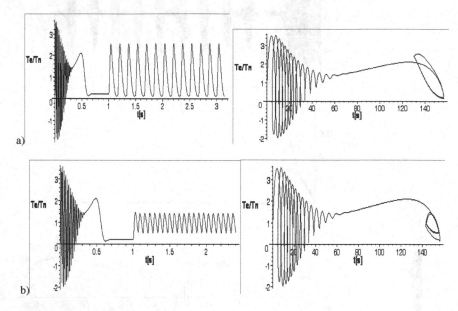

Fig. 3.16 Induction motor electromagnetic torque under periodic load changes for various frequency values: a) $f_l = 6$[Hz] b) $f_l = 15$[Hz]

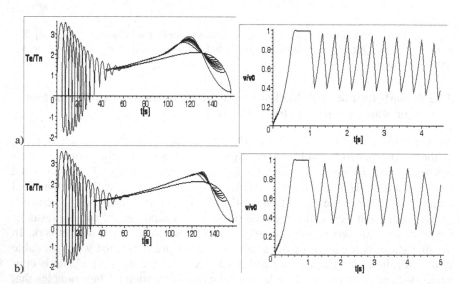

Fig. 3.17 Induction motor performance under periodic load changes close to the break torque loading: a) $f_l = 3$[Hz], $T_{lmax} = 5.2\,T_n$, $T_{lav} = 1.87\,T_n$, b) $f_l = 2$[Hz], $T_{lmax} = 3.5\,T_n$, $T_{lav} = 1.85\,T_n$

$$U_{s12} = U_m \sin(\omega_L t + \gamma) \qquad (3.100)$$

the soft-start of the motor follows for the phase angle:

$$\gamma = \chi_1 \pi \quad \chi_1 = 0.46...0.50 \ , \qquad (3.101)$$

while the connection of the remaining, third, clamp follows with a phase delay:

$$\delta = 2\pi/3 - \chi_2 \quad \chi_2 = 0.43...0.36 \qquad (3.102)$$

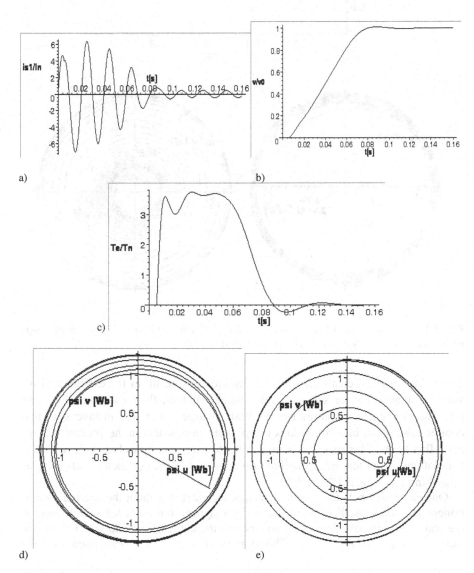

Fig. 3.18 Delay soft-start of the small power induction motor ($J = 3J_s$): a) phase current b) angular speed c) electromagnetic torque d) stator flux trajectory e) and rotor flux trajectory

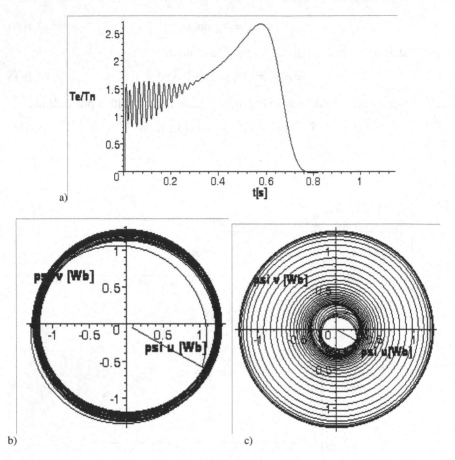

Fig. 3.19 Delay soft-start of the medium power induction motor ($J = 2J_s$): a) electromagnetic torque b) stator flux trajectory c) rotor flux trajectory

The values of coefficients χ_1, χ_2 are determined as a result of the calculations involving simulations for selected squirrel-cage motors; their extreme values are relevant with regard to motors from a small to large power. The instances of such connections, result in a virtual lack of aperiodic component in the generated magnetic flux Ψ_s, Ψ_r. This issue has been illustrated using examples based on computer simulations for motors from small to large power, and the obtained results are presented in Figs. 3.18, 3.19 and 3.20.

One can easily note the smooth curve of the current without the aperiodic component and small oscillations of the electromagnetic torque at the initial stage of the start-up. This comes as a consequence of the curve of trajectories of magnetic fluxes presented in the figures. This type of start-up, that is associated with the

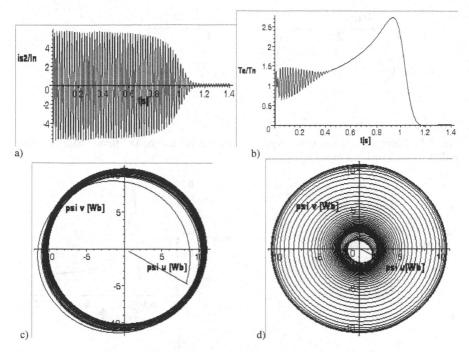

Fig. 3.20 Delay soft-start of the high power induction motor ($J = J_s$): a) phase current b) electromagnetic torque c) stator flux trajectory d) and rotor flux trajectory

need to apply power semiconductor switches, has considerably more advantages than direct connection, and this can be concluded from a comparison between the above illustrations and results presented in Figs. 3.7, 3.8, 3.9 for the same motors.

3.2.2.6 DC Braking of an Induction Motor

Braking using direct current involves DC supply to the suitably connected stator's windings in such a way that enables the potentially high constant magnetic flux in which the rotor is put in motion. The current produced by rotor windings as a consequence of induction combines with the magnetic field thus producing braking torque, which approaches idle run for a DC supply, i.e. the condition when the rotor is stalled. Fig.3.21 presents two typical layouts from among the list of the possible connections between the stator's windings for braking.

The modeling is based on equations (3.76-3.77) for an induction motor for untransformed currents of stator's windings. For a three-phase system of connections (Fig. 3.21a), we directly apply equations (3.76-3.77) by assuming that:

$$u_{s12} = U_{DC} \quad u_{s23} = 0 \quad i_{s3} = -i_{s1} - i_{s2} \tag{3.103}$$

For two-phase power supply during braking (Fig. 3.21b) the following constraints are applicable:

$$i_{s2} = -i_{s1} \quad u_{s12} = U_{DC} \tag{3.104}$$

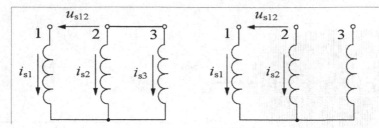

Fig. 3.21 Connection of induction motor stator windings for the DC three-phase breaking and the two-phase breaking

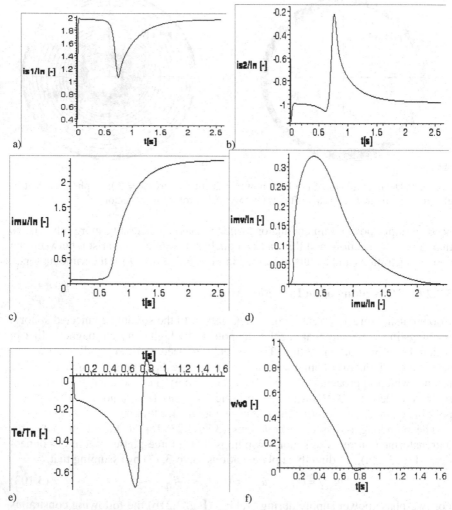

Fig. 3.22 3-phase DC breaking of the small power induction motor with $J = 5J_s$, $i_{DC} = 2I_n$:
a) stator current i_{s1} b) stator current i_{s2} c) magnetizing current i_m d) MMF trajectory i_{mu} / i_{mv} e) electromagnetic torque f) rotational speed

Consequently, the stator windings connected in a star have a single degree of freedom, for which we assume the variable i_{s1}. Using (3.104) and after elimination of the latter of equations (3.76), we obtain the model for this type of braking:

$$
\begin{bmatrix} U_{DC} \\ 0 \\ 0 \end{bmatrix} = \begin{bmatrix} R_{s1}+R_{s2} & 0 & 0 \\ -p\dot\theta_r L_m \frac{\sqrt2}{2} & R_r & p\dot\theta_r L_r \\ -p\dot\theta_r L_m \sqrt{\frac{3}{2}} & -p\dot\theta_r L_r & R_r \end{bmatrix} \begin{bmatrix} i_{s1} \\ i_{ru} \\ i_{rv} \end{bmatrix} +
$$

$$
+ \begin{bmatrix} 2L_s & L_m\sqrt{\frac{3}{2}} & -L_m\frac{\sqrt2}{2} \\ L_m\sqrt{\frac{3}{2}} & L_r & 0 \\ -L_m\frac{\sqrt2}{2} & 0 & L_r \end{bmatrix} \frac{d}{dt}\begin{bmatrix} i_{s1} \\ i_{ru} \\ i_{rv} \end{bmatrix}
$$

(3.105)

$$
T_e = -pL_m\sqrt2\, i_{s1}(\tfrac{1}{2}i_{ru} + \sqrt{\tfrac{3}{2}}\, i_{rv})
$$

On the basis of the obtained versions of the mathematical model, simulations were conducted for braking of a small and large power motors (3.95, 3.98) for a braking current, which in the steady state is equal to $i_{DC} = 2I_n$. The characteristic waveforms are presented in Figs. 3.22 to 3.25.

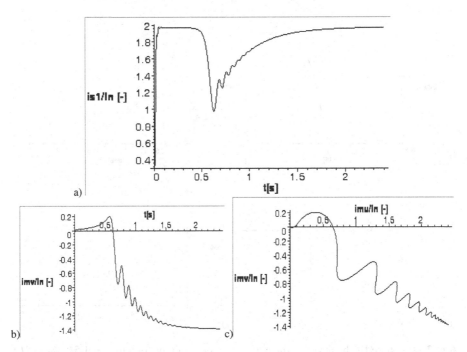

a)

b)   c)

Fig. 3.23 2-phase DC breaking of the small power induction motor with $J = 5J_s$, $i_{DC} = 2I_n$.: a) stator current i_{s1} b) magnetizing current i_m c) MMF trajectory i_{mu} / i_{mv} d) electromagnetic torque e) rotational speed

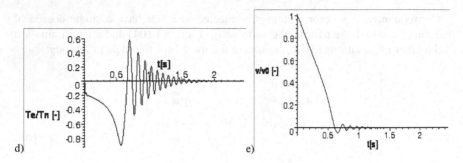

Fig. 3.23 (*continued*)

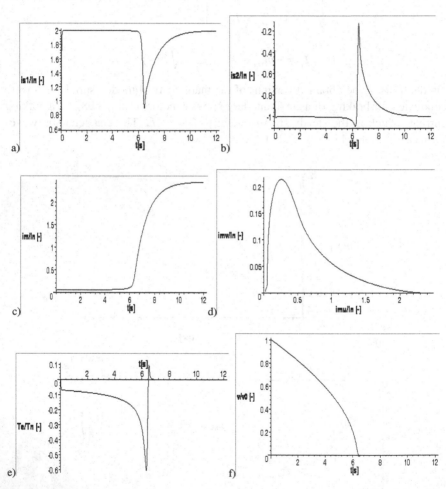

Fig. 3.24 3-phase DC breaking of the high power induction motor with $J = 1.5J_s$, $i_{DC} = 2I_n$:
a) stator current i_{s1} b) stator current i_{s2} c) magnetizing current i_m d) MMF trajectory i_{mu} / i_{mv}
e) electromagnetic torque f) rotational speed

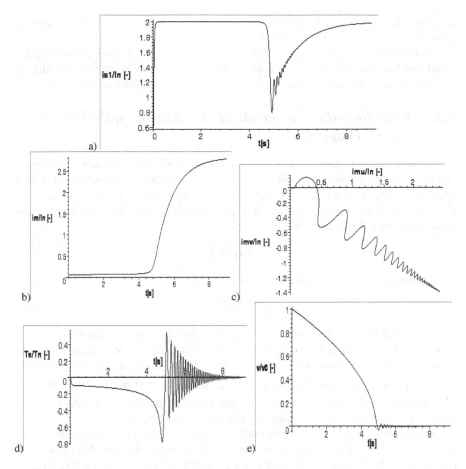

Fig. 3.25 2-phase DC breaking of the high power induction motor with $J = 1.5J_s$, $i_{DC} = 2I_n$:
a) stator current i_{s1} b) magnetizing current i_m c) MMF trajectory i_{mu} / i_{mv} d) electromagnetic torque e) rotational speed

The closer familiarity with the results of calculations for DC breaking leads to the following general conclusions:

- 2-phase braking is considerably more effective than 3-phase braking with direct current; however, its characteristics include oscillations of torque and speed in the final phase of braking. This results from the lack of damping of the clamped circuits in the windings in phase 2 and 3
- magnetizing current during braking is quite small and is definitely smaller than the magnetizing current during symmetric motor regime. After the rotor is stalled the magnetizing current reaches the value of i_{DC}. Hence, the saturation of the magnetic circuit over the entire range of speeds during braking is similar to characteristics of motor regime of operation and the applied models with constant parameters remain in the same precision range as during motor regime. This

concerns the supply of the motor with the direct voltage, for which current i_{DC} does not exceed several times the rated current.

- considerable differences are absent from the dynamic curves of braking for small and large power motors. Smaller motors tend to brake more dynamically, in accordance with the larger value of the electromechanical constant c_m (3.93).

3.2.3 Reduction of a Mathematical Model to an Equivalent Circuit Diagram

A dynamic system, such as electric drive described with ordinary differential equations for given initial conditions and input functions, is characterized with a specific trajectory of the motion. This trajectory represents the history of all variables in a system. The steady state of such a system occurs when the trajectory is represented by a fixed point, that is

$$\left\{\varphi^t(\mathbf{q})\right\} = \left\{\mathbf{q}\right\}$$

or by a periodic function with the period of T, when

$$\varphi^{t+T} = \varphi^T \tag{3.106}$$

For an electric drive this occurs when variables in a system forming the vector of generalized coordinates $\mathbf{q}$ are either constant functions or periodically variable ones. In a induction motor drive we can assume in an idealized way that the steady state occurs when the angular speed is constant, i.e. $\Omega_r = \dot{\theta}_r = const$ and the electric currents which supply the windings are periodic functions with the period in conformity with the voltages enforcing the flow of the currents.

One can note that the history of both the supply voltages and the resulting currents is relative to the transformation of the co-ordinates of the system, as presented in the models of the motor in α,β, d,q or x,y axes (3.61 – 3.65). In a x,y system rotating with the speed $\omega_c = \omega_s = p\omega_f$ the symmetric system of sinusoidal voltages supplying phase windings as a result of transformation (3.37) is reduced to constant voltages. In such coordinate system the steady state literally means a fixed point on the trajectories of all variables. The situation will be different for a steady state in the case of asymmetry of the supply voltages or cyclically variable load torque. In such a case the steady state will be characterized by periodically variable waveforms of electric currents and angular speed, while in the speed waveform the constant component will form the predominant element. The acquaintance with steady states is relevant for the design and exploitation of a drive since it provides information regarding its operating conditions and, hence, forms the basis for the development of strategies regarding methods of drive control. The familiarity with the steady states makes it possible to determine the characteristics of the drive, i.e. functional relations between variables that form the sets of constant points on a trajectory and ones that are time invariable. For the reasons given here the steady state of the induction motor drive can be conveniently described in axial coordinates x,y. Therefore, we will take as the starting point the transformed equations (3.60) in current coordinates, which after the substitution $\omega_c = \omega_s$ gives:

$$\begin{bmatrix} \mathbf{u}_{sxy} \\ 0 \end{bmatrix} = \begin{bmatrix} R_s & 0 \\ 0 & R_r \end{bmatrix} \begin{bmatrix} \mathbf{i}_{sxy} \\ \mathbf{i}_{rxy} \end{bmatrix} +$$
$$+ \frac{d}{dt} \begin{bmatrix} L_s & L_m \\ L_m & L_r \end{bmatrix} \begin{bmatrix} \mathbf{i}_{sxy} \\ \mathbf{i}_{rxy} \end{bmatrix} - \begin{bmatrix} X_s & X_m \\ sX_m & sX_r \end{bmatrix} \begin{bmatrix} \mathbf{A}_2 \mathbf{i}_{sxy} \\ \mathbf{A}_2 \mathbf{i}_{rxy} \end{bmatrix}$$
(3.107)

where: $X_s = \omega_s L_s \quad X_r = \omega_s L_r \quad X_m = \omega_s L_m$

while $s = \dfrac{\omega_s - p\Omega_r}{\omega_s}$ (3.108)

- is the slip of the rotor speed in relation to the rotating magnetic field (see 3.63). We assume that the steady state forms the fixed point of the trajectory $\varphi'(\mathbf{q}) = \mathbf{q}$, hence, it denotes the constant angular speed $\Omega_r = const$ and the constant slip $s = const$. This condition is possible due to the constant values of currents $\mathbf{i}_{sxy}, \mathbf{i}_{rxy}$, and, as a result, the constant electromagnetic torque T_e. This requires the constant supply voltages after the transformation of x,y, which take the following form in accordance with (3.37):

$$\mathbf{u}_{sxy} = \begin{bmatrix} u_{sx} \\ u_{sy} \end{bmatrix} = \sqrt{\frac{3}{2}} \sqrt{2} U_{ph} \begin{bmatrix} \cos\gamma \\ \sin\gamma \end{bmatrix}$$
(3.109)

The form of voltages (3.109) suggests the introduction of complex values:

$$\underline{U}_s = u_{sx} + ju_{sy} = \sqrt{3} U_{ph} e^{j\gamma} = \sqrt{3} \underline{U}_{ph}$$
(3.110)

where: U_{ph} - is the RMS value of the voltage supplying the phase of the motor. Subsequently, we can substitute:

$$\mathbf{i}_s = \underline{I}_s = i_{sx} + ji_{sy} \qquad \mathbf{i}_r = \underline{I}_r = i_{rx} + ji_{ry}$$
(3.111)

In the following transformations of equations (3.107) the latter of the equations in each pair is multiplied by the imaginary unit j and is added to the first of the equations, thus giving the equations for a complex variable. For the stator we obtain:

$$\underline{U}_s = R_s \underline{I}_s + jX_s \underline{I}_s + jX_m \underline{I}_r$$
(3.112)

Here we have applied: $\dfrac{d}{dt} \underline{I}_s = 0$, $\dfrac{d}{dt} \underline{I}_r = 0$, which results from the steady state and

$$\mathbf{A}_2 \mathbf{i}_{sxy} = \begin{bmatrix} & 1 \\ -1 & \end{bmatrix} \begin{bmatrix} i_{sx} \\ ji_{sy} \end{bmatrix} = \begin{bmatrix} ji_{sy} \\ -i_{sx} \end{bmatrix}$$

which after addition of row vectors leads to:

$$\mathbf{A}_2 \mathbf{i}_{sxy} \Rightarrow -j\underline{I}_s .$$

As we perform similar operations for the other pair of equations, i.e. rotor's equations, and dividing this equation by slip s, we obtain:

$$0 = \frac{R_r}{s}\underline{I}_r + jX_m\underline{I}_s + jX_r\underline{I}_r \qquad (3.113)$$

This makes it possible to develop an equivalent circuit for an induction motor in the steady state as a result of merging equations (3.112, 3.113) in the form of a two port, using a common magnetizing reactance term jX_m. The equivalent circuit in the form in Fig. 3.26, beside the voltage and current relations presented in every two port, also realizes in an undisturbed manner the energetic relations occurring in the steady state. This comes as a result of the application of orthogonal transformations that preserve scalar product and quadratic forms in the transformation of equations.

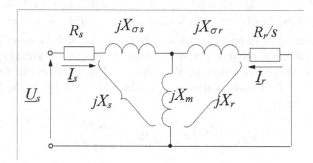

Fig. 3.26 Equivalent circuit of an induction motor for the steady state

In this circuit we have to do with a resistance term R_r/s, which realizes in the energetic sense both Joule's losses in the rotor windings and the mechanical output of the drive transferred via the machine's shaft as the product of torque T_e and the angular speed of the shaft Ω_r. Hence the resistance term can be divided into two terms: R_r, $R_r(1-s)/s$, which realize the losses of the power in the stator's windings and mechanical power P_m, as it is presented in Fig. 3.27. The following components of the electric power are encountered in the equivalent diagram:

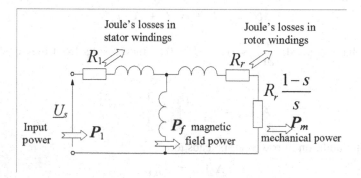

Fig. 3.27 Equivalent circuit of induction motor with physical interpretation of electric power components

$$P_1 = |\underline{U}_s||\underline{I}_s|\cos\varphi_s \qquad \text{input power}$$

$$P_{els} = \underline{I}_s^2 R_s \qquad \text{Joule's losses in stator windings}$$

$$P_f = \underline{I}_r^2 R_r / s \qquad \text{air gap field power} \qquad (3.114)$$

$$P_{elr} = \underline{I}_r^2 R_r \qquad \text{Joule's losses in rotor windings}$$

$$P_m = \underline{I}_r^2 R_r \frac{1-s}{s} \qquad \text{mechanical power}$$

$$P_1 = P_{els} + P_f$$

The energy balance for a 3-phase machine is preserved due to the fact that $\underline{U}_s = \sqrt{3}\underline{U}_{ph}$, hence, the transformed power is three times higher than the power of a single phase. In the analysis of the expression for the mechanical power output of an induction motor drive we can distinguish the following areas of operation:

1. for $0 < s < 1$ $\quad P_f > 0, \quad P_m > 0 \quad$ - motor regime
2. for $s < 0$ $\quad P_f < 0, \quad P_m < 0 \quad$ - generating regime
3. for $s > 1$ $\quad P_f > 0, \quad P_m < 0 \quad$ - braking regime
4. for $s = 1$ $\quad P_f > 0, \quad P_m = 0 \quad$ - stall of the motor
5. for $s = 0$ $\quad P_f = 0, \quad P_m = 0 \quad$ - idle run

From the expression for the mechanical power we can calculate the motor's torque in the steady state:

$$T_e = \frac{P_m}{\Omega_r} = \frac{p}{\omega_s} I_r^2 R_r \frac{1}{s} \qquad (3.115)$$

The equivalent circuit can additionally be useful in the calculation of the stator and motor currents:

$$\underline{I}_s = \frac{\underline{U}_s}{R_s + jX_s + \dfrac{X_m^2}{R_r / s + jX_r}} \qquad (3.116)$$

$$\underline{I}_r = -\underline{I}_s \frac{jX_m}{R_r / s + jX_r} \qquad (3.117)$$

It would be valuable to present the currents in the standardized parameters (3.57) since as a consequence of such presentation it is possible to depart from the particular design of an induction motor. The standardized parameters assume values in the ranges presented in Table 3.1. In this case the relations (3.116-3.117) take the form:

$$\underline{I}_s = \frac{\omega_s}{X_s} \frac{\underline{U}_s}{\alpha_s + j\omega_s + \frac{s\omega_s^2(1-\sigma)}{s\omega_s - j\alpha_r}}$$

$$\underline{I}_r = -\underline{I}_s k_r \frac{js\omega_s}{js\omega_s + j\alpha_r}$$

(3.118)

Currents $\underline{I}_r, \underline{I}_s$ represent symbolic values of stator and rotor currents for steady state sine curves. The electromagnetic torque in the steady state can be derived from relation (3.45)

$$T_e = pL_m(i_{sy}i_{rx} - i_{sx}i_{ry}) = pL_m \, \mathrm{Im}(\underline{I}_s \breve{\underline{I}}_r)$$

(3.119)

Using relations (3.116, 3.117) presenting stator and rotor currents we obtain:

$$T_e = \frac{pU_s^2}{\omega_s} \cdot \frac{\dfrac{R_r}{s}}{(\dfrac{R_s}{X_m}\dfrac{R_r}{s} - \dfrac{1}{k_r}X_z)^2 + (\dfrac{1}{k_s}\dfrac{R_r}{s} + \dfrac{1}{k_r}R_s)^2}$$

(3.120)

where: $X_z = \sigma X_s$ (3.121)

- is a blocked-rotor reactance.

The electromagnetic torque can also be presented using standardized parameters (3.59), and takes this form:

$$T_e = \frac{pU_s^2(1-\sigma)}{X_s} \frac{\dfrac{\alpha_r}{s}}{(\dfrac{\alpha_s}{\omega_s}\dfrac{\alpha_r}{s} - \sigma\omega_s)^2 + (\dfrac{\alpha_r}{s} + \alpha_s)^2}$$

(3.122)

The expressions (3.120), (3.122) representing electromagnetic torque relative to supply voltages and motor parameters are frequently subjected to certain simplifications in order to simplify the analysis of these expressions. The basic procedure applies disregarding of the resistance of the stator winding R_s and, subsequently, α_s in some or all terms of this expression. A detailed analysis of this type of simplification will be conducted later on during the determination of the characteristics of the drive regime.

3.2.4 Static Characteristics of an Induction Motor

Static characteristics concern the steady state of a drive and give in an analytic or graphic form the functional relations between the parameters characterizing motor regimes. Typical static characteristics can for instance indicate the relations between electromagnetic torque, current and the capacity of a motor or between efficiency and the slip, voltage supply and the power output of the drive and the like. One can note that static characteristics constitute a set of constant points along a trajectory $\{q_i \in \mathbf{q}\}$ for selected variables of the state q_i or their functions, that illustrate the values that are interesting from the point of view of the specific regimes of a machine, for example electromagnetic torque T_e. Static characteristics collect the end points of trajectories for which the system reaches a steady state. They do not provide information regarding the transfer from a specific point on the characteristics to another one, how much time it will take and whether it is attainable. Hence, in static characteristics we do not have to do with such parameters as moment of inertia J, and the electromagnetic torque T_e and load torque T_l are equal since the drive is in the state of equilibrium, i.e. it does not accelerate or brake (see 3.12). For example, very relevant characteristics are presented using functions (3.120, 3.122). They illustrate the electromagnetic torque for an induction machine depending on a number of parameters. A typical task involves the study of the relation between the characteristics of the electromagnetic torque and the slip $T_e(s)$ for constant remaining parameters, since it informs of the driving capabilities of the motor in the steady state. The relation between machine's torque and slip $T_e(s)$ gives the maximum of this function for two slip values called break torque slip or pull-out slip.

$$s_b = \pm \frac{k_r}{k_s} \frac{R_r}{\sigma X_s} \underbrace{\sqrt{\frac{\left(\dfrac{R_s}{X_s}\right)^2 + 1}{\left(\dfrac{R_s}{\sigma X_s}\right)^2 + 1}}}_{\chi} \tag{3.123}$$

or in standardized parameters

$$s_b = \pm \frac{\alpha_r}{\omega_s \sigma} \underbrace{\sqrt{\frac{\alpha_s^2 + \omega_s^2}{(\alpha_s / \sigma)^2 + \omega_s^2}}}_{\chi} \tag{3.124}$$

The root term χ in formulae (3.123, 3.124) is the factor for correction of the value of the break torque slip as a result of the of stator windings resistance R_s influence. Since leakage coefficient is $\sigma = 0.08...0.03$ (see Table 3.1) the following inequality is fulfilled

$$\left(\frac{R_s}{\sigma X_s}\right) >> \left(\frac{R_s}{X_s}\right) \tag{3.125}$$

In addition, these relations are inversely proportional to the square root of the frequency of the supply source. Two degrees of simplification that are applicable in the development of static characteristics of an induction motor result from the presented estimates. The first of them is not very far-reaching and involves disregarding of resistance R_s in the terms denoting torque (3.120, 3.122) and break torque slip (3.123, 3.124), in which this effect is smaller in accordance with the estimation in (3.125). In this case we obtain:

$$T_e = \frac{pU_s^2}{\omega_s} \frac{\dfrac{R_r}{s}}{(\dfrac{1}{k_r}\sigma X_s)^2 + (\dfrac{1}{k_s}\dfrac{R_r}{s} + \dfrac{1}{k_r}R_s)^2}$$

$$s_b = \pm \frac{k_r}{k_s} \frac{R_r}{\sigma X_s} \frac{1}{\sqrt{\left(\dfrac{R_s}{\sigma X_s}\right)^2 + 1}}$$

(3.126)

The most extensive simplification concerns the case when the resistance of the stator windings is completely disregarded, i.e. $R_s = 0$. In this case we obtain:

$$T_e = \frac{pU_s^2}{\omega_s} \frac{\dfrac{R_r}{s}}{(\dfrac{1}{k_r}\sigma X_s)^2 + (\dfrac{1}{k_s}\dfrac{R_r}{s})^2} \quad \text{and} \quad s_b = \pm\frac{k_r}{k_s}\frac{R_r}{\sigma X_s}$$

(3.127)

Fig. 3.28 presents static characteristics of the motor's torque in the function of the slip for a small power induction motor for the three examined variants of simplification regarding resistance R_s. One can note the small difference between the curve for the torque marked with solid line (i.e. the one presenting relations without simplifications (3.120,3.122)), and dotted line (i.e. the one presenting the result of calculations on the basis of formulae (3.126) involving the first degree of simplification). However, when the resistance of stator windings is totally disregarded ($R_s = 0$) in accordance with formulae (3.127), the error in the characteristics of torque T_e is considerable, as the relative involvement of resistance R_s in the stall impedance of small power motor is meaningful.

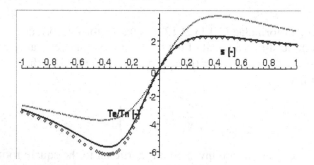

Fig. 3.28 Torque-slip characteristics for the small power induction motor illustrating simplifications concerning stator resistance R_s: _____ R_s taken into consideration completely, according to (3.120, 3.122); ▫▫▫▫▫▫▫ into consideration taken only the most significant component containing R_s, according to (3.126); -------- R_s totally disregarded (3.127)

It is noteworthy that for $R_s = 0$ the characteristic of motor torque becomes an odd function of the slip s, so it is symmetrical in relation to the point of the idle run $s = 0$. Accounting for resistance R_s torque waveform on the side of the motor regime ($s>0$) is considerably smaller in terms of absolute values than for the case of generating regime, i.e. for $s<0$. In addition, on the side of the generator regime the effect of the first degree of simplification accounting for resistance R_s is more clearly discernible than for the case of the motor regime, which can be simply interpreted by analyzing relations (3.120, 3.122). The presented effect of the resistance of stator windings on the characteristics of the torque increases along with the reduction of the pulsation of the supply voltage ω_s and becomes very high for small frequencies. This subject will be covered in more detail later. This effect is graphically presented in Fig. 3.29 in the range of the supply frequencies $1 < f_s \le 50$ [Hz].

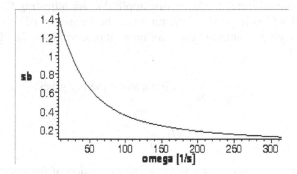

Fig. 3.29 Characteristic of the break-torque slip s_b versus pulsation of the supply voltage ω_s for the small power motor

The formulae for the break-torque slip and motor torque accounting for simplifications concerning the resistance can be additionally presented in formulae containing standardized parameters. The equivalent of the formulae (3.126) takes the form:

$$T_e = \frac{pU_s^2(1-\sigma)}{X_s} \frac{\dfrac{\alpha_r}{s}}{(\sigma\omega_s)^2 + (\dfrac{\alpha_r}{s} + \alpha_s)^2} \quad \text{and}$$

$$s_b = \pm\frac{\alpha_r}{\omega_s\sigma}\sqrt{\frac{\omega_s^2}{(\alpha_s/\sigma)^2 + \omega_s^2}} \tag{3.128}$$

Concurrently, formulae (3.127) are replaced with the form which disregards resistance R_s, by introducing $\alpha_s = 0$:

$$T_e = \frac{pU_s^2 k_s k_r}{X_s} \cdot \frac{\frac{\alpha_r}{s}}{(\sigma\omega_s)^2 + (\frac{\alpha_r}{s})^2} \qquad \text{and}$$

$$s_b = \pm\frac{\alpha_r}{\omega_s\sigma} \qquad\qquad (3.129)$$

In the latter case it is easy to calculate the value of the break-torque:

$$T_e(s_b) = T_b = \frac{pU_s^2 k_s k_r}{2\underbrace{\sigma X_s}_{X_z}\omega_s} = \frac{pk_s k_r}{2\underbrace{\sigma L_s}_{L_z}}\left(\frac{U_s}{\omega_s}\right)^2 \qquad (3.130)$$

Formula (3.130) constitutes the basic rule applicable for adjusting the RMS value of sinusoidal supply voltage U_s of the motor to the frequency of this voltage f_s in such a manner, that guarantees a constant break-torque value of T_b. Hence, the relation takes the form:

$$\frac{U_s}{\omega_s} = \frac{U_n}{\omega_{sn}}, \text{ which, subsequently gives:}$$

$$U_s = U_n\frac{f_n}{f_s} \qquad\qquad (3.131)$$

During the course of action that follows in the discussion of frequency based control of motor's rotational speed it will become evident that this rule is completely insufficient within the range of small supply frequencies. This is so due to the rising share of the resistance R_s in the impedance of the motor stall along with the decrease in the frequency of supply. The relation denoting the break-torque without simplifications, in which resistance R_s is not disregarded, is much more complex than the one in (3.130). The greater complexity of the relation results from the substitution of the break-torque slip s_b (3.123) in the expression denoting the electromagnetic torque of the motor. As a result we obtain:

$$T_b = \frac{sign(\omega_s)pU_s^2 k_s k_r}{\sigma L_s} \cdot \frac{1}{\chi\left[\left(\frac{\alpha_s}{\chi} - \omega_s\right)^2 + \left(\frac{\alpha_s}{\sigma} + \frac{\omega_s}{\chi}\right)^2\right]} \qquad (3.132)$$

where coefficient χ results from (3.123) and is given by the relation:

$$\chi = \sqrt{\frac{\left(\frac{R_s}{X_s}\right)^2 + 1}{\left(\frac{R_s}{\sigma X_s}\right)^2 + 1}} \qquad\qquad (3.133)$$

Under the simplifying assumption that $R_s = 0$, we have $\alpha_s = 0$ and $\chi = 1$ and, as a consequence, break-torque expression (3.132) is reduced to this form (3.130). The relation in (3.133) is applied to indicate the effect of resistance R_s on the break-torque T_b more clearly. The following illustrations in Fig. 3.30 show voltage-frequency relations required to provide constant value of nominal break-torque T_{bn} in the function of stator voltage pulsation ω_s. For the motor regime of operation the required voltage is clearly higher than for the generator regime. From Fig. 3.30 we can also see that smaller motors, within low frequency range, require much higher supply voltages than large motors to sustain the nominal level of T_{bn}. A close inspection of Fig. 3.30b indicates that for higher pulsations ω_s the differences between motors disappear, but still there is constant discrepancy between the symmetrical 'ideal' V-line for $\alpha_s = 0$ and the curves, for which stator resistance R_s was accounted for. For the motor operation the required voltages are higher while for generator operation they are lower in comparison to the 'ideal' V-line. One might say that the actual V-line for which resistance R_s is included is shifted in the direction of lower pulsations ω_s in respect to the 'ideal' V-line for which R_s is completely ignored.

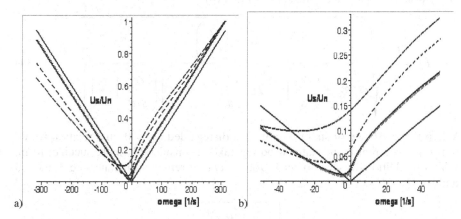

Fig. 3.30 Voltage-pulsation curves indicating the a stator voltage level required to sustain a nominal break-torque T_{bn} while ω_s pulsation changes. The curves are presented for different induction motors with $\alpha_s = 18.8, 5.4, 1.7, 1.2, 0.0$: a) for full range of stator voltage pulsation ω_s, b) range of ω_s limited to low values

Subsequently, Fig. 3.31 presents the characteristics of the motors in the function of the slip in two versions: completely accounting for parameter R_s - smaller characteristic in each pair, and the one totally disregarding resistance, i.e. for $R_s = 0$ - with the above presented characteristic. For nominal value of $\omega_s = 2\pi f_s$, the distinctive difference between the two versions take place for the small power motor.

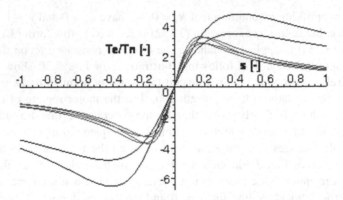

Fig. 3.31 Torque-slip curves (relative values) for the three induction motors: small, medium and high power. The effect of $R_s = 0$ simplification is illustrated for $f_s = 50$ [Hz]

Subsequently, Fig. 3.32 presents the characteristics of stator current for the three motors accounting for resistance R_s. The relation (3.134) is applied in this case, which comes as a consequence of (3.118):

$$I_s = \frac{U_s}{X_s} \sqrt{\frac{\omega_s^2 + \left(\dfrac{\alpha_r}{s}\right)^2}{\sigma^2 \omega_s^2 + \left(\dfrac{\alpha_r}{s}\right)^2 + 2\alpha_s \dfrac{\alpha_r}{s} k_s k_r + \alpha_s^2 \left[\left(\dfrac{\alpha_r}{s\omega_s}\right)^2 + 1\right]}} \qquad (3.134)$$

When the resistance of stator windings is disregarded ($\alpha_s = 0$), the relation which defines the current in the stator windings takes a considerably more succinct form, which is additionally easy to verify for the two extreme motor states, i.e. for $s = 0$ and $s = \infty$.

$$I_s = \frac{U_s}{X_s} \sqrt{\frac{\omega_s^2 + \left(\dfrac{\alpha_r}{s}\right)^2}{\sigma^2 \omega_s^2 + \left(\dfrac{\alpha_r}{s}\right)^2}} \qquad (3.135)$$

Self reactance of the stator windings X_s is encountered in a multitude of relations concerning induction motors. The value of this parameter can be easily determined from calculations or manufacturers' data for idle run. From the equivalent diagram (Fig. 3.26) of the motor it results that

$$\frac{\underline{U}_0}{\underline{I}_0} = \underline{Z}_0 = R_s + jX_s \qquad X_s = \underline{Z}_0 \sin \varphi_0 \qquad (3.136)$$

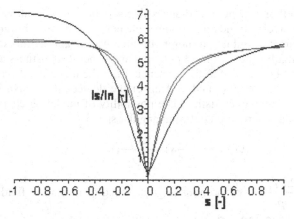

Fig. 3.32 Stator current-slip curves (relative values) for the three exemplary motors according to (3.134) presented in relative values, for $f_s = 50$ [Hz]

where $\underline{U}_0, \underline{I}_0, \underline{Z}_0, \varphi_0$ denote voltage, current, impedance and phase angle for the idle run of the motor. If the phase angle during idle run is not familiar, it is possible to use assessment relevant for the rated frequency: $R_s \ll X_s$ and calculate in an approximated way:

$$X_s = \frac{U_0}{I_0} \tag{3.137}$$

3.3 Methods and Devices for Forming Characteristics of an Induction Motor

By its very nature resulting from solid rotor windings and lack of power supply to its windings, an induction motor is most suitable for operation under steady conditions and with a small slip. In such a case the angular speed results from the frequency of the supply to the stator windings, number of pole pairs and value of the slip. Traditionally, it was applied in drives in which neither frequent changes of speed nor variable control were required (examples of such devices include pumps, blowers, compressors, belt conveyors, cranes, industrial hoists). There was virtually no possibility of controlling induction motors within wide range of speeds while concurrently preserving high energetic efficiency until 1970s. Drives in which the control of speed was necessary most frequently applied slip ring induction motors, in which it is possible to control rotational speed as a result of use of external elements. However, such systems are either complex, costly and problematic in control due to the use of cascaded systems. Alternatively, they have lower energetic efficiency due to additional resistance in the rotor's circuit. In addition, the start-up properties of an induction motor under direct connection to the network are adverse due to the initial period of oscillations of electromagnetic torque with a high amplitude and high value of the start-up current. Despite these drawbacks the induction motor has become the most common machine in electric drive systems due to the fundamental advantages including long service life and reliability as well as low price and accessible supply source. Following the development of power electronics and control elements enabling arbitrary shaping of voltages and currents, induction motors became widely applied in complex drives due to a new angular speed control potential and general robustness at heavy duty. This section will be devoted to the presentation of the methods of forming characteristics of induction motors and will cover the devices that make it possible to realize the required characteristics. The possibility of modeling characteristics results directly from the relation defining angular speed

$$\Omega_r = \omega_f (1-s) = \frac{2\pi f_s}{p}(1-s) \tag{3.138}$$

where: $\omega_f = \dfrac{2\pi f_s}{p}$ - synchronous angular speed of a rotating field. Each of the

values in relation (3.138) offers the possibility of modeling mechanical characteristics: number of pole pairs p, slip s as well as the frequency f_s of the supply voltages. The control of slip s is possible to a large extent as a result of the external interference in the rotor circuit and also voltage changes but within a small range of rotational speeds. The presentation of methods used for modeling characteristics associated with rotor slip changes will follow in the subsequent sections. Concurrently, a separate section will be devoted to an extensive presentation of control as a result of modifying the frequencies of the supply voltages. The application of the various number of pole pairs p for changing motor speed appears to be most straightforward to explain. A series of synchronous speeds ω_f for a given supply frequency consists of a discreet values. For the successive number of pole pairs p = 1,2,3,4,5,6,... and for example for the frequency of the supply f_s = 50 [Hz] they are, approximately:

$$\omega_f = 314.16, \quad 157.1, \quad 104.7, \quad 78.5, \quad 62.8, \quad 52.4,...$$

This finds application in multi-pole motors, in which the windings can be switched to two or three synchronous speeds, which leads to a stepwise change of rotor speed. This type of drive is applied in cranes and industrial hoists mainly with two speeds – transit speed with a higher value and a slower approach speed.

3.3.1 Control of Supply Voltage

The control of the supply voltage can offer only limited possibility of adjusting rotational speed of an induction motor. This results from the basic mechanical characteristic of the motor (Fig. 3.28) which indicates that the slip under a given load can be increased up to the limit of $s < s_b$, which means it has to keep below the break-torque slip beyond which a loss of the stability occurs and the motor stops. In addition, this type of control is achieved at the expense of efficiency loss since under a constant load the losses in the motor are $\Delta P > P_f s$. This comes as a consequence of the increase of the current and losses in the motor windings. At the same time, the control of the supply voltage is currently used in order to reduce the start-up current and perform a soft-start. This is realized with the use of an electronic device called a soft-starter. A diagram of such a device is found in Fig. 3.33.

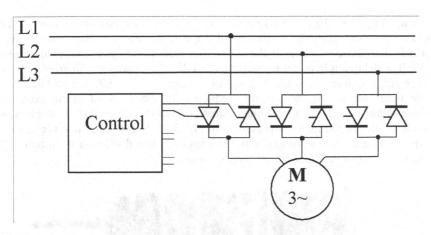

Fig. 3.33 Basic diagram of a soft-starter for an induction motor

The introduction of semiconductor elements (SCRs, IGBTs, GTOs, MOSFETs etc.) for the two directions of current flow for each line supplying the motor windings makes it possible to employ current flow with a selected delay angle α in relation to the zero crossing of the supply voltage curve. As a result, at some expense of altering the current and voltage from sinusoidal shape, it is possible to control the value of voltage and synchronize the motor with the network at the instant of connecting the particular motor phase windings during start-up. Soft-starters may, accordingly, realize the following functions related to the start-up and stopping of an induction motor:

- synchronization of the connection of particular phase winding to the network and thus enabling the reduction of the variable component of the torque (see 3.2.2.5)
- limitation of the start-up current in a selected range,
- braking with the use of direct current (see section 3.2.2.6) and conduct controlled stop of a drive.

Not all of the above functions have to be realized by a single type of soft starter. In the most economic versions designed for smaller drives, a soft starter sometimes contains switches in the two supply lines, which only leads to limitation of the start-up currents and does not provide symmetry of the supply voltages. The following Figs. 3.34-3.38 present the examples of application of a soft-starter for an medium power induction motor with a delay angle $\alpha = 40°$ and the basic value of the moment of inertia $J = J_s$. The figures present a comparison between start-up versions without synchronization during the connection of phases to the network and the one with synchronization involving the connection of line L_1, L_2 for phase angle $\delta_{1,2} = 0.48\pi$ [rad] and a later connection of the third supply line L_3 for angle: $\delta_3 = a + \delta_{1,2} - 0.1$ [rad]. As a result, we obtain a very soft starting curve during the initial stage of the start-up of the motor (Fig. 3.34) accompanied by a very favorable torque waveform (Fig. 3.36). The synchronized connection for such a large delay angle $\alpha = 40°$ also results in the reduction of the duration of the start-up (Fig. 3.35, Fig. 3.36) since the value of the constant component of the motor torque increases during the initial stage of the start-up. The current waveform in the phase winding of the motor for such a supply is presented in Fig. 3.37. The delay angle in the range of around 40° is virtually the sharpest one for which it is possible to conduct start-up of the motor during idle run within a sensible time, due to the considerable reduction of the value of electromagnetic torque of the motor. The approximate illustration of the effect of delay angle α on characteristics of the motor is presented in Fig. 3.38. Soft-starters find application in drives with an easy start-up due to the considerable reduction of the torque following the fall of the value of the supply voltage.

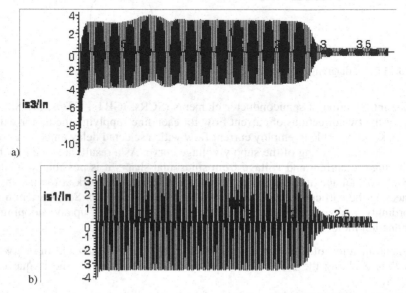

Fig. 3.34 Line current of the medium power induction motor during free acceleration with a soft starter ($\alpha = 40°$): a) without synchronization b) with synchronization: $\delta=0.48\pi$; $\delta_3 = a + \delta_{1,2} - 0.1$

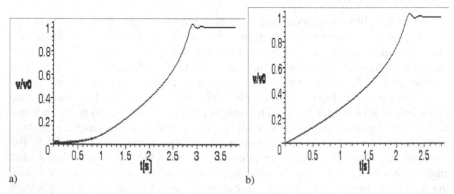

Fig. 3.35 Relative velocity curve for the medium power motor during the soft-start free acceleration, under conditions like in Fig. 3.34

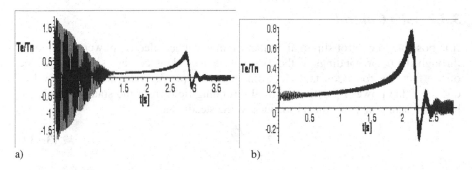

a)   b)

Fig. 3.36 Electromagnetic torque curve for the medium power motor during the soft-start free acceleration, under conditions like in Fig. 3.34

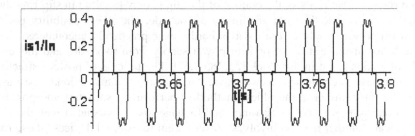

Fig. 3.37 Steady-state line current for the medium power motor during the soft-start free acceleration, under conditions like in Fig. 3.34

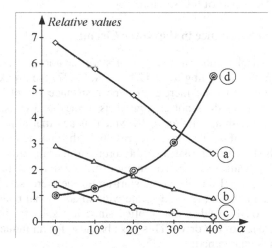

Fig. 3.38 Induction motor starting characteristics (relative values) for the medium power motor during the soft-start free acceleration in relation to delay angle α: a) starting current b) break torque c) starting torque d) idle run free-acceleration time

3.3.2 Slip Control

It is possible to control slip in an induction motor when electric power is delivered through the rotor windings to the external devices. This comes as a consequence of the fact that for a constant electromagnetic torque T_e and constant supply frequency f_s the power P_ψ delivered by the rotating field from the stator to the rotor has to remain constant. This is so since in the steady state

$$T_e = \frac{P_\psi}{\omega_f} \qquad (3.139)$$

After some power is extracted from the rotor windings, the electric power $P_{el} = P_\psi s$ increases and the mechanical power $P_m = P_\psi (1-s)$ decreases, which is possible as a result of an increase of slip s, i.e. the reduction of the rotational speed of the rotor. As we can see, the control of the slip is only possible in slip-ring motors, due to access to the rotor windings from outside. The other possibility associated with power supply to the rotor is hardly ever practically encountered. It is possible for instance in a motor with power supply from two sides and this case will not be discussed in this book [25,28,62,67]. The process of power extraction from windings is conducted in two ways. An inclusion of an additional resistance R_d in the rotor circuit is the oldest method of performing soft-start and possibly speed control; however, it is accompanied by huge losses associated with the produced heat. Another method involves power output to external devices whose role is to transform the power to useful forms, for instance its return to the supply network. Such devices, which used to be electromechanical, now predominantly are power electronic ones are called cascades. One of them is the Scherbius drive, and is a subject in the latter part of this section.

3.3.2.1 Additional Resistance in the Rotor Circuit

This method of control results in changes of static characteristics of the torque presented in Fig. 3.39. According to (3.123, 3.124, 3.127) the break-torque slip s_b increases proportionally to the increase of the resistance of rotor windings R_r, while the break-torque T_b does not change. This comes as a consequence of the maintenance of the constant relation α_r /s, which means that slip s rises proportionally to the increase of α_r. This in a way results in the change of the scale of the slip which extends the characteristics of the torque in the direction of higher values of the slip. This leads to an improvement of motor start-up since the start-up current is reduced for $s = 1$ and the static start-up torque increases. Unfortunately, the operation of the motor in the steady state with an additional resistance in order to reduce the rotational speed is not applied since it results in the reduction of the energetic efficiency of the drive. This is so because for an induction motor the following relation is maintained:

$$\eta < 1 - s \qquad (3.140)$$

which means that for instance the reduction of rotor speed Ω_r to reach the half of the synchronic speed ω_f with the use of resistance based control leads to the reduction of the efficiency to $\eta < 0.5$, which is unacceptable.

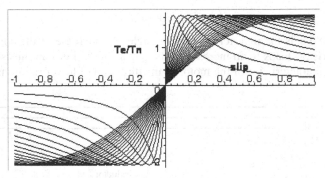

Fig. 3.39 Torque-slip characteristics of a wound induction motor with additional external resistance in the rotor for $\alpha_{rd} = \alpha_r \cdot (1...20)$; α_{rd} includes additional resistance R_d connected to the slip-rings

3.3.2.2 Scherbius Drive

As a result of the application of Scherbius drive it is possible to control the speed of a slip-ring induction motor as a consequence of electric power extraction from rotor windings and its return after the desired transformation into the network. In its historical model the Scherbius drive contains an electromechanical frequency converter connected on one side to the rotor of an induction motor and the other one to the supply network into which the power returns. In a modern solution of the Scherbius drive (Fig. 3.40), the currents of the rotor windings are rectified in a 3-phase rectifying bridge and subsequently supply a converter which returns the energy into the network via an adapting transformer. Between the two bridges there is an inductor that smoothens the flow of the current and whose role is to secure the continuity of current flow across the rotor even for small mechanical load of the motor shaft. The control parameter is delay angle α of the thyristor bridge, which for the desired inverter mode is contained in the range:

$$90° < \alpha < 180 - \mu$$

where μ is the emergency angle which prevents the inverter back-feed and has to be bigger then the maximum calculated commutation angle. Such control corresponds to the feeding of voltage $U_d(\alpha)$ into the rotor, which offers a possibility of controlling the slip of an idle run s_0. The slip during idle run s_0 corresponds to the theoretical idle run of an induction motor in which there is an equilibrium between the mean values of electromotive forces E_r and voltage at the output of the inverter $U_d(\alpha)$.

Since $\qquad E_r = U_{r0} s_0 \dfrac{3\sqrt{2}}{\pi} \qquad U_d(\alpha) = -U_L' \cos\alpha \dfrac{3\sqrt{2}}{\pi}$,

then: $\qquad\qquad\qquad\qquad s_0 = -\dfrac{U_L'}{U_{r0}} \cos\alpha \qquad\qquad\qquad\qquad$ (3.141)

For the case of an adequately selected transformer rate supplying the inverter the following is fulfilled: $U_L' = U_{r0}'$ and, as a consequence:

$$s_0 = -\cos\alpha = \cos\beta$$

where $\beta = 180-\alpha$ is the inverter advancing angle. An increase of the control angle α of the thyristor, leads to an increase of the slip s_0 of the idle run thus shifting the mechanical characteristics of the motor in the direction of lower rotational speeds.

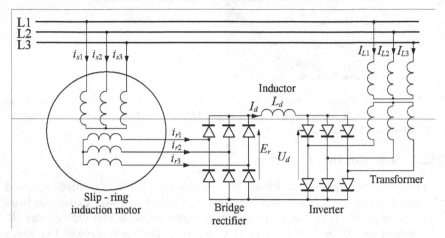

Fig. 3.40 Diagram of a semiconductor Scherbius drive

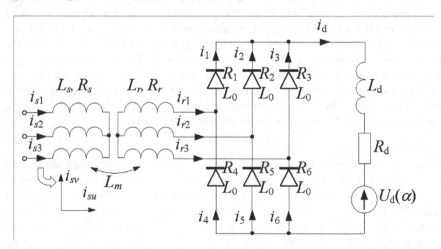

Fig. 3.41 Diagram of the simplified Scherbius drive for mathematical modeling

The control of the semiconductor cascaded drive has been modeled in a simplified form (Fig 3.41), where an inverter is reduced to lumped elements L_d, R_d, $U_d(\alpha)$. An induction motor is modeled so that the electric variables of the stator are transformed into orthogonal axes u,v while we have to do with natural variables i_{r1}, i_{r2}, i_{r3} in the rotor windings. This model is discussed in subsection 3.6.2 and is described by the system of equations in (3.88-3.89). The combination of the model of a slip-ring motor with a bridge on the side of the rotor and a circuit of direct current for an inverter is described by a system of equations with $s_e = 7$ electrical degrees of freedom and a single $s_m = 1$ degree of freedom for the rotational motion of the rotor. The generalized coordinates for the electric component of the model are the transformed currents of the stator windings i_{su}, i_{sv}, phase currents of the rotor windings i_{r1}, i_{r3} currents of the rectifier bridge i_1, i_3 and the current i_d in the inverter circuit.

This model for electric circuits can take the form of a matrix equation:

$$
\begin{bmatrix} u_{su} \\ u_{sv} \\ 0 \\ 0 \\ 0 \\ 0 \\ u_d(\alpha) \end{bmatrix} =
\begin{bmatrix}
R_s & 0 & v(c\tau-\sqrt{3}s\tau) & 2vc\tau \\
0 & R_s & v(s\tau+\sqrt{3}c\tau) & 2vs\tau \\
v(c\tau-\sqrt{3}s\tau) & v(s\tau+\sqrt{3}c\tau) & 2R_r+R_4+R_5 & R_r+R_5 \\
2vc\tau & 2vs\tau & R_r+R_5 & 2R_r+R_5+R_6 \\
0 & 0 & -R_4-R_5 & -R_5 \\
0 & 0 & -R_5 & -R_5-R_6 \\
0 & 0 & R_5 & R_5
\end{bmatrix}
$$

$$
\begin{bmatrix}
0 & 0 & 0 \\
0 & 0 & 0 \\
-R_4-R_5 & -R_5 & R_5 \\
-R_5 & -R_5-R_6 & R_5 \\
R_1+R_2+R_4+R_5 & R_2+R_5 & -R_2-R_5 \\
R_2+R_5 & R_2+R_3+R_5+R_6 & -R_2-R_5 \\
R_2+R_5 & -R_2-R_5 & R_d+R_2+R_5
\end{bmatrix}
\begin{bmatrix} i_{su} \\ i_{sv} \\ i_{r1} \\ i_{r3} \\ i_1 \\ i_3 \\ i_d \end{bmatrix} +
$$

$$
+\begin{bmatrix}
L_s & 0 & x(s\tau+\sqrt{3}c\tau) & 2xs\tau \\
0 & L_s & x(-c\tau+\sqrt{3}s\tau) & -2xc\tau \\
x(s\tau+\sqrt{3}c\tau) & x(-c\tau+\sqrt{3}s\tau) & 2L_0+L_r & L_0+L_r \\
2xs\tau & -2xc\tau & L_0+L_r & 2L_0+L_r \\
0 & 0 & -2L_0 & L_0 \\
0 & 0 & -L_0 & -2L_0 \\
0 & 0 & L_0 & L_0
\end{bmatrix}
$$

$$
\begin{bmatrix}
0 & 0 & 0 \\
0 & 0 & 0 \\
-2L_0 & -L_0 & L_0 \\
-L_0 & -2L_0 & L_0 \\
4L_0 & 2L_0 & -2L_0 \\
2L_0 & 4L_0 & -2L_0 \\
-2L_0 & -2L_0 & 2L_0+L_d
\end{bmatrix}
\frac{d}{dt}
\begin{bmatrix} i_{su} \\ i_{sv} \\ i_{r1} \\ i_{r3} \\ i_1 \\ i_3 \\ i_d \end{bmatrix}
\qquad (3.142)
$$

where:

$$v = \frac{\sqrt{2}}{2} p\dot{\theta}_r L_m \qquad x = \frac{\sqrt{2}}{2} L_m \qquad\qquad s\tau = \sin(p\theta_r) \quad c\tau = \cos(p\theta_r)$$

The term which expresses the electromagnetic torque takes the form:

$$T_e = \frac{\sqrt{2}}{2} pL_m *$$
$$* \left[c\tau \left(i_{r1}(\sqrt{3}i_{sv} + i_{su}) + 2i_{r3}i_{su} \right) + s\tau \left(i_{r1}(i_{sv} - \sqrt{3}i_{su}) + 2i_{r3}i_{sv} \right) \right] \tag{3.143}$$

On the basis of this model calculations were conducted with an aim of finding electric and mechanical transients of the drive and static characteristics as well, for various states of control of the inverter bridge. The calculations were carried out for a medium power slip-ring motor with the following parameters:

$$U_n = 6000\,[V] \qquad I_n = 53\,[A] \qquad P_n = 450\,[kW] \qquad T_n = 2920\,[Nm]$$
$$J_s = 35\,[Nms^2] \qquad U_{2n} = 525\,[V] \qquad I_{2n} = 545\,[A] \qquad \cos\varphi_n = 0.88$$
$$p = 2 \qquad\qquad n_n = 1473\,[rev/\min] \qquad \eta_n = 0.93 \qquad J_s = 35\,[Nms^2]$$

The static characteristics of this motor are presented in Fig. 3.39. The parameters determined on the basis of the motor's ratings are:

$$R_s = 1.2\,[\Omega] \qquad R_r' = 0.989\,[\Omega]$$
$$L_m = 1.04\,[H] \qquad L_s = L_r = 1.0703\,[H]$$

and the parameters of the bridge are:

$$L_d = 0.3\,[H] \quad L_0 = 0.01\,[H] \qquad R_d = 0.2\,[\Omega]$$

The dynamic calculations were conducted for the drive's moment of inertia $J = 3J_s = 105\,[Nms^2]$. Fig 3.42 presents the start-up of the drive under a load $T_l = 0.15\,T_n$ and control angle α corresponding to $U''_d = 2400\,[V]$. Hence, the slip of the idle run, calculated according to (3.141), is equal to:

$$s_0 = \frac{\pi}{3\sqrt{2}} \frac{U''_d}{U_{sn}} \cong 0.296$$

where U''_d is the voltage of the inverter bridge expressed in stator windings voltage terms. In consequence the angular speed of the idle run is

$$\Omega_0 = \omega_f(1 - s_0) = \frac{2\pi f_s}{p}(1 - s_0) = 110.6\,[rad/s]$$

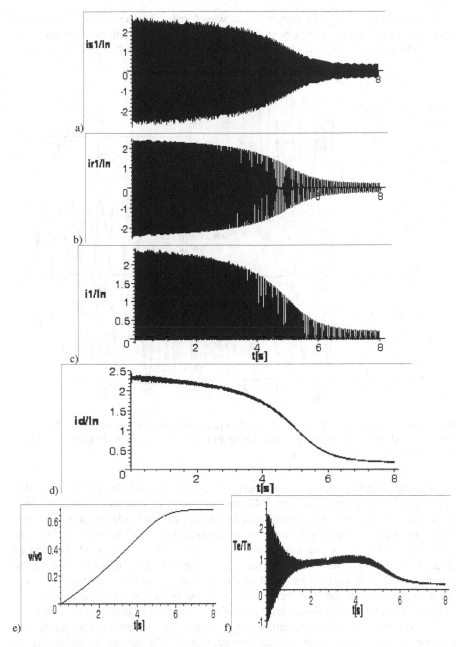

Fig. 3.42 Starting of the 450 [kW] Scherbius drive with $s_0 \approx 0.3$, $T_l = 0.15 T_n$, $J = 3 J_s$: a) stator current b) rotor current c) bridge current d) DC link current e) angular speed f) electromagnetic torque

Subsequently, Fig. 3.43 presents the waveforms of electric and mechanical variables of the drive in steady state for $U''_d = 1200$ [V], which corresponds to $s_0 = 0.148$ and $\Omega_0 = 133.8$ [rad/s], respectively for the load of $T_l = 0.5\ T_n$.

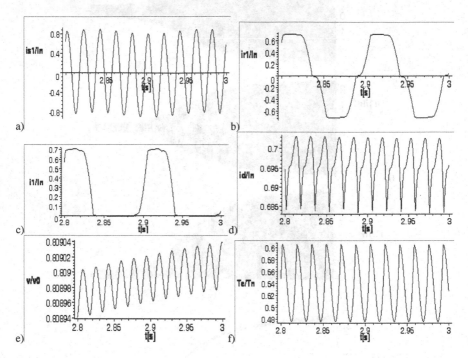

Fig. 3.43 Steady state time-curves of the 450 [kW] Scherbius drive with $s_0 \approx 0.15$, $T_l = 0.5T_n$, $J = 3J_s$: a) stator current b) rotor current c) bridge current d) DC link current e) relative speed f) electromagnetic torque

One can note that start-up of this drive does not normally occur in a cascaded system, as in the example in Fig. 3.42 but in system with an additional resistance R_d in the rotor circuit, which ensures a faster start-up with a larger motor torque. After this initial stage of start-up rotor is reconnected to the Scherbius drive system. As indicated by the comparison of static characteristics of the torque (Fig. 3.39) with the start-up characteristics in the Scherbius drive system, in the latter case the motor develops much smaller torque due to the deformations of the rotor currents from the sine curve accompanied by a considerable decrease of voltages associated with the components of semiconductor bridges in the rotor.

To give an illustration of a transient operation of the Scherbius drive a stepwise change in control of inverter voltage was introduced. The output voltage of the inverter changed abruptly from $U''_d = 1200$ [V] to $U''_d = 2400$ [V], without a change of the load being $T_l = 0.5\ T_n$. The drive has to slow down due to the change of the idle slip value from $s_0 \approx 0.15$ to $s_0 \approx 0.25$. The resulting transients are presented in Fig. 3.44.

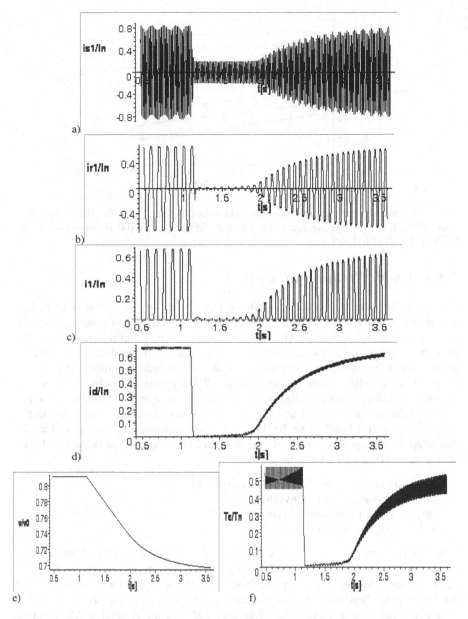

Fig. 3.44 Transient characteristics of the Scherbius drive during inverter control change from $s_0 \approx 0.15$ to $s_0 \approx 0.25$: a) stator phase current b) rotor phase current c) bridge current d) DC link current e) relative rotor speed f) relative electromagnetic torque

Static mechanical characteristics of the semiconductor Scherbius drive were determined on the basis of a series of calculations of the steady state for various inverter bridge control angles. An illustration is found in Fig. 3.45.

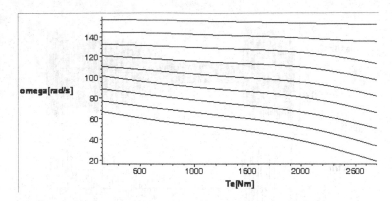

Fig. 3.45 Torque-speed characteristics the Scherbius drive for $U''_d = 0$, 600, 1200, 1800, 2400, 3000, 4200, 4800 [V] or equivalent idle run slip values: $s_0 = 0.0, 0.074, 0.148, 0.296, 0.370, 0.444, 0.518, 0.594$

3.3.3   Supply Frequency f_s Control

One of the fundamental methods applied for control of angular speed of an induction motor in accordance with (3.138) is based on changing the frequency f_s of the voltage supply to the stator's windings. Although it was difficult to execute in the past, this method has become widespread as a result of the application of various power electronic converters. It finds application in induction motor drives in the range of power ratings from a fraction of a [kW] to powerful machines exceeding 10 MW [10]. Depending on the power output and design it is possible to apply various solutions of inverters and various frequencies of energy conversion in the range from several hundred to 30 [kHz] for small and medium power devices. In this chapter an emphasis will be on the basic solutions applied in induction drives with power inverters, including:

- direct frequency converters – cycloconverters,
- two-level voltage source inverters,
- three-level voltage source inverters,
- PWM current source inverters.

This list of converter drives does not form the complete record of the applied drives – in particular with regard to large power drives but contains the most common ones. Moreover, resonance based current inverters and load commuted inverters are applied in addition to the listed ones. For each one of the systems it is possible to apply several methods of control realizing the various voltages waveforms and output currents. The issues thereof are very extensive and are widely discussed in the references [10,14,22,51,52,97].

3.3.3.1   Direct Frequency Converter–Cycloconverter

The role of a cycloconverter is the conversion of 3-phase alternating voltage and current of the supply network with the frequency f_L into single-phase voltage and current of a load with the frequency of f_s without conversion into direct current. In order to obtain a 3-phase system of voltages and currents supplying the motor on the output of a converter it is necessary to apply three separate conversion unit or two units in the economical versions of a converter [52]. Each of such units consists of 2 antiparallel groups of controlled rectifiers most common of which include 6-pulse rectifiers ($q = 6$). The necessity of using two antiparallel rectifying groups results from the need of symmetric conducting currents in two opposite directions. Thyristors (SCRs) are applied in the rectifying groups of the converter and hence in a typical frequency converter we have to do with current commutation. This comes as a consequence of the fact that the typical application of a converter is the controlled large power induction or synchronous motor drive with a capacity of up to a dozen MW. A SCR–Silicon Controlled Rectifier is a power electronic component with the highest operating voltages and high conduction currents; hence, it is used in large power converters. One of the standard applications of a frequency converter in a 3-phase load (ac motor) is presented in Fig. 3.46.

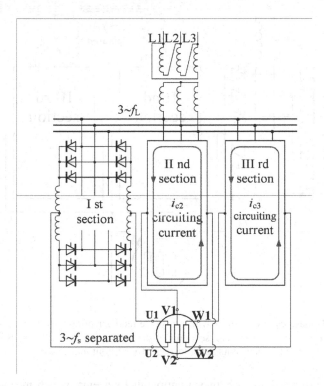

Fig. 3.46 A 3-phase cycloconverter with a 6-pulse rectifier bridges and separated outputs. The system with inductors limiting the circulating current

This is a converter with bridge rectifying units and separate phases of the load, which, however, is supplied from a transformer with a single secondary winding. Between the rectifiers supplying the windings of the motor there are inductors

limiting the impact of the equalizing current in the circuit of the antiparallel rectifier units since in this solution both of them are controlled by the delay angle over
the entire range of the operating conditions of the converter. This is a solution that
does not require detection of the instant of the load current crossing zero and a
subsequent separate control of the two rectifier units. Another solution of the
cycloconverter system is presented in Fig. 3.47. In this case the 3-phase load is
connected in a star, which leads to the supply of particular systems of the antiparallel converters from separate secondary windings of the transformer in order to
avoid shorts.

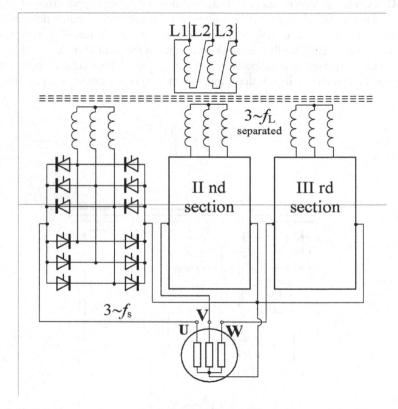

Fig. 3.47 A 3-phase cycloconverter with a 6-pulse rectifier bridges and the Y - connected
output, which requires 3 separate secondary windings of the supply transformer. The
system without circulating currents and consequently without inductors

Inductors are not applied between antiparallel systems, which means that such
units do not involve simultaneous control of both rectifying bridges. Each of them
feeds the current into one direction of conducted current that is singular for it. This
is associated with a need to apply more advanced control of the converter, which
involves the detection of the instant of a current flow direction change, and a short
break during the conduction of both bridges in this period to restore blocking
ability. The basic distinction in the applied control system involves selection the

control method of a cycloconverter with equalizing currents or without them and it is possible to use one or the other in every type of converter. However, the limitation of the equalizing current is associated with the need to use massive and expensive inductors designed for conducting currents with large values. This requirement leads to a tendency to apply a system without equalizing currents. The principle for the control of cycloconverters results from the adequate control of a 3-phase rectifier in such a way that ensures an output voltage whose basic harmonic is a sine waveform with the frequency of f_s. Since the output voltage and current originate from co-operation of the two rectifying units (bridges) forming the input of a single phase of a load, it is necessary to control both of them, simultaneously or in succession, in order to ensure that they supply uniform output voltage within the range of the basic harmonic:

$$U_{sm} \sin(\omega_s t) = U_{d0} \cos \alpha_1$$
$$-U_{sm} \sin(\omega_s t) = U_{d0} \cos \alpha_2$$

for the unit 1 and unit 2 respectively. Hence control angles α_1, α_2 are:

$$\alpha_1 = \arccos(m_a \sin(\omega_L m_f t))$$
$$\alpha_2 = \arccos(-m_a \sin(\omega_L m_f t))$$

(3.144)

or

$$\alpha_2 = \pi - \alpha_1$$

(3.145)

where: $m_a = \dfrac{U_{sm}}{U_{d0}}$ - amplitude modulation factor

$m_f = \dfrac{f_s}{f_L}$ - frequency modulation factor

U_{sm} - maximum value of the basic harmonic of the output voltage of a converter

$U_{d0} = \dfrac{3\sqrt{2}}{\pi} U_2$ - mean voltage of the rectifying unit ($q = 6$) for delay angle $\alpha = 0$

U_2 - RMS voltage of the secondary side of a transformer.

The delay angle waveform of the rectifier group 1 is presented in Fig. 3.48 in the function of the phase angle of the output voltage.

The figures that follow present how the output voltage is formed by the units 1 and 2 of antiparallel rectifiers. Rectifier unit 1 performs the descending section of the modulated voltage, while unit 2 is responsible for the ascending section of the voltage by application of 3-phase voltages of the supply network. Both voltages for these units of the converter generate an identical harmonic of the output voltage with the frequency of f_s, while in the range of the higher harmonics the waveforms are different. Hence, the equalizing current occurs in the antiparallel system of the rectifying units for the case of controlling both groups over the entire period of the output voltage.

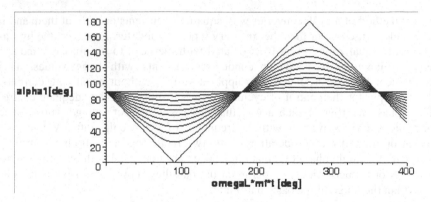

Fig. 3.48 α_1 control angle of a cycloconverter anode group as a function of the output voltage phase angle, for amplitude modulation factor $m_a = 0, 0.1,...0.9, 1.0$

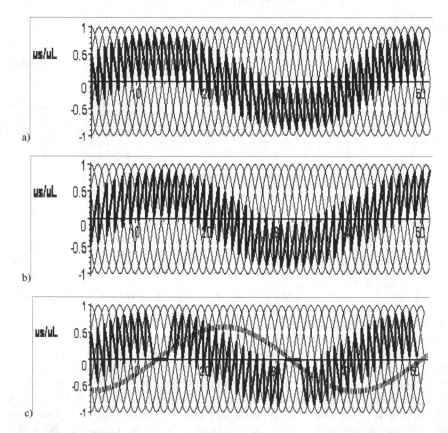

Fig. 3.49 Cycloconverter's output voltages for $m_a = 0.55$, $m_f = 0.1667$, $f_s = 10$ [Hz] and for $f_L = 60$ [Hz]: a) unit 1 voltages b) unit 2 voltages c) output voltage for separate control of rectifying units – control without circulating current

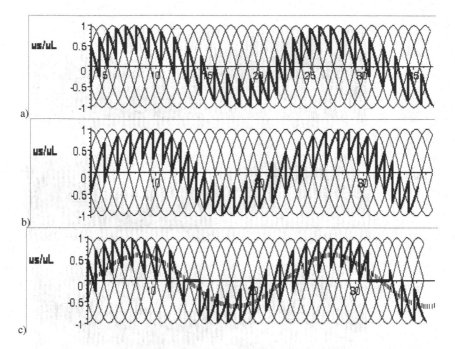

Fig. 3.50 Cycloconverter's output voltages for $m_a = 0.75$, $m_f = 0.3333$, $f_s = 20$ [Hz] and for $f_L = 60$ [Hz]: a) unit 1 voltages b) unit 2 voltages c) output voltage for separate control of rectifying units – control without circulating current

From the illustration of the transformed voltage curves, whose basic frequency is f_s it results that the content of higher harmonics in the output voltage is high and increases along with the decrease of the amplitude modulation factor m_a. The analysis of the output frequency indicates that the dominant part is occupied by harmonics with the frequencies of

$$f_{h,k} = qf_L \pm kf_s \qquad k = 0,1,2,\dots \qquad (3.146)$$

which for a low output frequency f_s means that the major higher harmonics have a frequency around qf_L. This means around 300 [Hz] for a converter with 6-pulse rectifying units and the frequency of the supply network of $f_L = 50$ [Hz].

Practical considerations lead to the limitation of the upper boundary of the output frequency to around $0.4 f_L$ and the adaptation of voltage U_2 of the transformer supplying the converter to this frequency. This comes as a consequence of the principle in (3.131), which defines the adaptation of the value of the supply voltage to the frequency, while preserving an adequate surplus of voltage. The aim of this is to apply a potentially high amplitude modulation factor m_a and, thus, the limitation of the amplitudes of higher harmonics of the output voltage.

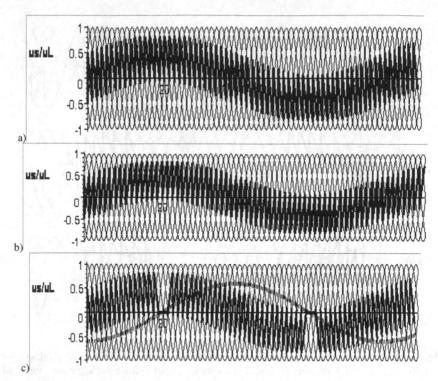

Fig. 3.51 Cycloconverter's output voltages for $m_a = 0.45$, $m_f = 0.1$, $f_s = 5$ [Hz] and for $f_L = 50$ [Hz]: a) unit 1 voltages b) unit 2 voltages c) output voltage for separate control of rectifying units – control without circulating current

3.3.3.2 Two-Level Voltage Source Inverter

Voltage Source Inverter – VSI is a power electronic device capable of transforming the DC voltage and direct current into voltage and alternating current with the desired characteristics. It is possible to design single- and multiphase inverters. 3-phase inverters are commonly applied for the supply of induction motors. The basic diagram of a 2-level voltage inverter is presented in Fig. 3.52.

This inverter is commonly referred to as VSI inverter since it forms the source with voltage characteristics and the voltage curve on the output (load) is not relative to the value of the load current in a wide range of operating conditions. This is made possible due to the powerful voltage source with small internal impedance additionally boosted by a capacitor with adequately large capacity C_s on the input of the inverter. A 3-phase inverter has 3 branches with two semiconductor switches and free-wheeling diodes presented in Fig. 3.52. The control switches apply IGBTs (Isolated Gate Bipolar Transistors) or GCTs (Gate Commuted Thyristors) or MOSFET (Metal-Oxide Semiconductor Field Effect Transistor) depending on required working conditions, firstly supply voltage, load current and switching frequency. An output voltage filter with capacitors with the capacity of

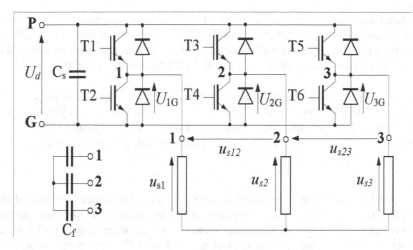

Fig. 3.52 Diagram of a 2-level 3-phase voltage source inverter (VSI)

C_f is connected to the load. The name for a two-level converter comes as a consequence of the fact that voltages U_{1G}, U_{2G}, U_{3G} can only assume two values, i.e. U_d or 0. Regardless of the specific manner of control of the inverter's gates at any instant only one of the semiconductor switches can conduct current in a specific branch. The commutation involves the process in which in a single branch one of the switches terminates the conduction while the other switch starts the conduction process only after an adequate break to prevent shorts. At any instant with the only exception of the commutation break in the inverter we have to do with conduction of three semiconductor switches, i.e. one in each branch. The desired waveform of the output voltage is gained as a result of an adequate control of inverter semiconductor switches. For the supply of an induction motor drive it is desirable to have 3-phase voltage with a sine waveform having controllable frequency and amplitude. The basic method applied for the modeling of the output voltage consists in Pulse Width Modulation–(PWM) [3,6,39,41,44,59,85]. Under standard conditions it involves adequate switching of the potential U and potential 0 at the output by semiconductor switches from the inverter branches in a short intervals corresponding to a small fraction of the period of the output voltage. Concurrently, there is a large number of modulation methods, some of which will be discussed in this section. Every modulation method should result in output voltages close to 3-phase symmetrical sinusoidal system with small content of higher harmonics. The other postulate regards the application of possibly small number of switchings between the control semiconductor elements corresponding to a single cycle of the output voltage. Every commutation in an inverter branch is associated with resistive losses in the power electronic switches, hence the effort to make transition period short. Also the commutation is associated with losses in the dielectric in the windings' insulation leading to wear of the insulation layer. Hence, the postulate of the limitation of the number of connections follows. The discussion here will focus on sinusoidal PWM (SPWM) modulation in which the triangular carrier signal is modulated with the sinusoidal waveform as well as several varieties of the space vector modulation (SVM) [50,55,70,71,72,73,87]. The first of the listed methods has its origin in the analogue technique of control and was thus undertaken in this area, while the other one corresponds to the digital technique of control.

3.3.3.2.1 Sinusoidal Pulse Width Modulation (SPWM). Sinusoidal pulse width modulation (SPWM–Sinusoidal PWM) involves appropriate employment of the crossing points between saw carrier signal and sinusoidal modulation wave. When modulation voltages u_{m1}, u_{m2}, u_{m3} are higher then the voltages of the carrier wave.

$$u_{mi} > u_{saw} \qquad (3.147)$$

the potential of the output semiconductor switch $i = 1,2,3$ assumes the value of the supply U_d. In the opposite case this potential has the value of 0 since the ground semiconductor switch of the adequate branch of the inverter is in the ON state. The formation of the carrier wave is presented in Fig. 3.53. The output voltage results from the difference of the potential between the appropriate pairs of output points between inverter's branches $i = 1,2,3$ like:

$$U_{12} = U_{1G} - U_{2G} \qquad (3.148)$$

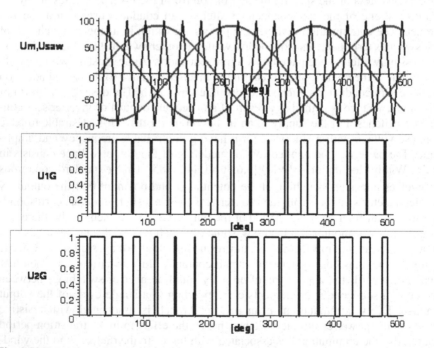

Fig. 3.53 Formation of an output voltage by the Sinusoidal PWM

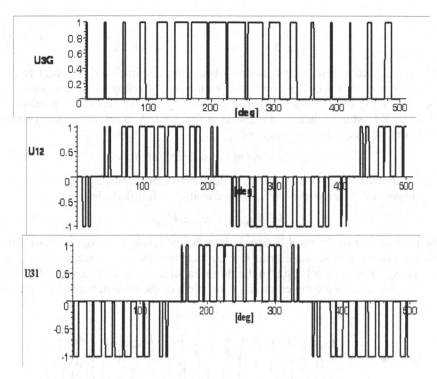

Fig. 3.53 (*continued*)

In the description of the inverter's mode of operation we apply amplitude and frequency modulation factors

$$m_a = \frac{U_{mi}}{U_{saw}} \qquad m_f = \frac{f_s}{f_{saw}} \qquad (3.149)$$

When the carrier frequency f_{saw} is an integer multiple of the output frequency f_s we have to do with synchronic modulation. The case when $m_a < 1$ is called proper modulation and in that case the frequency of switching is

$$f_{swch} = f_{saw} = f_s / m_f \qquad (3.150)$$

while for $m_a > 1$ we have to do with overmodulation, the switching frequency is smaller than it results from (3.150), and the voltage waveform is distorted. The highest attainable RMS value of the output voltage basic harmonic with the frequency of f_s is equal to [10,52]

$$U_L = 0.612 U_d \quad \text{for} \quad m_a = 1 \qquad (3.151)$$

which indicates a relatively small application of the supply voltage. Concurrently, higher harmonic orders in this curve for the synchronic modulation [10] amount to

$$k / m_f \pm 2 \qquad k = 2n - 1$$
$$l / m_f \pm 1 \qquad l = 2n \qquad n = 1,2,3,\ldots \tag{3.152}$$

and do not contain low order harmonics being the multiple of f_s. In order to increase the inverter's scope of application of the supply voltage a method with the injection of the third harmonic of the modulating voltage has been developed, [34,39,61,89], whose illustration is found in Fig. 3.54. As a result, the saw carrier wave is modulated with the voltage of

$$u_m = u_{m1} \sin(\omega_s t) + u_{m3} \sin(3\omega_s t) \tag{3.153}$$

in a manner that ensures the wave u_m of the modulating voltage does not exceed the voltage of the carrier wave U_{saw}. This condition is fulfilled when

$$u_{m3} \le 0.4\, u_{m1} \qquad u_{m1} \le U_{saw} \tag{3.154}$$

The introduction of the third harmonic into the modulating voltage results in the distortion of the voltage waveforms in relation to the reference potential U_{1G}, U_{2G}, U_{3G}. However, it does not result in the distortion of the output voltages in the load U_{12}, U_{23}, U_{31} since the compensation of the effect of the third harmonic according to (3.148).

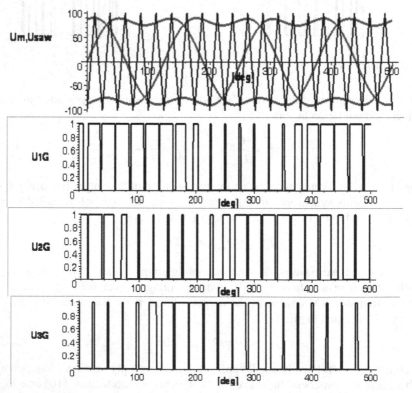

Fig. 3.54 Forming an output voltage by the SPWM method with a 3rd harmonic injection

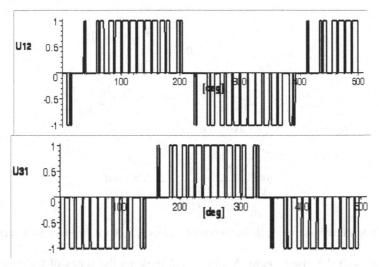

Fig. 3.54 (*continued*)

In the result we have to do with an increase of the output voltage to

$$U_L = 0.707 U_d \tag{3.155}$$

which means that it is over 15% more comparing with (3.151).

3.3.3.2.2 Space Vector Modulation (SVM). Two-level voltage inverter has three branches, each of which is in one of conduction states. The conduction states of an inverter can be determined as:

$$S_k \in 1,0 \qquad k = 1,2,3 \tag{3.156}$$

For example the state $S_1 = 1$ means that in the first branch the upper semiconductor switch is in the conduction state and the output potential is equal to $U_{1G} = U_d$. Concurrently, $S_2 = 0$ means that the ground semiconductor switch in the second branch is 'ON' and then $U_{2G} = 0$. In this method the inverter's output short time (T_p) averaged voltage vector $\mathbf{V}_s$ could be defined and effectively constructed by use of a concept of the space vector (complexor) $\mathbf{V}_k$, which is determined by the basic harmonic U_{1ph} of the required output voltage and the states of the particular branches S_1, S_2, S_3:

$$\mathbf{V}_k = (S_1 + S_2 e^{j120} + S_3 e^{-j120}) U_{1ph} \tag{3.157}$$

Since there are 3 branches, each of which can be in either of two states, the instantaneous outputs $\mathbf{V}_k$ from the inverter can assume any of 8 states illustrated graphically in Fig. 3.55, in accordance with (3.157).

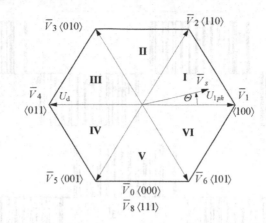

Fig. 3.55 Space vector plane with the switching states $\langle S_1, S_2, S_3 \rangle$ defined for each vector

Along with the given vector $\mathbf{V}_k$, Fig. 3.55 presents the states of the inverter's branches for which either one occurs. There are also two zero states of the inverter outputs $\mathbf{V}_0$, $\mathbf{V}_8$ for which semiconductor switches in all branches connect the load either to the ground (G) or positive (P) supply bar. In result the output voltages are equal to zero. Directly from the states of the branches it is possible to determine output voltages

$$\begin{bmatrix} u_{s12} \\ u_{s23} \\ u_{s31} \end{bmatrix} = U_d \begin{bmatrix} 1 & -1 & 0 \\ 0 & 1 & -1 \\ -1 & 0 & 1 \end{bmatrix} \begin{bmatrix} S_1 \\ S_2 \\ S_3 \end{bmatrix} \tag{3.158}$$

In a similar manner, we can establish phase voltages:

$$\begin{bmatrix} u_{s1} \\ u_{s2} \\ u_{s3} \end{bmatrix} = \frac{U_d}{3} \begin{bmatrix} 2 & -1 & -1 \\ -1 & 2 & -1 \\ -1 & -1 & 2 \end{bmatrix} \begin{bmatrix} S_1 \\ S_2 \\ S_3 \end{bmatrix} \tag{3.159}$$

The relation (3.159) illustrates voltage waveform presented in Fig. 3.56. The first harmonic has the amplitude of

$$U_{1ph} = \frac{2}{\pi} U_d \tag{3.160}$$

which corresponds to the length of vector $\mathbf{V}_k$ of the voltage star presented in Fig. 3.55. The amplitude of this value is achievable only for phase angles $\theta = k\pi/3$. The instantaneous position of vector $\mathbf{V}_s$ is determined by a phase angle θ. The method of modulation using space vectors SVM will be presented on the example of the synthesis of vector $\mathbf{V}_s$ situated in the first sector of the voltage star of the inverter. In the other sectors the situation is similar, as a value of phase angle θ could be reduced to the range of the first sector. This vector is synthesized by adequately selected switching times of the states that determine vectors $\mathbf{V}_1$ and $\mathbf{V}_2$ as well as zero vectors $\mathbf{V}_0$ and $\mathbf{V}_8$. This is illustrated in Fig 3.57.

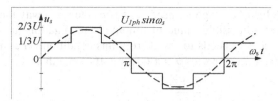

Fig. 3.56 Phase voltage of the two-level inverter, illustrating (3.159)

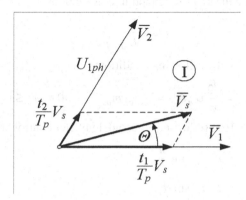

Fig. 3.57 A method of synthesis of the $\mathbf{V}_s$ vector in the first segment of the space vector plane

The construction of short time averaged vector $\mathbf{V}_s$ involves the fact that within sufficiently short pulsation time T_p, which corresponds to a fraction of the total cycle, voltages $\mathbf{V}_1$, $\mathbf{V}_2$ and $\mathbf{V}_0$ or $\mathbf{V}_8$ are switched on for the selected duration t_x, t_y, t_n. These intervals are obviously relative to the instant position of vector $\mathbf{V}_s$ determined by angle θ. As a matter of simplification it is assumed that within a single pulsation time the angle θ is invariable. The determination of time intervals t_x, t_y, t_n is performed using the relation:

$$T_p\mathbf{V}_s = t_x\mathbf{V}_1 + t_y\mathbf{V}_2 + t_n\mathbf{V}_n$$
$$T_p = t_x + t_y + t_n$$

(3.161)

where:
$$\mathbf{V}_s = U_s e^{j\theta} \qquad \mathbf{V}_1 = \frac{2}{\pi}U_d e^{j0} \qquad \mathbf{V}_2 = \frac{2}{\pi}U_d e^{j\frac{\pi}{3}}$$

t_n - is a sum of the intervals of the occurrence of zero vectors $\mathbf{V}_0$, $\mathbf{V}_8$.
After solving (3.161), we obtain:

$$t_x = T_p \frac{U_s}{U_d}\frac{\pi}{\sqrt{3}}\sin\left(\frac{\pi}{3} - \theta\right)$$

$$t_y = T_p \frac{U_s}{U_d}\frac{\pi}{\sqrt{3}}\sin\theta$$

(3.162)

$$t_n = T_p - t_x - t_y$$

The maximum value of voltage U_s needs to be selected in a way that $t_n \geq 0$. In addition, the highest admissible value of amplitude modulation factor for a linear range of inverter control needs to result from this condition. The maximum value of time intervals $t_x = t_y$ occurs for $\theta = \pi/6$ (see 3.166) and then:

$$\frac{t_x + t_y}{T_p} = \frac{U_s}{U_d} \frac{\pi}{\sqrt{3}} \leq 1 \tag{3.163}$$

Defining for this case the amplitude modulation factor as

$$m_a = \frac{U_s}{U_d} \frac{\pi}{\sqrt{3}} \tag{3.164}$$

from (3.163) we obtain the following condition:

$$U_s = \frac{\sqrt{3}}{\pi} U_d m_a = U_{s\max} m_a \qquad 0 < m_a \leq 1 \tag{3.165}$$

The time intervals (3.163) reflecting sinusoidal inverter control are equal to:

$$t_x = T_p m_a \sin\left(\frac{\pi}{3} - \theta\right)$$

$$t_y = T_p m_a \sin\theta \tag{3.166}$$

$$t_n = T_p - t_x - t_y \qquad \frac{t_n}{T_p} = 1 - m_a \cos\left(\frac{\pi}{6} - \theta\right)$$

Output frequency results from control of the angle $\theta = \omega_s t = 2\pi f_s t$, while the amplitude of the sinusoidal waveform results from an adequate selection of factor m_a. The maximum amplitude of the sinusoidal voltage for $m_a = 1$ is $U_{s\max}$ according to (3.165). Alternatively it could be calculated as:

$$U_{s\max} = U_{1ph} \frac{\sqrt{3}}{2} = \frac{2}{\pi} \frac{\sqrt{3}}{2} U_d = \frac{\sqrt{3}}{\pi} U_d \approx 0.551 U_d \tag{3.167}$$

- and it is the value of the amplitude of the phase voltage, while

$$U_{1L} = \sqrt{3} U_{s\max} = \frac{3}{\pi} U_d \approx 0.955 U_d$$

- is amplitude value of line-to-line voltage, and finally

$$U_L = \frac{U_{1L}}{\sqrt{2}} \approx 0.675 U_d$$

- is the RMS value of the output voltage, which is over 10% higher than (3.151) for the case of control using sinusoidal pulse width modulation (SPWM).

From the above relations it results that for a motor with the rated voltage U_{sn}, the voltage U_d delivered to the inverter should be, respectively:

$$\begin{array}{ll} U_{sn} = 400 \ [V] & U_d = 600 \ [V] \\ U_{sn} = 6000 \ [V] & U_d = 9000 \ [V] \end{array} \tag{3.168}$$

- that is $U_d \approx 1.5 U_n$.

3.3.3.2.3 Switching Sequence in the Pulsation Cycle. The basic requirements to be met by an inverter's control system are the following:

- wide range of linear operation,
- small number of switchings per cycle of output voltage aimed at reducing energy losses during commutation,
- limitation of the amplitudes of the harmonics of voltage and current of the load associated with switching frequency,
- limitation the incidence and amplitude of lower harmonics of the output voltage associated with load frequency f_s.

6-pulse switching sequence

A standard solution involves 6-pulse switching sequence for a single pulsation cycle. The duration t_n of the zero vectors $\mathbf{V}_0$ or $\mathbf{V}_8$ in this cycle is divided into three parts: ¼ at the beginning and the end of a cycle and ½ in the mid section of the cycle. The characteristic property in this case is that the change of inverter state occurring between the individual pulses requires only a single switching. An example of the waveform for a single pulsation cycle in 6-pulse sequence is presented in Fig. 3.58. As one can note, for this kind switching sequence the beginning and termination of the cycle come with the $\mathbf{V}_0$ <000> state. In addition, it involves only a single switching for each successive pulse and there is symmetry in relation to the mid period of pulsation. For the entire period of the output voltage of the inverter the 6-pulse switching sequence is presented in Table 3.2.

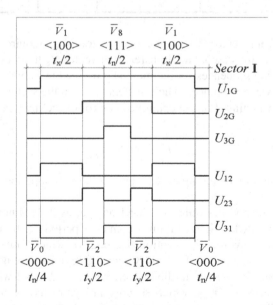

Fig. 3.58 Exemplary 6-pulse sequence in a pulse cycle in the I-st sector of the inverter's operation

Table 3.2 Six-pulse switching sequence with minimum number of commutations

Sector	Switched vector with the duration of a pulse						
	$t_n/4$			$t_n/2$			$t_n/4$
I	$\mathbf{V}_0$	$\mathbf{V}_1$ $t_x/2$	$\mathbf{V}_2$ $t_y/2$	$\mathbf{V}_8$	$\mathbf{V}_2$ $t_y/2$	$\mathbf{V}_1$ $t_x/2$	$\mathbf{V}_0$
II	$\mathbf{V}_0$	$\mathbf{V}_3$ $t_y/2$	$\mathbf{V}_2$ $t_x/2$	$\mathbf{V}_8$	$\mathbf{V}_2$ $t_x/2$	$\mathbf{V}_3$ $t_y/2$	$\mathbf{V}_0$
III	$\mathbf{V}_0$	$\mathbf{V}_3$ $t_x/2$	$\mathbf{V}_4$ $t_y/2$	$\mathbf{V}_8$	$\mathbf{V}_4$ $t_y/2$	$\mathbf{V}_3$ $t_x/2$	$\mathbf{V}_0$
IV	$\mathbf{V}_0$	$\mathbf{V}_5$ $t_y/2$	$\mathbf{V}_4$ $t_x/2$	$\mathbf{V}_8$	$\mathbf{V}_4$ $t_x/2$	$\mathbf{V}_5$ $t_y/2$	$\mathbf{V}_0$
V	$\mathbf{V}_0$	$\mathbf{V}_5$ $t_x/2$	$\mathbf{V}_6$ $t_y/2$	$\mathbf{V}_8$	$\mathbf{V}_6$ $t_y/2$	$\mathbf{V}_5$ $t_x/2$	$\mathbf{V}_0$
VI	$\mathbf{V}_0$	$\mathbf{V}_1$ $t_y/2$	$\mathbf{V}_6$ $t_x/2$	$\mathbf{V}_8$	$\mathbf{V}_6$ $t_x/2$	$\mathbf{V}_1$ $t_y/2$	$\mathbf{V}_0$

The transfer from vector $\mathbf{V}_0$ to any odd vector $\mathbf{V}_{2k-1}$ requires just a single switching, just as in the case of the transfer from vector $\mathbf{V}_8$ to an even vector $\mathbf{V}_{2k}$. Hence, the switching sequences for the method presented in Table 3.2 require only one switching between the pulses. The transfers between the sectors occur without the necessity of switching. For the examined control of the inverter the frequency of switching is

$$f_{sw} = 6 \cdot N \cdot f_s \qquad (3.169)$$

where: N - is number of switching cycles corresponding to a period of output voltage.

The presented method has only one disadvantage, as it does not fulfill one of the postulates presented earlier. Within a complete period T_s the output voltage does not form an antisymmetric function, i.e. one for which $f(\omega t + T_s/2) = -f(\omega t)$ since in the opposite sectors e.g. I-IV, III-VI etc. the segments are not identical but have opposite signs. One can note this by referring to Table 3.2 where for active vectors the relation $\mathbf{V}_i \rightarrow \mathbf{V}_{i+3}$, which secures antisymmetry, is not fulfilled for sectors $I \rightarrow I+3$. This is graphically presented in Fig. 3.59 for sector IV.

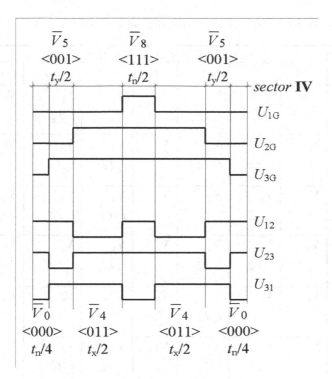

Fig. 3.59 Six-pulse sequence in a pulse cycle of the IV sector

From the comparison between the sequence of the pulses in sector I (Fig. 3.58) and sector IV (Fig. 3.59) and the shape of the output voltage it is clear that the signals corresponding to the voltages have equal values with opposite signs; however, their phase shifts are different. This results in originating even order harmonics in the output voltage with frequencies being the multiple of f_s. The amplitudes of these harmonics are not considerable and are acceptable in small and medium power drives; however, in high power drives they cannot be accepted due to exceeding the requirements of the standard values of current deformations. For this reason, there is a number of 6-pulse control cycles ensuring the absence of even numbered harmonics in the voltages generated by the inverter. One of such cycles will be presented in the section that follows.

6-pulse switching sequence eliminating even harmonics

In this kind of control an increased number of switchings in a sequence comes as a result of the elimination of the even harmonics in the output voltage. This elimination most easily occurs as a result of the adjustment of pulses in the opposite sectors in a way that ensures that they are realized by opposite vectors in the voltage star. This is illustrated in Table 3.3.

Table 3.3 Six-pulse switching sequence with elimination of even harmonics

Sector	Switched vector with the duration of a pulse						
	$t_n/4$			$t_n/2$			$t_n/4$
I	V_0	V_1 $t_x/2$	V_2 $t_y/2$	V_8	V_2 $t_y/2$	V_1 $t_x/2$	V_0
II	V_0	V_3 $t_y/2$	V_2 $t_x/2$	V_8	V_2 $t_x/2$	V_3 $t_y/2$	V_0
III	V_0	V_3 $t_x/2$	V_4 $t_y/2$	V_8	V_4 $t_y/2$	V_3 $t_x/2$	V_0
IV	V_8	V_4 $t_y/2$	V_5 $t_x/2$	V_0	V_5 $t_x/2$	V_4 $t_y/2$	V_8
V	V_8	V_6 $t_x/2$	V_5 $t_y/2$	V_0	V_5 $t_y/2$	V_6 $t_x/2$	V_8
VI	V_8	V_6 $t_y/2$	V_1 $t_x/2$	V_0	V_1 $t_x/2$	V_6 $t_y/2$	V_8

The operating principle in this method is presented in Fig. 3.60 on the basis of the example of the sequence of pulses in sectors II and V. The presented courses ensure that the resulting voltage waveform is an odd function; hence, it does not contain even harmonics. An increase of the number of swithings occurs during the transfer from sector III to IV and from sector VI to sector I since there is a change of the zero vector from V_0 to V_8 and the reverse. This leads to the increase of the switching frequency by

$$\Delta f_{sw} = 6 \cdot f_s \qquad (3.170)$$

4-pulse switching sequence (DSVM)

This method of inverter control can lead to a further limitation of the number of switchings. During a single cycle of switching one of the nodes of the inverter does not change the state and the switching occurs in the two remaining ones. For this reason this mode of control is called discontinuous SVM (DSVM). We can distinguish two types of switching cycles: type A – when a branch that is not in-volved in switching remains at the level of potential G and type B when the branch that does not switch is connected to a high potential P. This is illustrated in

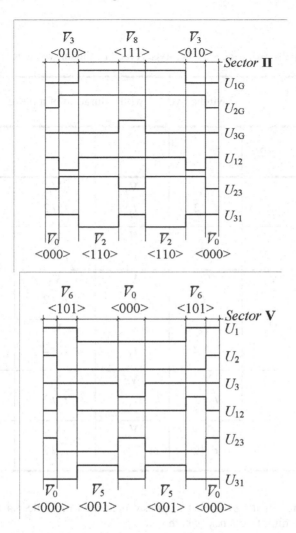

Fig. 3.60 Six pulse switching sequence with elimination of even order harmonics, demonstrated for the II and V sector

Fig. 3.61. Both types of switching sequences equally form the basic harmonic of the output voltage while its shape is different in the range of higher harmonics since in each type of sequence a different sector is divided into two time sections. One can also note that the negation of the switching sequence type A in sector I is transformed into the sequence of switchings (DSVM) type B in sector IV, etc. Table 3.4 contains a summary of the switching sequences for the full period of the output voltage.

Table 3.4 Four-pulse switching sequence–type A for a minimum switching frequency

Sector	Switched vector with a duration of a pulse				
	$t_n/2$				$t_n/2$
I	V_0	V_1 $t_x/2$	V_2 t_y	V_1 $t_x/2$	V_0
II	V_0	V_3 $t_y/2$	V_2 t_x	V_3 $t_y/2$	V_0
III	V_0	V_3 $t_x/2$	V_4 t_y	V_3 $t_x/2$	V_0
IV	V_0	V_5 $t_y/2$	V_4 t_x	V_5 $t_y/2$	V_0
V	V_0	V_5 $t_x/2$	V_6 t_y	V_5 $t_x/2$	V_0
VI	V_0	V_1 $t_y/2$	V_6 t_x	V_1 $t_y/2$	V_0

The advantage of the switching sequence presented in Table 3.4 is the limitation of the switching frequency, which is

$$f_{sw} = 4 \cdot N \cdot f_s \qquad (3.171)$$

However, the disadvantage regards the shape of the voltage, which does not form an antisymmetric function and results in the origin of even harmonics type $2k\, f_s$. The lack of the antisymmetry of the voltage curve is clearly visible in Table 3.4 as s result of the comparison between the voltages in the opposite sectors, e.g. I-IV, II-V, or III-VI. This drawback can be eliminated by the application of the type B sequences e.g. in sectors IV to VI. This is summarized in Table 3.5.

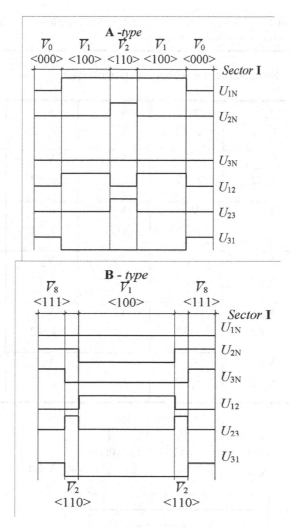

Fig. 3.61 Four-pulse switching sequence of the A - and B - type, presented for the I-st sector

The application of the switching sequence presented in Table 3.5 results in an increase of the number of inverter switchings per period of the output voltage. Following the transfers from sectors III → IV and VI → I we have to do with switching regarding zero vectors, i.e. $V_0 → V_8$ and $V_8 → V_0$, respectively. This results in an increase of frequency expressed by (3.170). Fig. 3.62 illustrates the application of 4-pulse switching sequence with elimination of even harmonics of the voltage, for voltages generated in sectors III and VI of the phase plane.

Table 3.5 Four-pulse switching sequence – type A and B combined, for eliminating even harmonics of $2kf_s$ order in an output voltage

Section	Switched vector with a duration of a pulse				
I	$\mathbf{V}_0$ $t_n/2$	$\mathbf{V}_1$ $t_x/2$	$\mathbf{V}_2$ t_y	$\mathbf{V}_1$ $t_x/2$	$\mathbf{V}_0$ $t_n/2$
II	$\mathbf{V}_0$ $t_n/2$	$\mathbf{V}_3$ $t_y/2$	$\mathbf{V}_2$ t_x	$\mathbf{V}_3$ $t_y/2$	$\mathbf{V}_0$ $t_n/2$
III	$\mathbf{V}_0$ $t_n/2$	$\mathbf{V}_3$ $t_x/2$	$\mathbf{V}_4$ t_y	$\mathbf{V}_3$ $t_x/2$	$\mathbf{V}_0$ $t_n/2$
IV	$\mathbf{V}_8$ $t_n/2$	$\mathbf{V}_4$ $t_x/2$	$\mathbf{V}_5$ t_y	$\mathbf{V}_4$ $t_x/2$	$\mathbf{V}_8$ $t_n/2$
V	$\mathbf{V}_8$ $t_n/2$	$\mathbf{V}_6$ $t_y/2$	$\mathbf{V}_5$ t_x	$\mathbf{V}_6$ $t_y/2$	$\mathbf{V}_8$ $t_n/2$
VI	$\mathbf{V}_8$ $t_n/2$	$\mathbf{V}_6$ $t_x/2$	$\mathbf{V}_1$ t_y	$\mathbf{V}_6$ $t_x/2$	$\mathbf{V}_8$ $t_n/2$

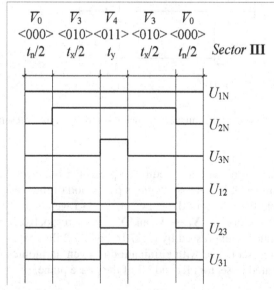

Fig. 3.62 Four pulse switching sequence with elimination of even numbered harmonics, demonstrated for the opposite sectors III and VI

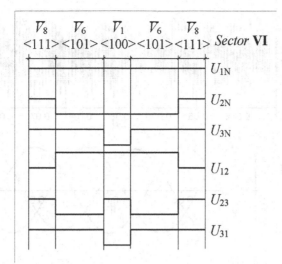

Fig. 3.62 (*continued*)

3.3.3.2.4 Exemplary Curves of Inverter Currents. The output waveforms of a 3-phase load current supplied from an inverter depend the control method, i.e. the number N of pulse sequences per period of the curve, amplitude modulation factor m_a and the load. The examples of the waveforms under a load type R, L and the output frequency of the inverter $f_s = 50$ [Hz] are presented in the figures that follow in this way illustrating the effect of the number N of the sequences of pulses and modulation factor m_a.

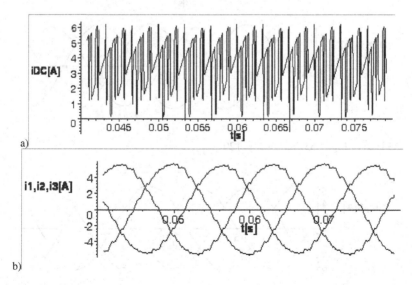

Fig. 3.63 Performance of a two-level VSI inverter with DSVM and even harmonics elimination control for $N = 18$, $f_s = 50$ [Hz], $m_a = 0.85$: a) DC source current b) 3-phase load currents

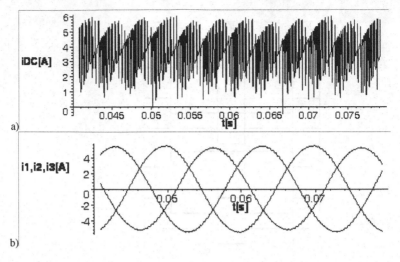

a)

b)

Fig. 3.64 Like in Fig. 3.63, but for $N = 48$

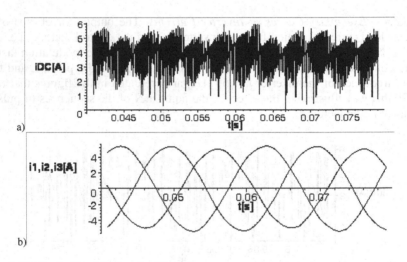

a)

b)

Fig. 3.65 Like in Fig. 3.63, but for $N = 120$

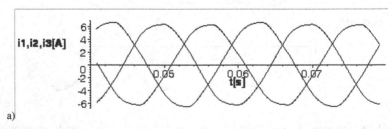

a)

Fig. 3.66 Performance of two-level VSI inverter for the over-modulation control area. Comparison of 3-phase load currents for: a) $m_a = 1.05$ b) $m_a = 1.25$ c) $m_a = 2.0$

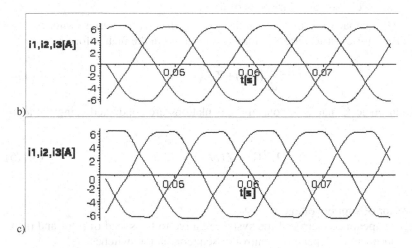

b)

c)

Fig. 3.66 (*continued*)

3.3.3.3 Induction Motor Supplied from 2-Level Voltage Inverter

The mathematical model for the simulations of the operation of a VSI inverter – induction motor drive is based on the diagram presented in Fig. 3.52 and on the system of equations with untransformed variables of the electric circuits of the stator (3.76-3.77). However, some changes have been implemented. In the place of the filter C_f connected to the load there are capacitors C placed parallel to every branch of the inverter. In addition, inductance L_{DC} and resistance R_{DC} are introduced to play the role of residual parameters of the DC voltage source. Moreover, residual inductance L_d between the branches of the inverter bridge has been introduced. The controlled switches present in the inverter branches have been simulated with resistance with a controlled value: a fraction of an ohm in the conduction state and several $k\Omega$ in the blocking state, so that it approximately corresponds to the actual operating conditions of an inverter. This mathematical model of a drive system is described using 16 state variables, 15 of which are electrical ones. Together, they form a vector of the state variables:

$$\mathbf{Y} = \left[i_{DC}, Q_2, Q_3, i_4, i_5, i_6, i_7, i_{ru}, i_{rv}, \omega_e, Q_{11}, ..., Q_{16}\right]^T \qquad (3.172)$$

where:

i_{DC} - DC source current

$Q_2 - Q_3$ - charge of the main capacitor C_g

i_4, i_5, i_6, i_7 - mesh current in the inverter's branches selected so, that the currents of the motor's stator are equal to:

$$i_{s1} = i_5 - i_4 \qquad i_{s3} = i_6 - i_7 \qquad (3.173)$$

i_{ru}, i_{rv} - transformed axial currents of the rotor

$\omega_e = p\Omega_r$ - electric angular speed of the rotor

$Q_{11},...Q_{16}$ - capacitor C charges on the six branches with inverter switches.
The model for electric variables may be presented in the matrix form:

$$\mathbf{U} = \frac{d}{dt}(\mathbf{LY}) + \mathbf{Z}(\omega_e)\mathbf{Y} \qquad (3.174)$$

Since inductance $\mathbf{L}$ matrix is constant, calculations are conducted using the algorithm

$$\dot{\mathbf{Y}} = \mathbf{L}^{-1}(\mathbf{U} - \mathbf{Z}(\omega_e)\mathbf{Y}) \qquad (3.175)$$

where:

$\mathbf{U}$ - vector of supply voltages
$\mathbf{Z}(\omega_e)$ - impedance matrix of the system relative to the speed of rotor and resistances, whose values depend on control of semiconductor switches
$\mathbf{L}$ -constant parameters' inductance matrix of the system.
These equations are further supplemented with an equation for the mechanical motion

$$\dot{\omega}_e = (T_e - T_l - D\,\omega_e)/J \qquad (3.176)$$

where:

$$T_e = pL_m\left((-\frac{\sqrt{2}}{2}i_{s1} - \sqrt{2}i_{s3})i_{ru} - \sqrt{\frac{3}{2}}i_{s1}i_{rv}\right) \qquad (3.177)$$

- denotes electromagnetic torque of the motor
T_l - is a load torque
D - is a coefficient of viscous damping
J - is a moment of inertia of the rotating parts of the drive.

This mathematical model was applied for simulation calculations of the drive in operation. The testing involved several methods of inverter control presented earlier in the book. The exemplary waveforms presented in figures regard control with 4-pulse switching sequence (DVSM) and elimination of harmonics with even numbers (Fig. 3.62). The term N in the description of the figures denotes the number of the sequences of pulses per period of the output voltage of an inverter.

During the simulation calculations the residual inductances L_{DC}, L_d assume the values in the range of 0.5% - 1% of the inductance of the block-rotor state $L_z = \sigma L_s$, (3.121), while blocking resistance R_b from 4 [kΩ] to 20 [kΩ], depending on the rated voltage of the motor. The voltage U_{DC} supplying the inverter corresponds to the voltages determined from (3.168). Below is a presentation of some results of simulations for small and medium power motors.

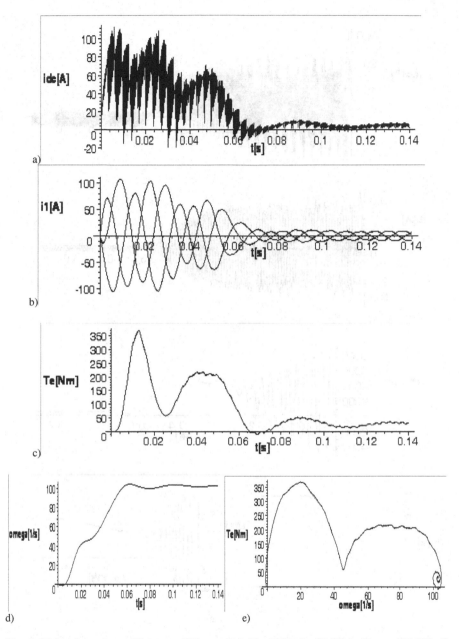

Fig. 3.67 Starting of a small power (S1) motor fed by VSI; f_s= 50 [Hz], U_{DC} = 600 [V], T_l = 0.5 T_n , J = 2J_s, N = 120: a) DC source current b) stator currents c) electromagnetic torque d) angular speed e) torque-speed trajectory

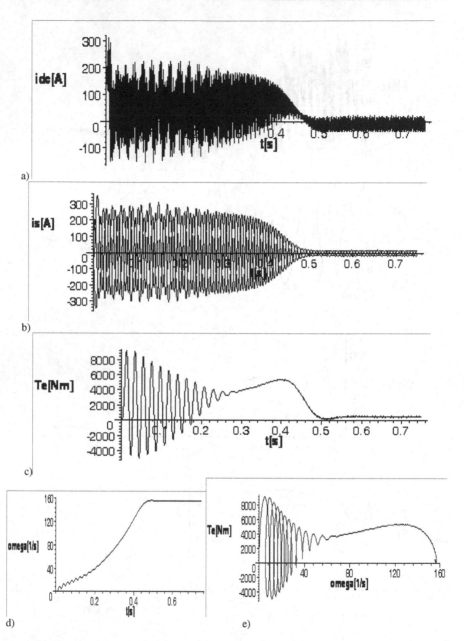

Fig. 3.68 Direct starting of a medium power (S2H) motor fed by VSI; f_s= 50 [Hz], U_{DC} = 9000 [V], T_l = 0.2, J = 1.5 J_s, N = 120: a) DC source current b) stator currents c) electromagnetic torque d) angular speed ω_e e) torque-speed trajectory

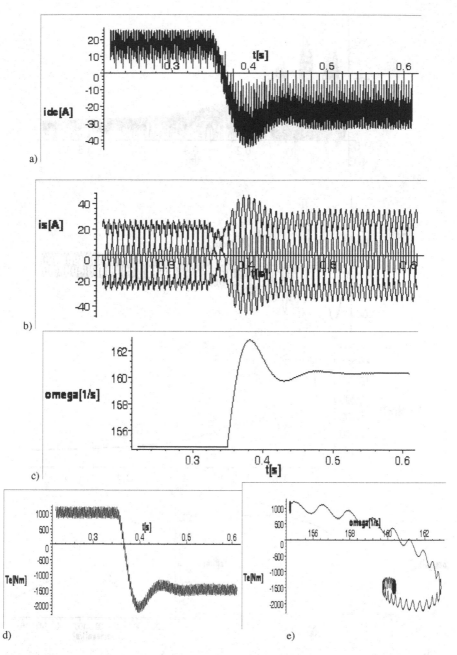

Fig. 3.69 Medium power (S2H) motor – transition to generating mode; f_s= 50 [Hz], U_{DC} = 9000 [V], T_l =0.5T_n → - 0.75T_n, J = 1.5 J_s, N = 120: a) DC source current b) stator currents c) angular speed d) electromagnetic torque e) torque-speed trajectory

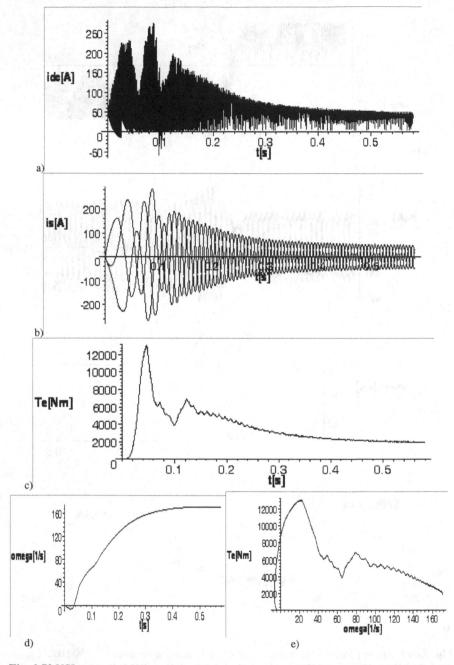

Fig. 3.70 VSI controlled U/f starting of the medium power (S2H) motor. $T_l = T_n, f_s=2...50$ [Hz], $k_u = 0.1...0.95, J = 1.5 J_s, N = 120$: a) DC source current b) stator currents c) electromagnetic torque d) angular speed e) torque-speed trajectory f) steady-state DC current g) steady state stator currents

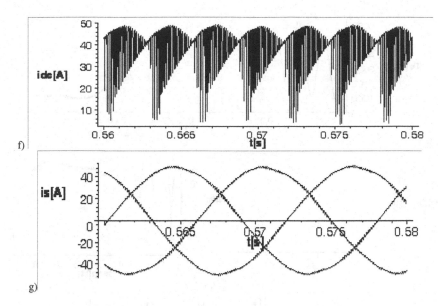

f)

g)

Fig. 3.70 (*continued*)

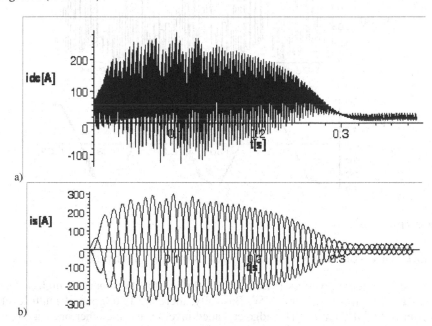

a)

b)

Fig. 3.71 VSI controlled U/f starting of the medium power (S2H) motor. $T_l = 0.5\ T_n$, $f_s =$ 25...50 [Hz], $k_u = 0.15...0.95$, $J = 1.5\ J_s$, $N = 60$: a) DC source current b) stator currents c) electromagnetic torque d) angular speed e) torque-speed trajectory f) steady-state DC current g) steady state stator current

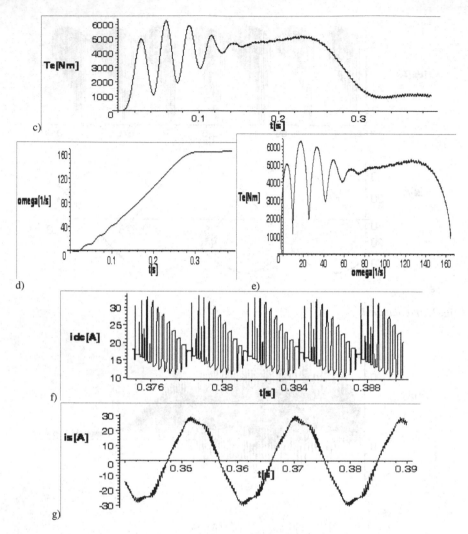

Fig. 3.71 (*continued*)

3.3.3.4 Three-Level Diode Neutral Point Clamped VSI Inverter

Multi-level voltage inverters [81,86,93] are applied for the supply of high power drives ranging around several MW. Basically, there are two types of multi-level inverters. One of them is a H-bridge cascaded inverter and the other one - a diode clamped inverter [10,33,56,64]. The advantage of the application of the latter type is associated with the possibility of high output voltage without serial connection of the semiconductor switches, lower values of the voltage dV/dt derivative for switchings as well as lower total harmonic distortion (THD) level, which determines the degree of waveform distortion resulting from higher order harmonics. It is possible to apply inverters with diode clamped neutral point in the form of three as well as multi-level ones [8,88], however, only the systems based on three-level design have found common applications in high-power drives. The name for the three-level inverter comes as a consequence of the fact that the output clamp 1, 2 or 3 can be connected to the positive potential of the source P, to the neutral point G or to the negative potential of the supply N. The neutral potential G is gained as a result of capacitor-based division of the supply voltage U_{DC} and appropriate control of switches in a way that balances the fluctuations of the potential at this point.

3.3.3.4.1 Structure and Operating Principle of a Three-Level Inverter. The diagram of a three-level inverter with diode clamped neutral points G is presented in Fig. 3.72.

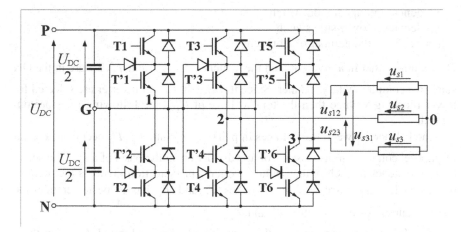

Fig. 3.72 Three-level diode clamped VSI

The particular branches contain four IGBT or GTO (Gate Torn Off) switches per each branch connected in a series. The neutral point G is formed by the upper and lower pairs of switches in branches being connected via diodes to the mid part of a capacity voltage divider. The state of a particular branch of an inverter is determined as a result of the algebraic description of operation using a respective variable S_1, S_2, S_3 corresponding to the branch number. Under normal operating conditions of an inverter there are possible three states such that variables S_1, S_2, S_3 may assume the values in the set <1,0,-1>. A branch assumes state '1' when the two upper switches in a branch are in the ON state – in that case the clamp of the load is connected to point P with the positive potential of the source. The '0' state of the branch means that the two middle switches in a specific branch are in the ON state – in this case the output clamp is connected to the neutral G using a upper or lower clamping diode depending on the direction of the current flow. A branch variable assumes the value of '-1' for the case when both lower switches in a branch are in ON state and, as a consequence, the load clamp is connected to the negative rail of the source with potential N. This is illustrated in Table 3.6.

Table 3.6 Possible states of the branch in three-level VSI inverter with a diode clamped neutral point

	$S_i = 1$	$S_i = 0$	$S_i = -1$
T_x	on	off	off
T_x'	on	on	off
T_y'	off	on	on
T_y	off	off	on

where:

x - denotes an upper side switch
y - denotes a lower side switch
$i = 1,2,3$ - is the number of a branch

One can note that in a given branch the switches T_x, T_y' and T_x', T_y are in the ON state in a complementary manner. Since the switches in an inverter are selected for lower blocking voltage then the total voltage of U_{DC}, the following rules are observed:

- the balance between the voltages in pairs T_x, T_x' and T_y, T_y' occurs as a result of parallel connection of resistor voltage divider or on the basis of a specific selection of switches in each branch of an inverter. This selection involves the requirement that leakage current of the switches T_x and T_y has to be lower then the leakage current of the switches T_x' and T_y'

- the direct switching between the states 1 and −1 is prohibited. Between these states the inverter must pass through zero state.

Since the inverter has 3 branches and each of them can assume any of three states, the total number of inverter states is equal to

$$LS = 3^3 = 27 \tag{3.178}$$

Each of these states can be described using a vector $\langle S_1, S_2, S_3 \rangle$. This involves the following states:

a)   zero states described with vectors:

$$\mathbf{V}_0 = \langle 0,0,0 \rangle$$
$$\mathbf{V}_{0P} = \langle 1, 1, 1 \rangle \tag{3.179}$$
$$\mathbf{V}_{0N} = \langle -1,-1,-1 \rangle$$

b)   6 states for the major star pattern voltages:

$$
\begin{aligned}
&\mathbf{V}_{1L} = \langle 1,-1,-1 \rangle && \mathbf{V}_{2L} = \langle 1, 1,-1 \rangle \\
&\mathbf{V}_{3L} = \langle -1, 1,-1 \rangle && \mathbf{V}_{4L} = \langle -1, 1, 1 \rangle \\
&\mathbf{V}_{5L} = \langle -1,-1, 1 \rangle && \mathbf{V}_{6L} = \langle 1,-1, 1 \rangle
\end{aligned}
\tag{3.180}
$$

c)   6 states for the middle star pattern voltages:

$$
\begin{aligned}
&\mathbf{V}_{1M} = \langle 1,0,-1 \rangle && \mathbf{V}_{2M} = \langle 0, 1,-1 \rangle \\
&\mathbf{V}_{3M} = \langle -1, 1,0 \rangle && \mathbf{V}_{4M} = \langle -1, 0,1 \rangle \\
&\mathbf{V}_{5M} = \langle 0,-1, 1 \rangle && \mathbf{V}_{6M} = \langle 1,-1,0 \rangle
\end{aligned}
\tag{3.181}
$$

d)   6 states for the minor star pattern voltages type P:

$$
\begin{aligned}
&\mathbf{V}_{1SP} = \langle 1,0,0 \rangle && \mathbf{V}_{2SP} = \langle 1,1,0 \rangle \\
&\mathbf{V}_{3SP} = \langle 0,1,0 \rangle && \mathbf{V}_{4SP} = \langle 0, 1, 1 \rangle \\
&\mathbf{V}_{5SP} = \langle 0,0, 1 \rangle && \mathbf{V}_{6SP} = \langle 1,0, 1 \rangle
\end{aligned}
\tag{3.182}
$$

e)   6 states for the minor star pattern voltages type N:

$$
\begin{aligned}
&\mathbf{V}_{1SN} = \langle 0,-1,-1 \rangle && \mathbf{V}_{2SN} = \langle 0,0,-1 \rangle \\
&\mathbf{V}_{3SN} = \langle -1,0,-1 \rangle && \mathbf{V}_{4SN} = \langle -1,0,0 \rangle \\
&\mathbf{V}_{5SN} = \langle -1,-1,0 \rangle && \mathbf{V}_{6SN} = \langle 0,-1,0 \rangle
\end{aligned}
\tag{3.183}
$$

The graphical representation of the vector forming these star patterns is based on the relation (3.157), where U_{1ph} is the phase voltage of the minor star pattern voltages. This is illustrated in Fig. 3.73.

The angular (and time) diagrams for the instantaneous output voltages are obtained from the following relation:

$$
\begin{bmatrix} u_{s12} \\ u_{s23} \\ u_{s31} \end{bmatrix} = \frac{U_{DC}}{2} \begin{bmatrix} 1 & -1 & 0 \\ 0 & 1 & -1 \\ -1 & 0 & 1 \end{bmatrix} \begin{bmatrix} S_1 \\ S_2 \\ S_3 \end{bmatrix}
\tag{3.184}
$$

Similarly, for phase voltages of the 3-phase load:

$$
\begin{bmatrix} u_{s1} \\ u_{s2} \\ u_{s3} \end{bmatrix} = \frac{U_{DC}}{6} \begin{bmatrix} 2 & -1 & -1 \\ -1 & 2 & -1 \\ -1 & -1 & 2 \end{bmatrix} \begin{bmatrix} S_1 \\ S_2 \\ S_3 \end{bmatrix}
\tag{3.185}
$$

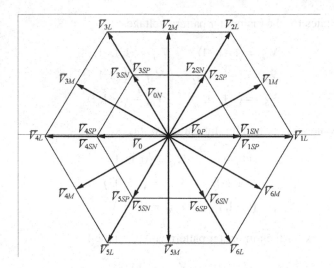

Fig. 3.73 Space vector plane for a three level VSI inverter

The application of the relations (3.184) and (3.185) with regard to vectors (3.180), which determine major pattern voltages leads to the voltage waveforms (Fig. 3.74) for such a cycle of operation. The switching sequence is performed every $\omega_s t = \pi/3$. A similar course of action undertaken for middle pattern voltages (3.181) gives voltage curves presented in Fig. 3.75. For the two sets of vectors (3.182), (3.183) that determine states of minor pattern voltages the voltage curves are identical in terms of the shape and form to the waveforms in major pattern voltages (Fig. 3.74); however, the amplitudes of these voltages are reduced by a half. The summary of the characteristics presenting three configuration models for a 3-phase inverter is presented in Table 3.7.

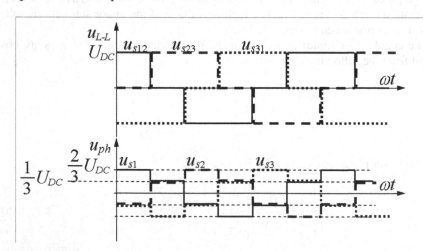

Fig. 3.74 Voltage curves for the major pattern switching of a three-level VSI: line-to-line voltages (3-step) and phase voltages (4-step)

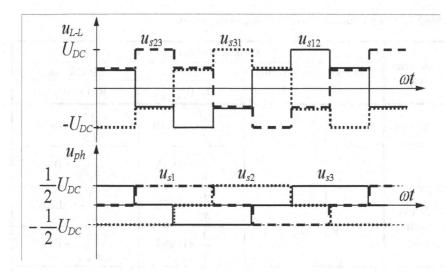

Fig. 3.75 Voltage curves for the middle pattern switching of a three-level VSI: line–to–line voltages (4-step) and phase voltages (3-step)

As it results from data in Table 3.7, the relations between the voltages for the particular switching models are:

$$U_{middle}/U_{major}(1h, RMS) = \frac{\sqrt{3}}{2}$$

$$U_{minor}/U_{major}(\max,1h, RMS) = \frac{1}{2} \qquad (3.186)$$

3.3.3.4.2   SVM Control of 3-Level VSI Voltage Inverter. As presented in Fig. 3.73, for the control of a three-level inverter with space vector modulation (SVM) we have 24 active vectors and 3 zero ones when compared to the total of 6 active and 2 zero vectors in a two-level inverter. The general rules regarding the control are similar to the previously considered case:

- there is a single pair of switching at a time (see Table 3.6), direct transition from state '1' to '-1' and reverse is forbidden

- transfer of the vector from a switching sequence to another one, both within a single control region and during the transfer between the neighboring regions, should not require more than a single pair of switching,

- switching within a control sequence is configured in a way that limits the fluctuations of the potential of the neutral point G to a maximum degree. This issue is not encountered in two-level inverters and occurs in three-level ones due to the capacitor based setting of the neutral level of the potential of point G.

Table 3.7 Output voltages for two- and three-level inverters

Inverter type	shape	maximum $*U_d$	1h * $\quad$ 1h $(V_s)_{max}$	RMS - step $\quad$ RMS $(V_s)_{max}$
			Output voltage U_L/U_d	
3-level major star	3 step	1	$\dfrac{2\sqrt{3}}{\pi} \approx 1.10$	$\sqrt{\dfrac{2}{3}} \approx 0.816$
			$\dfrac{3}{\pi} \approx 0.955$	$\dfrac{3}{\pi\sqrt{2}} \approx 0.675$
3-level medium star	4 step	1	$\dfrac{3}{\pi} \approx 0.955$	$\dfrac{1}{\sqrt{2}} \approx 0.707$
			$\dfrac{3}{\pi} \approx 0.955$	$\dfrac{3}{\pi\sqrt{2}} \approx 0.675$
3-level minor star	3 step	1/2	$\dfrac{\sqrt{3}}{\pi} \approx 0.551$	$\dfrac{1}{\sqrt{6}} \approx 0.408$
			$\dfrac{3}{2\pi} \approx 0.478$	$\dfrac{3}{2\pi\sqrt{2}} \approx 0.338$
2 level	3 step	1	like for the major star	
			Output voltage U_{ph}/U_d	
3-level major star	4 step	2/3	$\dfrac{2}{\pi} \approx 0.637$	$\dfrac{\sqrt{2}}{3} \approx 0.471$
			$\dfrac{\sqrt{3}}{\pi} \approx 0.551$	$\sqrt{\dfrac{3}{2}}\dfrac{1}{\pi} \approx 0.390$
3-level medium star	3 step	1/2	$\dfrac{\sqrt{3}}{\pi} \approx 0.551$	$\dfrac{1}{\sqrt{6}} \approx 0.408$
			$\dfrac{\sqrt{3}}{\pi} \approx 0.551$	$\sqrt{\dfrac{3}{2}}\dfrac{1}{\pi} \approx 0.390$
3-level minor star	4 step	1/3	$\dfrac{1}{\pi} \approx 0.318$	$\dfrac{1}{3\sqrt{2}} \approx 0.236$
			$\dfrac{\sqrt{3}}{2\pi} \approx 0.276$	$\sqrt{\dfrac{3}{2}}\dfrac{1}{2\pi} \approx 0.195$
2 level	4 step	2/3	like for the major star	

The situation of the synthesis of the short time averaged (T_p) output voltage $\mathbf{V}_s$ using vectors present in the voltage stars for the sector I of a complex plane is presented in Fig. 3.76.

For the case of the 3-level inverter the control sector in the complex plane is divided into 4 smaller triangles called regions and defined as a,b,c,d. In each of the regions there are 3 constitutive vectors and the total number of them in each sector is equal to 5. In the sector I they are: $\mathbf{V}_{1S}$, $\mathbf{V}_{1L}$, $\mathbf{V}_{1M}$, $\mathbf{V}_{2S}$, $\mathbf{V}_{2L}$ and in addition the zero vectors. The indices S, L, M denote the small, large and medium voltage star, respectively. The vectors in the small star may occur either of two versions N or P; for instance: $\mathbf{V}_{2S(P)} = <1,1,0>$; $\mathbf{V}_{2S(N)} = <0,0,-1>$.

Calculation of time intervals for SVM inverter control

Time intervals t_x, t_y, t_z of the duration of particular vectors in the pulse sequence lasting T_p are relative to angle θ of the vector $\mathbf{V}_s$ position and its magnitude. In order to maintain the postulate of the control by means of a single switching pair it is possible to apply only the vectors pointing the apexes of the triangle forming a region. The switchings between the 1,-1 states can occur only in two stages by switching through the zero state. The method used for the calculation of time intervals t_x, t_y, t_z will be presented for vector $\mathbf{V}_s$ situated in region d as in Fig. 3.76. The time interval t_x regards the vector with the minimum value of angle θ, t_y with the medium value, and t_z for the highest value of angle θ.

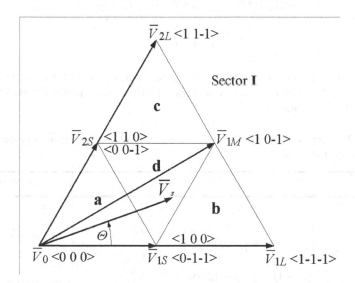

Fig. 3.76 Division of a sector I into 4 regions a, b, c, d as an illustration for constitution of an output voltage vector $\mathbf{V}_s$

The sum of the time intervals

$$t_x + t_y + t_z = T_p \tag{3.187}$$

is equal to the period of the pulsation sequence T_p. For instance for region d the following vector equation is valid:

$$\mathbf{V}_s e^{j\theta} T_p = \mathbf{V}_{1S} e^{j0} t_x + \mathbf{V}_{1M} e^{j\pi/6} t_y + \mathbf{V}_{2S} e^{j\pi/3} t_z$$

Regarding $\mathbf{V}_s$ as a phase vector

$$V_{1S} = \frac{1}{\pi} U_d \quad V_{2S} = \frac{1}{\pi} U_d \quad V_{1M} = \frac{\sqrt{3}}{\pi} U_d \tag{3.188}$$

the vector equation assumes the form:

$$\frac{\pi}{\sqrt{3}} T_p U_s (\cos\theta + j\sin\theta) =$$

$$= \frac{1}{\sqrt{3}} U_d T_x + U_d T_y (\frac{\sqrt{3}}{2} + j\frac{1}{2}) + \frac{1}{\sqrt{3}} U_d T_z (\frac{1}{2} + j\frac{\sqrt{3}}{2}) \tag{3.189}$$

After the introduction of amplitude modulation coefficient (3.164): $m_a = \dfrac{U_s}{U_d}\dfrac{\pi}{\sqrt{3}}$

and separation of (3.189) into the real and imaginary part, we obtain:

$$m_a T_p \cos\theta = \frac{1}{\sqrt{3}} T_x + T_y \frac{\sqrt{3}}{2} + T_z \frac{1}{2\sqrt{3}}$$

$$m_a T_p \sin\theta = \frac{1}{2} T_y + \frac{1}{2} T_z \tag{190}$$

The solution of the system (3.190), (3.187) makes it possible to determine time intervals t_x, t_y, t_z for region d. The summary of the time intervals for the entire Sector I is presented in Table 3.8.

Table 3.8 Time intervals t_x, t_y, t_z for SVM inverter control in a,b,c,d regions of the I-st sector of the 3-level inverter

Region	t_x	t_y	t_z	formulae
a	$\langle \mathbf{V}_{1S} \rangle \rightarrow$ $\rightarrow T_p w_1$	$\langle \mathbf{V}_0 \rangle \rightarrow$ $\rightarrow T_p (1 - w_3)$	$\langle \mathbf{V}_{2S} \rangle \rightarrow$ $\rightarrow T_p w_2$	$w_1 =$ $2m_a \sin(\dfrac{\pi}{3} - \theta)$
b	$\langle \mathbf{V}_{1L} \rangle \rightarrow$ $\rightarrow T_p (w_1 - 1)$	$\langle \mathbf{V}_{1S} \rangle \rightarrow$ $\rightarrow T_p (2 - w_3)$	$\langle \mathbf{V}_{1M} \rangle \rightarrow$ $\rightarrow T_p w_2$	$w_2 =$ $2m_a \sin(\theta)$
c	$\langle \mathbf{V}_{1M} \rangle \rightarrow$ $\rightarrow T_p w_1$	$\langle \mathbf{V}_{2S} \rangle \rightarrow$ $\rightarrow T_p (2 - w_3)$	$\langle \mathbf{V}_{2L} \rangle \rightarrow$ $\rightarrow T_p (w_2 - 1)$	$w_3 =$ $2m_a \sin(\dfrac{\pi}{3} + \theta)$
d	$\langle \mathbf{V}_{1S} \rangle \rightarrow$ $\rightarrow T_p (1 - w_2)$	$\langle \mathbf{V}_{1M} \rangle \rightarrow$ $\rightarrow T_p (w_3 - 1)$	$\langle \mathbf{V}_{2S} \rangle \rightarrow$ $\rightarrow T_p (1 - w_1)$	

If a vector $\mathbf{V}_s$ is situated in one of the latter regions, angle θ has to be reduced to a value within a range of $0 < \theta \leq \pi/3$ and formulae from Table 3.8 should be applied for an appropriate region, changing the numbers of indexes in $\mathbf{V}$ vectors, accordingly.

Switching sequence over a pulsation period

The matters in this case is are more complex in comparison to the two-level inverter since due to the necessity of limiting the fluctuations of the potentials of the neutral point G it is necessary to balance the time intervals of the duration of vector in the minor star in version N as well as in P. This occurs differently in regions b and c, where there is a single constitutive vector of the minor star than in regions a and d, where there are two constitutive vectors of the minor star. An example of 6-pulse control in the pulsation sequence for region b, i.e. one in which there is one constitutive vector of the minor star, is presented in Fig. 3.77. The balanced realization of vector $\mathbf{V}_{1S}$ is ensured as a result of performing the switching sequence <1,0,0> and <0,0,-1> with the identical summarized duration. The situation is different for regions a and d, where there are two constitutive vectors forming the minor voltage star. In this case the region divides into two symmetrical parts and for each of them the realization of P and N is balanced for the vector of the minor star that is dominant in each of the halves in a region. Concurrently, the other vector of the minor star is unbalanced in the half of the region in which it plays a less important role. This is presented in Fig. 3.78 for Sector I, subregion $d2$, i.e. for the half of the region d that lies closer to vector $\mathbf{V}_{2S}$.

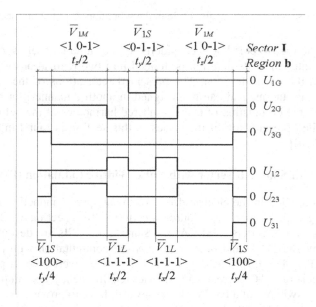

Fig. 3.77 Six-pulse sequence of pulse cycle in the region b of the first sector

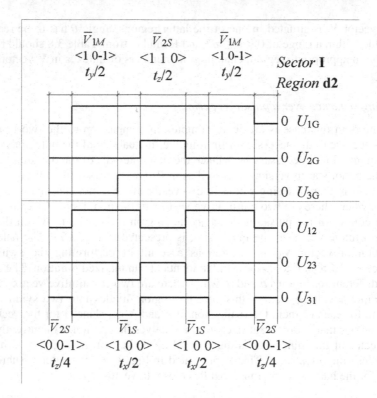

Fig. 3.78 Six-pulse sequence of a pulse cycle in the subregion $d2$ of the first sector

In this subregion the vector $\mathbf{V}_{2S}$ is considered in the same manner as zero vectors and the time interval of its duration is divided into two realizations: $<1,0,0>$ and $<0,0,-1>$ with the equal duration. The presented realizations of the sequence of switching are not unique and can be designed in another manner, for instance by accounting for the elimination of even numbered harmonics of the output voltage. A more detailed presentation of the issues in this section is found in references, e.g. [10,38,39,70].

3.3.3.5 Current Source Inverter with Pulse Width Modulation (PWM)

The inverters applied in the electric drive for the supply of induction motors can be generally grouped into voltage source inverters (VSI – section 3.3.2 and 3.3.3) and current source inverters (CSI). At the same time, CSIs are designed as load commuted inverter (LCI) and ones with imposed commutation with pulse width modulation (PWM) [106]. Further on, current source inverters with pulse width modulation generate AC current with the desired frequency and amplitude. The semiconductor switches of a PWM current inverter have to provide a possibility of current shut-off and have to have reverse voltage blocking capacity. The current models apply reverse blocking Gate Commuted Thyristors (GCT) while formerly they applied GTO thyristors. PWM current source inverters display a number of advantages and are therefore commonly applied in industrial drives for several reasons:

- in contrast to VSI inverters they do not use high frequency switched voltage to the motor's windings but tend to supply AC with a smooth waveform to the windings; hence, the problem with the high value of *du/dt* derivative is absent,

- they have a considerably simple engineering structure and the switches do not need the application of freewheeling diodes,

- due to the necessity of applying an in-series inductor there is no hazard of short-circuiting on inverter or motor clamps.

The drawbacks of this solution include:

- necessity of applying in-series inductor designed for a load current value and a capacitor filter at the output,

- relatively slow reaction time in response to control due to the presence of an inductor in the DC circuit.

This type of inverter (CSI-PWM) is successfully applied in medium and large power drives where we have to do with considerable load over the entire cycle of operation and a relatively fast control of the motor's angular speed is not required.

3.3.3.5.1 Structure and operating principle of CSI – PWM inverter. The diagram of CSI current inverter with PWM control is presented in Fig. 3.79. A silicon controlled rectifier (SCR) with an adequate current output, an inductor L_d with

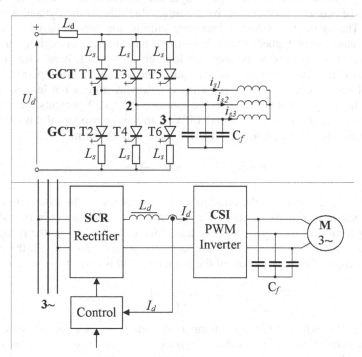

Fig. 3.79 Diagram of a PWM current source inverter (CSI): inverter motor connections and overall control system

adequate inductance and a controller that regulates the current form the current source for the inverter. PWM-CSI inverter serves merely for the switching of the current to energize particular windings while the capacitor filter assembly C_f provides the current in the phase windings which are not supplied from the CSI inverter in a given time period. This procedure is followed in such a manner since at any instant except for the commutation period, only one switch of the positive group (anode group of the inverter) and one in the ground group (cathode one) is in the ON state. For the current source supply, the situation when two switches in a group are in the ON state, results in the loss of the ability to control in the sense of current diffusion into the two windings. In turn, if only one switch in the entire CSI inverter were in the ON state, the current flow forced by the source will result in a very strong overvoltage resulting in its failure. For this reason, the adequate control of the CSI inverter has to secure continuity of current flow through an inverter and motor in a way that only two switches are in the conduction state, each one in a different branch and another group of switches.

3.3.3.5.2 *Control of CSI – PWM Inverter.*

Most common are 3 methods of inverter control [4,10,39]: trapezoidal pulse-width-modulation (TPWM), control using selective harmonics elimination (SHE) and space vector modulation (SVM). Each of the methods has a number of advantages and drawbacks to them. The trapezoidal PWM modulation can be realized in real time or in the form of a look-up table. In addition, it displays good dynamic characteristics; however, it does not give the possibility of current by-passing, which limits the dynamics of drive control. The method of selected harmonic elimination offers the best results in terms of suppressing higher order harmonics of the current; although it must operate on the basis of previously prepared look-up tables and, hence, the dynamic properties deteriorate. The method of space vector modulation operates in real time and has excellent dynamic properties; however, it does not lead to a small level of higher order harmonics in the current. Fig. 3.80 presents the principle governing the control using trapezoidal PWM. In the description of this control we apply amplitude and frequency modulation factors in the form:

$$m_a = \frac{U_s}{U_{saw}} \qquad m_f = \frac{f_s}{f_{saw}} \tag{3.191}$$

where: U_{saw}, f_{saw} - denotes the amplitude and frequency of the saw carrier wave. As indicated in Fig. 3.80, the switch in a branch of the inverter is in the ON state for a flat section of the top base of the trapezoid as well as for its the rising slope when it is higher than the signal of the saw carrier wave. As a result, the number N_p pulses corresponding to a half of the current period is equal to:

$$N_p = \frac{f_{saw}}{3f_s} + 1 \tag{3.192}$$

It is beneficial when the integer number of periods of saw carrier wave corresponds to one of the slopes of the trapezoid, i.e. when: $1/m_f = 6n$, $n = 1,2,3....$ This case illustrates a symmetrical and synchronic switching and hence the major harmonics [10] of the output current have the frequencies of

$$f_h = (3(N_p - 1) \pm 1)f_s = f_{saw} \pm f_s$$
$$f_h = (3(N_p - 1) \pm 5)f_s = f_{saw} \pm 5f_s$$

(3.193)

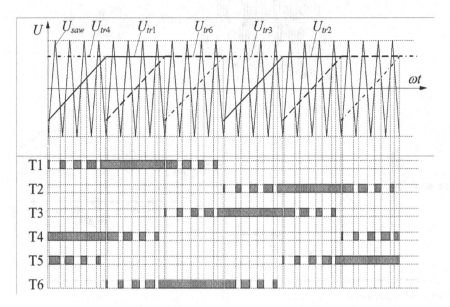

Fig. 3.80 Illustration of a trapezoidal pulse-width modulation (PWM)

3.3.3.5.3 Exemplary curves of induction motor drive supplied from CSI-PWM. The illustrations that follow present the start-up curves of a medium-power induction motor (S2L) with the application of CSI current inverter controlled by means of PWM modulation. The induction motor supplied from CSI inverter with small inertia and low load torque as well as from capacitors with low value of capacity tends to operate in an unstable way. An illustration of this is found in Fig. 3.81.

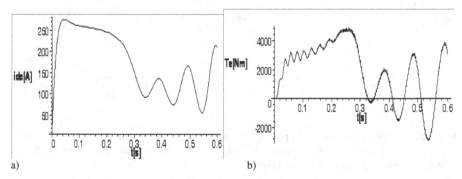

a)   b)

Fig. 3.81 Starting of the CSI PWM fed induction drive without any feedback, with $J = J_s$, $T/T_n = 0.2$, $C_f = 100$ [μF] for medium power motor (3.96): a) DC current b) electromagnetic torque c) stator currents d) angular speed e) torque-speed trajectory

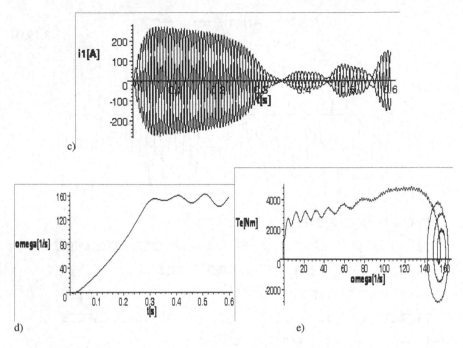

c)

d)   e)

Fig. 3.81 (*continued*)

The higher the inertia, capacity of a filter capacitor C_f and the negative feed-back in relation to the speed, the more stabilized is the operation of the drive after start-up – see Fig. 3.82.

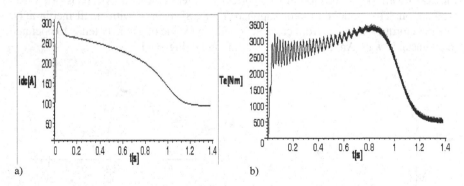

a)   b)

Fig. 3.82 Starting of the medium power (3.96) CSI PWM drive with $J = 3J_s$, $T/T_n = 0.2$, $C_f = 150$ [μF]: a) DC current b) electromagnetic torque c) stator currents d) angular speed e) torque-speed trajectory

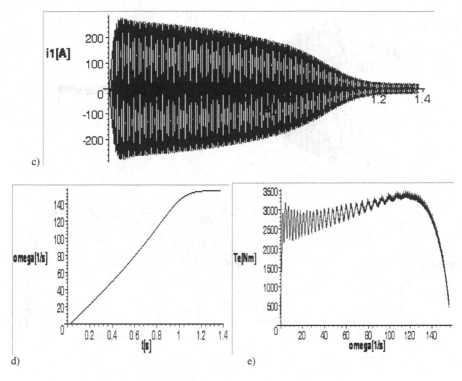

Fig. 3.82 (*continued*)

The examples presented in Figs. 3.82 and 3.83 indicate that the appropriate selection of the motor's parameters and feedback lead to the stabilization of the drive even for relatively small load $T_l/T_n = 0.25$. The current $I_{DCr} = 98$ [A] is the input value for stabilization of the inverter supply current after start-up.

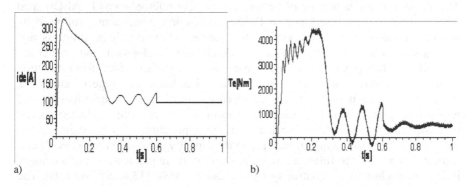

Fig. 3.83 Starting of the medium power (3.96) CSI PWM drive with DC current stabilization after start-up with $J = 3J_s$, $T_l/T_n = 0.2$, $C_f = 150$ [μF], $I_{DCr} = 98$ [A]: a) DC current b) electromagnetic torque c) stator currents d)) angular speed e) torque-speed trajectory

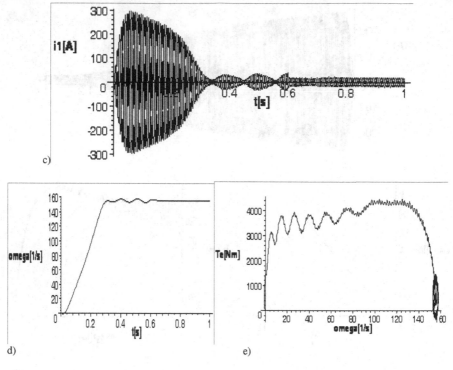

Fig. 3.83 (*continued*)

3.4 Control of Induction Machine Drive

3.4.1 Vector Control

The theoretical fundamentals of vector control, also referred to as Field Oriented Control (FOC), were developed in 1970s and 80s and subsequently followed by attempts conducted on induction machine drives. However, huge progress and wide application of the system was preceded by the development and greater accessibility of fast power electronic semiconductor switches which form the basis of power converters in engineering. On the other hand, this type of control is strictly relative to the use of fast microprocessors to process numerical data of mathematical models and measured data from sensors recording selected signals (variables) in a drive. As a consequence, the application of these methods has largely expanded following an increase in the capacity of signal processors. Therefore, it was possible to fulfill the prerequisites for the implementation of advanced induction machine drive control systems as late as 1990s [13,42,68,76,87,99]. The idea of vector control forms a response to the difficulties associated with the control of a 3-phase squirrel cage induction machine, which has a number of advantages despite not being susceptible to control by means of easily accessible methods such as control of supply voltage and frequency in a manner that offers quick response, oscillation free and agreement with the designed pattern in terms of torque, acceleration and angular speed. The principle governing vector control is the following: from the physical values with sine spatial distribution along the air gap, which can be presented in the form of rotating vectors (complexors), the flux linkage with rotor windings Ψ_r is identified and along this direction the axis x_ρ of the rotating two-axial system is situated. The axis y_ρ for this system, which is established perpendicular to this direction and the angle of orientation ρ, determines the position of this specific system with reference to a selected two axial system $(u,v,0)$. Strictly speaking, this forms a particular case of the $(x,y,0)$ system rotating with the speed of the magnetic field, oriented in the space so that the rotor's flux linkage Ψ_r is presented in the x_ρ axis (Fig. 3.84), hence the name for Field Oriented Control originated. In this new coordinate system the complexor of the stator's current $\mathbf{i}_s$ is made up of the terms $i_{x\rho}$, $i_{y\rho}$. The physical relevance of the entire undertaking is the following: the term $i_{x\rho}$ of the stator current forms the magnetizing current and directly affects the value of the rotor's flux $|\Psi_r|$.

Concurrently, the term $i_{y\rho}$ of the stator current that is perpendicular to it directly affects the value of the motor's electromagnetic torque T_e. As a result, the control of the value of the flux in an induction motor is decoupled from control of torque in a way that follows the model of a separately excited DC machine. In this analogy the term $i_{x\rho}$ of the stator current corresponds to the excitation current of i_f of the DC machine, while the term $i_{y\rho}$ corresponds to the armature current i_a. In a DC commutator machine the control procedure is conducted in an easy way since the orientation is a result of the specific design of a machine. The axes of the excitation and armature windings remain perpendicular as a result of the positioning of pairs of brushes collecting the armature current perpendicular to the axis of the excitation winding, which is wound at the machine stator's salient poles.

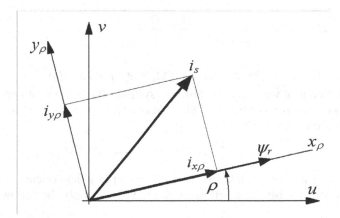

Fig. 3.84 Rotor magnetic flux Ψ_r and stator current $\mathbf{i}_s$ orientation in the field oriented control method

The cost to be incurred with the simplified control of DC current machine is associated with the overcoming problems with commutation, brushes, possibility of sparking at the commutator and necessity of regular inspecting such devices. On the other hand, in an induction machine there are difficulties associated with measurements and control. The steps to follow include: identification, orientation of the rotating system in space in the sense of determination of the orientation angle ρ followed by adequate decisions regarding the opening of the converter's switches with three-phase output in a way that this corresponds to the decoupled control by means of $i_{x\rho}$, $i_{y\rho}$ currents in a two-phase x_ρ, y_ρ system. This occurs in a quite complex control system that is currently applied in numerous versions differing in terms of specific values measured in the system and calculated on-line. The two basic varieties of applying field vector orientation method called direct vector control and indirect vector control will be discussed later in this chapter.

3.4.1.1 Mathematical Model of Vector Control

The specific property of vector control is associated with the orientation of a system (Fig. 3.84) by establishing angle ρ, which determines the direction of the axis of the rotor flux Ψ_r. The focus of the considerations here is a system of equations for a motor expressed in mixed co-ordinates i_s, Ψ_r (3.56) rotating with the speed of the field, i.e. for $\omega_c = \omega_s = p\Omega_0$. It takes the following form:

$$\dot{\psi}_{rx} = -\alpha_r \psi_{rx} + L_m \alpha_r i_{sx} + p\Omega_0\, s\, \psi_{ry}$$

$$\dot{\psi}_{ry} = -\alpha_r \psi_{ry} + L_m \alpha_r i_{sy} - p\Omega_0\, s\, \psi_{ru}$$

$$i_{sx} = \beta(\alpha_r \psi_{rx} + p\Omega_r \psi_{ry}) - \gamma i_{sx} + p\Omega_0 i_{sy} + \frac{\beta}{k_r} U_{sx} \qquad (3.194)$$

$$i_{sy} = \beta(\alpha_r \psi_{ry} - p\Omega_r \psi_{rx}) - \gamma i_{sy} - p\Omega_0 i_{sx} + \frac{\beta}{k_r} U_{sy}$$

$$\dot{\Omega}_r = \underbrace{\mu(\psi_{rx} i_{sy} - \psi_{ry} i_{sx})}_{T_e} - \frac{T_l}{J} - \frac{D}{J}\Omega$$

where: - $\mu = pk_r/J$ $\qquad \beta = k_r/(L_s\sigma)$ $\qquad \Omega_0 = \omega_s/p = 2\pi f_s/p$.

For vector control we apply a coordinate system, which takes into account the angle of orientation ρ, rotor flux Ψ_r and, additionally, rotation angle θ_r of the rotor. A transformation is established:

$$(i_s, \Psi_r): \qquad \psi_{rx}, \psi_{ry}, i_{sx}, i_{sy}, \Omega_r \;\rightarrow\; \theta_r, \Omega_r, \rho, \psi_r, i_{x\rho}, i_{y\rho} \qquad (3.195)$$

where: - $i_{x\rho}$, $i_{y\rho}$ - are the components of stator current in a field oriented system x_ρ, y_ρ. In order to establish the transformation (3.195) we apply the following relations:

$$\rho = \arctan\left(\frac{\psi_{ry}}{\psi_{rx}}\right) \qquad (3.196)$$

$$\psi_r = \sqrt{\psi_{rx}^2 + \psi_{ry}^2} \qquad (3.197)$$

and

$$\begin{bmatrix} i_{x\rho} \\ i_{y\rho} \end{bmatrix} = \begin{bmatrix} \cos\rho & \sin\rho \\ -\sin\rho & \cos\rho \end{bmatrix} \begin{bmatrix} i_{sx} \\ i_{sy} \end{bmatrix} \tag{3.198}$$

After the calculation of derivatives and substitutions, we obtain:

$$\dot{\theta}_r = \Omega_r$$

$$\dot{\Omega}_r = \underbrace{\mu\psi_r i_{y\rho}}_{T_e} - \frac{1}{J}(T_l + D\Omega_r)$$

$$\dot{\rho} = L_m \alpha_r \frac{i_{y\rho}}{\psi_r} - p\Omega_0 s$$

$$\dot{\psi}_r = -\alpha_r \psi_r + L_m \alpha_r i_{x\rho} \tag{3.199}$$

$$i_{x\rho} = \frac{1}{\sigma L_s} U_{x\rho} + \alpha_r \beta \psi_r - \gamma i_{x\rho} + p\Omega_r i_{y\rho} + L_m \alpha_r \frac{i_{y\rho}^2}{\psi_r}$$

$$i_{y\rho} = \frac{1}{\sigma L_s} U_{y\rho} - p\beta\Omega_r \psi_r - \gamma i_{y\rho} - p\Omega_r i_{x\rho} - L_m \alpha_r \frac{i_{x\rho} i_{y\rho}}{\psi_r}$$

From the very form of the model (3.199) for the vector control it is possible to directly derive the following conclusions:

- control is possible only when $\Psi_r > 0$
- the term representing electromagnetic torque

$$T_e = \mu\psi_r i_{y\rho} \tag{3.200}$$

is very simple in the same way as the one for electromagnetic torque in a separately excited DC machine

- magnetic flux of the rotor Ψ_r is relative only to the current $i_{x\rho}$. There is a complete analogy with reference to the excitation flux and excitation current in a separately excited DC machine. In the steady state

$$\Psi_r = L_m i_{x\rho}, \tag{3.201}$$

- and for $\dot{\rho} = 0$, machine's slip is equal to

$$s = \frac{L_m \alpha_r}{p\Omega_0} \frac{i_{y\rho}}{\Psi_r}, \tag{3.202}$$

- voltages $U_{x\rho}$, $U_{y\rho}$ are transformed in accordance with (3.198).

3.4.1.2 Realization of the Model of Vector Control

Vector control of an induction motor is realized on the basis of a mathematical model (3.199) which applies control signals from sensors and quantities calculated on-line in the signal processor. In the up-to-date solutions of vector control the measured signals include the currents and voltages supplying stator windings or, alternatively, rotor's angular speed or its angle of rotation measured by an encoder, which calculates the pulses corresponding to the units of the angle of rotation [53]. Among the calculated quantities there must be the angle of rotor orientation ρ, which secures the orientation of the system x_ρ, y_ρ in accordance with the rotor's flux linkage Ψ_r. The assigned values in the control process include the rotor's flux $\Psi_{r\,ref}$ and the reference torque $T_{e\,ref}$, while the subscript *ref* denotes the reference value.

3.4.1.2.1 Direct Vector Control System. The diagram of such a control system is presented in Fig. 3.85. The control procedure involves the measurement of the values of stator current $i_{s1,2,3}$ and voltage $u_{s1,2,3}$. On the basis of this and applying the relation

$$\dot{\Psi}_s = U_s - R_s i_s + p\Omega_0 A_2 \Psi_s$$
$$\Psi_r = \Psi_s - \sigma L_s i_s$$

(3.203)

we can calculate the terms Ψ_{rx}, Ψ_{ry} of the rotor flux. Subsequently, we can calculate orientation angle ρ, flux Ψ_r and three-phase currents are transformed to the constituent terms $i_{x\rho}$, $i_{y\rho}$ in a two-axis w field oriented system. This method is called direct field control since it involves the calculation of the flux Ψ_r.

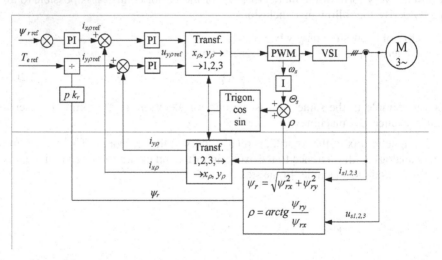

Fig. 3.85 Block diagram of a direct vector control system

3.4.1.2.2 Indirect Vector Control System. The diagram of this control manner is presented in Fig. 3.86. The control procedure involves the measurement of the values of stator current $i_{s1,2,3}$ and angular speed Ω_r. This system maps the derivative of the orientation angle $\dot{\rho}$, which is subsequently integrated and added to the phase angle of stator voltage θ_s and applied for the transformation of stator currents to two-axial field-oriented components $i_{x\rho}$, $i_{y\rho}$. The system does not involve the determination of flux Ψ_r and for this reason this version of the field control method is denoted as indirect.

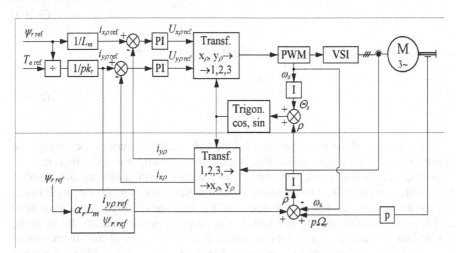

Fig. 3.86 Block diagram of an indirect vector control system

3.4.1.3 Formalized Models of Vector Control

3.4.1.3.1 Asymptotic Decoupling of Flux and Speed Control. If we name the particular variables in the model (3.199) in the following way:

$$(\theta_r, \Omega_r, \Psi_r, i_{x\rho}, i_{y\rho}, \rho) = (x_0, x_1, x_2, x_3, x_4, x_5) \qquad (3.204)$$

we can obtain the equations of the model in the form:

$$\dot{x}_0 = x_1$$
$$\dot{x}_1 = \mu x_2 x_4 - a$$
$$\dot{x}_2 = -\alpha_r x_2 + L_m \alpha_r x_3$$
$$\dot{x}_3 = -\gamma x_3 + v_x \qquad (3.205)$$
$$\dot{x}_4 = -\gamma x_4 + v_y$$
$$\dot{x}_5 = \alpha_r L_m \frac{x_4}{x_2} - p(\Omega_0 - x_1)$$

$$\mu = pk_r / J \qquad a = (T_l + Dx_1)/J$$

where:
$$v_x = \frac{1}{\sigma L_s} U_{x\rho} + p\Omega_r i_{y\rho} + \alpha_r \beta \Psi_r + \alpha_r L_m \frac{i_{y\rho}^2}{\Psi_r}$$

$$v_y = \frac{1}{\sigma L_s} U_{y\rho} - p\Omega_r i_{x\rho} - p\beta\Omega_r \Psi_r - \alpha_r L_m \frac{i_{x\rho} i_{y\rho}}{\Psi_r}$$

The mathematical model of the drive (3.205) can be broken down into two subsystems – one associated with the control of the flux with the output of $x_2 = \Psi_r$

$$\dot{x}_2 = -\alpha_r\, x_2 + L_m \alpha_r x_3$$
$$\dot{x}_3 = -\gamma x_3 + v_x \tag{3.206}$$

and the system of the control of rotor position or angular speed:

$$\dot{x}_0 = x_1$$
$$\dot{x}_1 = \mu\, x_2 x_4 - a \tag{3.207}$$
$$\dot{x}_4 = -\gamma x_4 + v_y$$

The decoupling of the two systems (3.206) and (3.207) occurs in the asymptotic sense, i.e. for a strictly controlled flux. As a result, we can assume that: $x_2 = x_{2fer}$. This situation fully resembles the control of excitation flux and speed in a separately excited DC motor. In the case of which, however, the decoupling occurs as a result of the engineering design of the machines and requires a commuter and brushes for armature. In contrast, the decoupling in this case results from the application of the equation for the orientation angle $x_5 = \rho$ and transformation of the variables. One can note here that in the presented system variable $x_5 = \rho$ is not followed and is uncontrolled. The familiarity with it is necessary for the transformation of physical quantities to the state variables and, hence, it needs to be conducted separately.

3.4.1.3.2 Input-Output Decoupling. In the system (3.199) we now will introduce new variables:

$$z_1 = \theta_r$$
$$z_2 = \Omega_r$$
$$z_3 = \mu \Psi_r i_{y\rho} - a$$
$$z_4 = \Psi_r \tag{3.208}$$
$$z_5 = -\alpha_r \Psi_r + \alpha_r L_m i_{x\rho}$$
$$z_6 = \rho$$

Since the flux is determined in polar coordinates, it is necessary that $\Psi_r > 0$, which secure the existence of an inverse transformation:

$$\theta_r = z_1$$
$$\Omega_r = z_2$$
$$i_{x\rho} = \frac{1}{\alpha_r L_m}(z_5 + \alpha_r z_4) \tag{3.209}$$
$$i_{y\rho} = \frac{1}{\mu z_4}(z_3 + a)$$
$$\rho = z_6$$

After the introduction of transformation (3.208) we obtain:

$$\dot{z}_1 = z_2$$
$$\dot{z}_2 = z_3$$
$$\dot{z}_3 = v_y$$
$$\dot{z}_4 = z_5$$
$$\dot{z}_5 = v_x$$
$$\dot{z}_6 = \alpha_r L_m (z_3 + a) - p(\Omega_0 - z_2)/(\mu z_4^2)$$

(3.210)

where:

$$v_x = -(\alpha_r + \gamma)z_5 + \alpha_r(\alpha_r \beta L_m - \gamma)z_4 + p z_2(z_5 + \alpha_r z_4) +$$

$$+ \left(\frac{\alpha_r L_m}{\mu}\right)^2 \frac{1}{z_4^3}(z_3 + a)^2 + \frac{k_s \alpha_r}{\sigma} U_{x\rho}$$

$$v_y = -(\alpha_r + \gamma)(z_3 + a) - z_2 z_4^2 \mu p \left(\beta + \left(\alpha_r + \frac{z_5}{z_4}\right)\frac{1}{\alpha_r L_m}\right) +$$

(3.211)

$$+ \frac{\mu}{k_r} \beta U_{y\rho} - b$$

$$a = (T_l + D x_1)/J \qquad b = \dot{a} \qquad (z_3 + a) = \frac{1}{J} T_e$$

In the presented system (3.210) we have to do with input-output decoupling and independent control of the two outputs: $z_1 = \theta_r$ and $z_4 = \Psi_r$. In this version of control input to state decoupling does not occur since variable $x_6 = \rho$ remains unobserved and uncontrolled. The control requires that $z_4 = \Psi_r > 0$. The control signals v_x, v_y are strictly relative to the variables of the state and their derivatives:

$$v_x, v_y = f(z_2, z_3, z_4, z_5, a, b) = f(\Psi_r, \dot{\Psi}_r, \Omega_r, \dot{\Omega}_r, T_e, \dot{T}_e)$$

Such relations tend to be very complex and involve quantities that are difficult to measure in the drive. They make it possible to derive control variables on the basis of measurements using state observers. The cost of the simple and linear control of the I/O system, as presented above, is associated with complex control signals and necessity of their combining from various sources (observers/estimators/sensors).

3.4.1.3.3 Input to State Linearization by Dynamic Feedback. The achievement of input to state linearization by dynamic feedback for an induction motor drive is associated with the need to apply other set of variables than those applied before, i.e. ones accounting for orientation angle ρ. Therefore, the currently applied primary variables (technically outputs) include:

$$y_1 = \Omega_r \qquad z_1 = \rho$$

(3.212)

These primary variables need to be differentiated until the latest of the derivatives presents the control inputs, which ensures the controllability the system. For the adopted variables (3.212) it will require the differentiation of each variable three times and consideration of the control voltage v_{yp} as an additional variable in the system:

$$x_6 = v_{yp} \qquad (3.213)$$

while its time derivative $\dot{x}_6 = \tilde{v}_{yp}$ will be the control value subjected to integration by dynamic feedback at the input of the system. This is symbolically illustrated in Fig. 3.87.

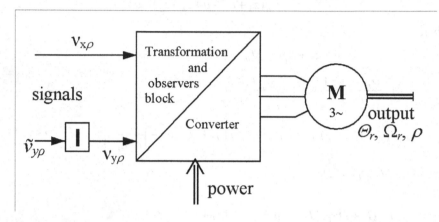

Fig. 3.87 Linearized and decoupled FOC system of an induction motor with an integrator of the $\tilde{v}_{yp}$ signal

For variables (3.212) and an additional $y_0 = \theta_r$, the decoupled and linearized FOC system of an induction motor takes the following form:

$$\dot{y}_0 = y_1$$
$$\dot{y}_1 = y_2$$
$$\dot{y}_2 = v_1 \quad [s^{-4}]$$
$$\dot{z}_1 = z_2 \qquad (3.214)$$
$$\dot{z}_2 = z_3$$
$$\dot{z}_3 = v_2 \quad [s^{-3}]$$

where in accordance with the definition of variables (3.212)

$$v_1 = \frac{d^3\Omega_r}{dt^3} = \frac{d^4\theta_r}{dt^4}$$
$$\qquad (3.215)$$
$$v_2 = \frac{d^3\rho}{dt^3}$$

In order to determine control quantities v_1, v_2 it is necessary to conduct subsequent differentiations in accordance with (3.215), by application of formulae (3.205) for

x_1, x_5. As a result, we obtain:

$$v_1 = \mu\alpha_r L_m(\alpha_r + 3\gamma)x_3 x_4 - \mu(2\alpha_r + \gamma)x_2 x_6 + 2\mu\alpha_r L_m x_3 x_6 +$$
$$+ 2\mu(\alpha_r + \gamma)^2 x_2 x_4 + \mu\alpha_r L_m x_4 v_x + \mu x_2 \tilde{v}_y - \dot{b}$$

$$v_2 = (\alpha_r L_m)^2 \frac{x_3 x_4}{x_2^2}\left[2\alpha_r L_m \frac{x_3}{x_2} - 3(\alpha_r - \gamma)\right] +$$

$$+ \alpha_r L_m(\alpha_r - \gamma)^2 \frac{x_4}{x_2} + 2\alpha_r L_m(\alpha_r - \gamma)^2 \frac{x_6}{x_2}\left(\alpha_r - \gamma - \alpha_r L_m \frac{x_3}{x_2}\right) + \tag{3.216}$$

$$+ p\mu[\alpha_r x_4(L_m x_3 - x_2) + x_2(x_6 - \gamma x_4)] - pb -$$

$$- v_x(\alpha_r L_m)^2 \frac{x_4}{x_2^2} + \tilde{v}_y \alpha_r L_m \frac{1}{x_2}$$

On the basis of (3.216) it is possible to calculate the control quantities:

$$v_x \, [A/s] \qquad \tilde{v}_y \, [A/s^2] \qquad x_6 = \int_0^t \tilde{v}_y d\tau \, [A/s] \tag{3.217}$$

The determination of the controls is quite complex and requires the familiarity of $x_2 = \Psi_r$, $x_3 = i_{x\rho}$, $x_4 = i_{y\rho}$ as well as the first and second derivative of the load torque, i.e. $b = \dot{a}$, $\dot{b}$. The prerequisite for the control is condition $\Psi_r > 0$, just as in the previous cases. This type of control is difficult to realize, however, due to the observability of the system and selection of the variable $z_1 = \rho$ it ensures the orientation of the system as a result of its control. It is to some degree easier to realize the linearized system with reduced dynamics of the system after realizing the stabilization of the motor excitation current, i.e. $i_{x\rho}$ current, which is responsible for flux Ψ_r, as presented in the following chapter.

3.4.1.3.4 Linearization of a Reduced System with a Stabilized Excitation Current.
We will examine a FOC control system in which a separate control is used for the stabilization of the excitation current

$$i_{x\rho} = i_{xref} \tag{3.218}$$

Then, the model of the drive (3.205) is reduced to:

$$(\theta_r, \Omega_r, \Psi_r, i_{y\rho}, \rho) = (x_0, x_1, x_2, x_3, x_4) \tag{3.219}$$

$$\begin{aligned}
\dot{x}_0 &= x_1 \\
\dot{x}_1 &= \mu x_2 x_3 - a \\
\dot{x}_2 &= -\alpha_r x_2 + L_m \alpha_r w_1 \\
\dot{x}_3 &= w_2 \\
\dot{x}_4 &= \alpha_r L_m \frac{x_2}{x_3} - p(\Omega_0 - x_1)
\end{aligned} \tag{3.220}$$

where: - $w_1 = i_{x\,ref}$, w_2 - are the control values. The system (3.220) involves new variables, such as (3.212):

$$y_1 = \Omega_r \qquad z_1 = \rho \tag{3.221}$$

As a result of calculations of time derivatives of variables (3.221), we obtain:

$$
\begin{aligned}
\dot{z}_1 &= z_2 \\
\dot{z}_2 &= -\mu\alpha_r x_2 x_3 - b + \mu\alpha_r L_m x_3 w_1 + \mu x_2 w_2 \\
\dot{y}_1 &= y_2 \\
\dot{y}_2 &= \alpha_r^2 L_m \frac{x_2}{x_3} + \mu p x_2 x_3 - a - (\alpha_r L_m)^2 \frac{x_3}{x_2^2} w_1 + \alpha_r L_m \frac{1}{x_2} w_2
\end{aligned} \tag{3.222}
$$

After the introduction of input quantities in the form:

$$
\begin{aligned}
u_1 &= \mu\alpha_r L_m x_3 w_1 + \mu x_2 w_2 - \mu\alpha_r x_2 x_3 - b \\
u_2 &= -(\alpha_r L_m)^2 \frac{x_3}{x_2^2} w_1 + \alpha_r L_m \frac{1}{x_2} w_2 + \alpha_r^2 L_m \frac{x_2}{x_3} + \mu p x_2 x_3 - a
\end{aligned} \tag{3.223}
$$

we obtain a decoupled and linearized drive control system of the reduced order:

$$
\begin{aligned}
\dot{z}_1 &= z_2 \\
\dot{z}_2 &= u_1 \quad [s^{-3}] \\
\dot{y}_1 &= y_2 \\
\dot{y}_2 &= u_2 \quad [s^{-2}]
\end{aligned} \tag{3.224}
$$

The input equations can take the following form:

$$
\begin{bmatrix} u_1 \\ u_2 \end{bmatrix} = \begin{bmatrix} \mu\alpha_r L_m x_3 & \mu x_2 \\ -(\alpha_r L_m)^2 \dfrac{x_3}{x_2^2} & \dfrac{\alpha_r L_m}{x_2} \end{bmatrix} \begin{bmatrix} w_1 \\ w_2 \end{bmatrix} + \begin{bmatrix} -\mu\alpha_r x_2 x_3 - b \\ \alpha_r^2 L_m \dfrac{x_3}{x_2} + \mu p x_2 x_3 - a \end{bmatrix} \tag{3.225}
$$

This equation (3.225) offers the possibility of calculating the controls:

$$
\begin{bmatrix} w_1 \\ w_2 \end{bmatrix} = \begin{bmatrix} i_{x\rho} \\ \dfrac{di_{y\rho}}{dt} \end{bmatrix} = \frac{1}{W(x)} \begin{bmatrix} \mu\alpha_r L_m x_3 & (\alpha_r L_m)^2 \dfrac{x_3}{x_2^2} \\ -\mu x_2 & \dfrac{\alpha_r L_m}{x_2} \end{bmatrix} *
$$

$$
* \begin{bmatrix} u_1 + \mu\alpha_r x_2 x_3 + b \\ u_2 - \alpha_r^2 L_m \dfrac{x_3}{x_2} - \mu p x_2 x_3 + a \end{bmatrix} \tag{3.226}
$$

where: - $W(x) = 2\mu(\alpha_r L_m)^2 \dfrac{x_3}{x_2} \qquad x_2 = \Psi_r > 0$

Despite more simple control method in this equation, it requires that the values of rotor flux $x_2 = \Psi_r$, transformed current $x_3 = i_{yp}$, load torque $a = (T_l + T_f)/J$ and its derivative $b = \dot{a}$ are familiar. The control of the variable $y_1 = \Omega_r$ should occur around the natural values of the angular speed resulting from the load and conditions of power supply.

3.4.2 Direct Torque Control (DTC)

3.4.2.1 Description of the Method

Direct Torque Control (DTC) forms a very effective and relatively simple method; hence, is more and more commonly applied in the control of induction motor drives [19,74,79,100]. The power device responsible for the control in this system is most often a two-level voltage inverter (Fig. 3.52). The output quantities include: stator flux vector Ψ_s and motor's electromagnetic torque T_e. Eight possible states of inverter form the pool of input vectors to execute control in accordance with relation (3.157) and Fig. 3.55. The basis of the analysis of DTC control method is the term for electromagnetic torque expressed in flux coordinates, which in accordance with model (3.59) takes the form:

$$T_e = p\beta(\psi_{sv}\psi_{ru} - \psi_{su}\psi_{rv}) =$$
$$= p\beta\Psi_s \times \Psi_r = p\beta|\Psi_s| \times |\Psi_r| \sin\underbrace{(\varphi_s - \varphi_r)}_{\varphi_T} \qquad (3.227)$$

where: $\varphi_T = \varphi_s - \varphi_r$ - is a field angle.

Direct Torque Control (DTC) involves the control of the stator flux module $|\Psi_s|$ and its position on the u,v plane. In this method it is assumed that the changes of stator flux occur considerably faster than the changes of rotor flux Ψ_r. On the basis of the first of the equations (3.53)

$$\dot{\Psi}_s = U_s - R_s i_s + \omega_c A_2 \Psi_s$$

and the second one of (3.55)

$$\dot{\Psi}_r = -\alpha_r \Psi_r + L_m \alpha_r i_s + (\omega_c - p\Omega_r)A_2 \Psi_r$$

we can conclude that the change of stator flux $\Delta\Psi_s$ occurs directly under the effect of applying an adequate stator voltage over a specific period $\Delta t U_s$ and this is an instant effect. Concurrently, the change of rotor field vector $\Delta\Psi_r$ occurs under the effect of the change of the stator current or rotor speed. The change of the rotor speed in the examined time scale of a single control pulse occurs totally unnoticeably while the change of the stator current occurs with a time constant of $1/\alpha_s = L_s/R_s$, which means a considerable delay (see Table 3.1). Hence, depending on the angular position of vector Ψ_s we can achieve the effect of changing stator field Ψ_s and increasing field angle φ_T (3.227) by switching on of one of the instantaneous vectors $V_0, V_1, \ldots V_8$ representing one of the output states of the voltage inverter. Thus, the motor torque T_e is effected quickly and directly. This is illustrated in

Fig 3.88 for counterclockwise direction of rotation. Fig. 3.88 presents also six sectors I,II,...VI into which the control plane u,v can be divided. The decision regarding the selection of a particular vector $\mathbf{V}_k$ for motor control is relative to the position of vector $\mathbf{\Psi}_s$ in a specific control sector and to whether the value (module) of stator field vector exceeds one of the threshold control values as well as whether the value of motor torque T_e exceeds or is below the prescribed values determining the admissible torque fluctuations by the band-band method. This is done with the aid of controllers with a hysteresis based control characteristics, as presented in Fig. 3.89.

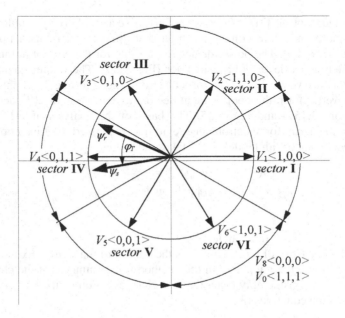

Fig. 3.88 Clarification of the Direct Torque Control method

In order to clarify selection method we will take into account the situation presented in Fig. 3.88, where the vector of stator flux $\mathbf{\Psi}_s$ is situated in sector IV and the direction of field rotation is counterclockwise. If the control system required an increase of the flux module $|\mathbf{\Psi}_s|$ and an increase of the torque T_e (which corresponds to an increase of φ_T), the state of the inverter switches corresponding to vector $\mathbf{V}_5$ would be in ON state. In order to reduce flux $|\mathbf{\Psi}_s|$ and increase torque it is necessary to switch on the state of the inverter corresponding to vector $\mathbf{V}_6$, while in order to decrease flux $|\mathbf{\Psi}_s|$ and torque T_e it would be necessary to switch on the state of the inverter corresponding to vector $\mathbf{V}_2$. The summary of the inverter's switching states for the particular sectors of the control plane and requirements regarding $|\mathbf{\Psi}_s|$ and T_e are found in Tables 3.9 and 3.10 for the counterclockwise and clockwise direction of field rotation, accordingly.

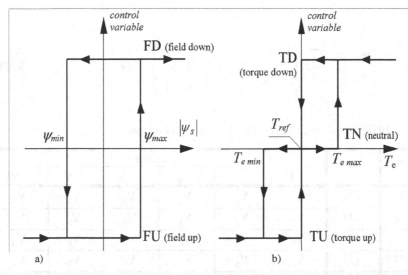

Fig. 3.89 Control characteristics: a) for a stator field $|\Psi_s|$ with a two positional hysteresis loop b) for a torque T_e with a three-positional hysteresis loop

Table 3.9 DTC switching table for the counterclockwise field rotation

$|\Psi_s|$	T_e	Sector					
		I	II	III	IV	V	VI
FU	TU	V_2	V_3	V_4	V_5	V_6	V_1
FU	TD	V_6	V_1	V_2	V_3	V_4	V_5
FU	TN	V_8	V_0	V_8	V_0	V_8	V_0
FD	TU	V_3	V_4	V_5	V_6	V_1	V_2
FD	TD	V_5	V_6	V_1	V_2	V_3	V_4
FD	TN	V_0	V_8	V_0	V_8	V_0	V_8

Index list:
FU – Flux $|\Psi_s|$Up; FD – Flux $|\Psi_s|$Down
TU – Torque T_e Up TD – Torque T_e Down; TN – Torque T_e Neutral

In the DTC control method the fundamental task is associated with the determination of stator flux vector $|\Psi_s|$ magnitude and its position on the u,v plane. This is done by the application of state observers as the tools based on calculations and measurements. This issue will be dealt with in section 3.5. For the purposes of mathematical modeling and research of motor drives the flux Ψ_s can be restated in terms of formulae (3.49) and subsequent transformations. Thus, we obtain:

$$\Psi_s = L_s \mathbf{i}_{suv} + L_m \mathbf{i}_{ruv}$$

or

$$\begin{bmatrix} \Psi_{su} \\ \Psi_{sv} \end{bmatrix} = \sqrt{\frac{2}{3}} L_s \begin{bmatrix} 1 & -\dfrac{1}{2} & -\dfrac{1}{2} \\ 0 & \dfrac{\sqrt{3}}{2} & -\dfrac{\sqrt{3}}{2} \end{bmatrix} \begin{bmatrix} 1 & 0 \\ -1 & -1 \\ 0 & 1 \end{bmatrix} \begin{bmatrix} i_{s1} \\ i_{s3} \end{bmatrix} + L_m \begin{bmatrix} i_{ru} \\ i_{rv} \end{bmatrix} \tag{3.228}$$

Table 3.10 DTC switching table for the clockwise field rotation

$|\Psi_s|$	T_e	Sector					
		I	II	III	IV	V	VI
FU	TU	V_6	V_1	V_2	V_3	V_4	V_5
FU	TD	V_2	V_3	V_4	V_5	V_6	V_1
FU	TN	V_8	V_0	V_8	V_0	V_8	V_0
FD	TU	V_5	V_6	V_1	V_2	V_3	V_4
FD	TD	V_3	V_4	V_5	V_6	V_1	V_2
FD	TN	V_0	V_8	V_0	V_8	V_0	V_8

This leads to the following result:

$$\begin{bmatrix} \Psi_{su} \\ \Psi_{sv} \end{bmatrix} = \begin{bmatrix} \sqrt{\dfrac{2}{3}} L_s i_{s1} + L_m i_{ru} \\ -\sqrt{2L_s}\left(\dfrac{1}{2} i_{s1} + i_{s3}\right) + L_m i_{rv} \end{bmatrix} \tag{3.229}$$

Thanks to this, it is possible to determine the module of stator flux

$$\Psi_s = \sqrt{\Psi_{su}^2 + \Psi_{sv}^2}$$

and the sector in which vector is actually located on the basis of the relations between its components Ψ_{su}, Ψ_{sv}. Similarly, in order to determine electromagnetic torque and field angle φ_T (3.227) it is necessary to determine the components of rotor's flux linkage:

$$\begin{bmatrix} \Psi_{ru} \\ \Psi_{rv} \end{bmatrix} = \begin{bmatrix} \sqrt{\dfrac{2}{3}} L_m i_{s1} + L_r i_{ru} \\ -\sqrt{2} L_m\left(\dfrac{1}{2} i_{s1} + i_{s3}\right) + L_r i_{rv} \end{bmatrix} \tag{3.230}$$

$$\Psi_r = \sqrt{\Psi_{ru}^2 + \Psi_{rv}^2}$$

The acquaintance with the components (3.229), (3.230) of stator and rotor field makes it possible to calculate field angle φ_T (Fig. 3.90)

$$\varphi_T = \arctan\frac{\Psi_{su}\Psi_{rv} - \Psi_{sv}\Psi_{ru}}{\Psi_{su}\Psi_{ru} + \Psi_{sv}\Psi_{rv}} \tag{3.231}$$

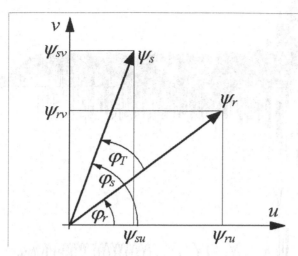

Fig. 3.90 Presentation of a field angle φ_T

3.4.2.2 Examples of Direct Torque Control (DTC) on the Basis of a Mathematical Model

DTC control was modeled with the aid of the relations (3.227-3.331) and Tables 3.9, 3.10 for a two-level voltage inverter as the power output device, the one whose operating principle is described in section 3.3.3.3. The modeling was conducted on the basis of an assumption that the determination of flux $\mathbf{\Psi}_s$ in the sense of the length of the vector and its position is error free, since in this model it is based on the relation in (3.229). The similar course of reasoning is assumed for electromagnetic torque T_e, whose calculation is based on the mathematical model in (3.227) thus providing error free result. This means that the conducted calculations and their results constitute the illustration of the operating principle of DTC method but do not reflect precisely the operation of the drive due to the assumption of idealized operating conditions, in particular in terms of determination of control quantities. The presented illustrations regard the start-up, braking, control of flux, etc. for DTC controlled and operated drive. The first set of illustrations in Figs. 3.91–3.93 presents the start-up of a high voltage medium-power motor (S2H) for a load of $T_l = 0.25T_n$, moment of inertia equal to $J = 1.5J_s$. The band limitations of DTC control involve the values of:

$$20.5 \le \left|\mathbf{\Psi}_s\right| \le 21.0 \quad [Wb]$$
$$3450 \le T_e \le 3550 \qquad T_{eref} = 3500\,[Nm] \tag{3.232}$$

Since the system does not apply speed control after the start-up the electromagnetic torque falls to reach the value resulting from the load and the speed reaches a steady value resulting from the operating conditions of the drive.

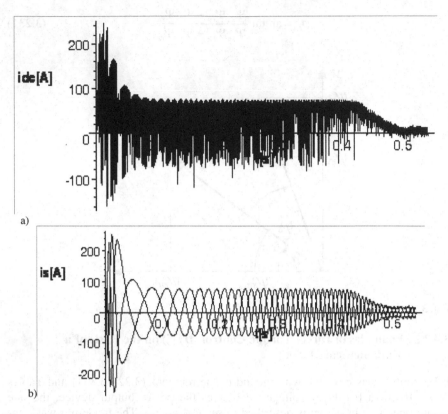

Fig. 3.91 Start-up of 400 [kW], 6 [kV] motor: a) DC source current b) stator currents; DTC parameters given by (3.232)

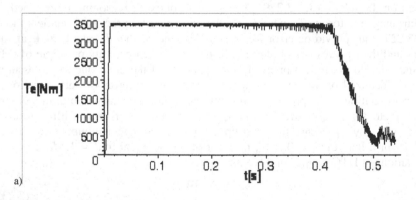

Fig. 3.92 Start-up of 400 [kW], 6 [kV] motor: a) electromagnetic torque b) angular speed: DTC parameters given by (3.232)

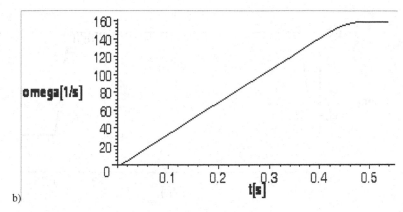

b)

Fig. 3.92 (*continued*)

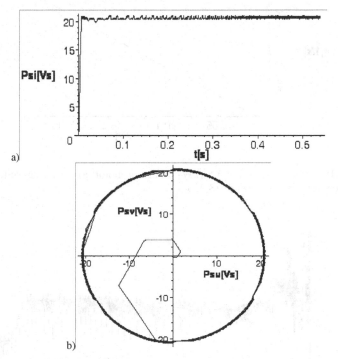

a)

b)

Fig. 3.93 Start-up of 400 [kW], 6 [kV] motor: a) stator magnetic flux |Ψ_s| time history b) *u,v* axis trajectory; DTC parameters given by (3.332)

The following set in Figs. 3.94–3.96 presents the start-up of a small induction motor with the given characteristics of the torque. Its waveform is shaped in a way that ensures a fast start-up and maintenance of the given speed after start for $J = 4J_s$, $T_l = 0.5T_n$. This system operates without any feedback, for the following DTC control limitations:

$$1.18 \leq |\Psi_s| \leq 1.22 \quad [Wb] \qquad \Delta T_e = 0.5 \quad [Nm] \qquad (3.233)$$

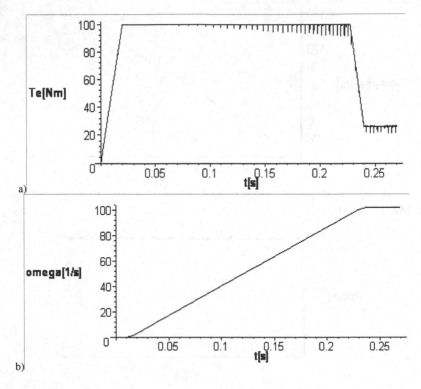

a)

b)

Fig. 3.94 DTC guided start of the 5.5 [kW] induction motor: a) electromagnetic torque b) angular speed. DTC limits given by (3.233)

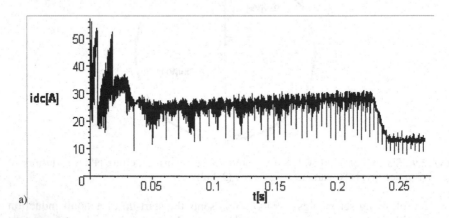

a)

Fig. 3.95 a) DC source current b) stator currents during the torque guided start up of the 5.5 [kW] induction motor. DTC limits given by (3.233)

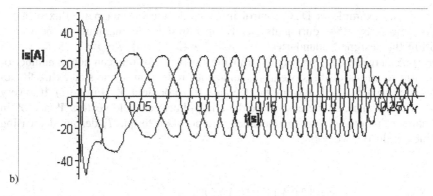

b)

Fig. 3.95 (*continued*)

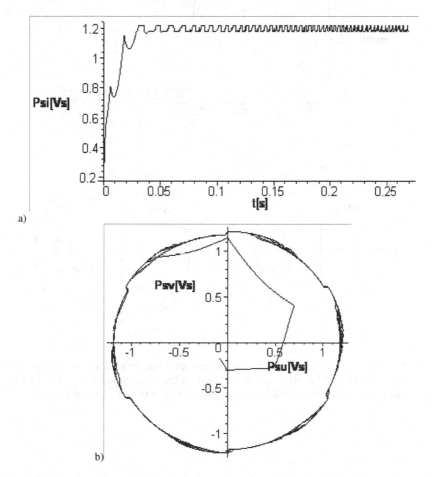

a)

b)

Fig. 3.96 a) stator magnetic flux $|\Psi_s|$ in time b) its *u,v* trajectory, during the torque guided start up of the 5.5 [kW] induction motor. DTC limits given by (3.233)

Another example of DTC control involves decreasing magnetic flux of a 5.5 [kW] motor by 25% during its steady operation while maintaining the torque within the designed boundaries, i.e. $38.8 \leq Te < 42.2$ [Nm]; $T_{e\,ref} = 0.75\ T_n = 41.5$ [Nm]. As a result of observing the transitory waveforms one can note a number of specific reactions of the system: instantaneous reduction of the stator flux $\mathbf{\Psi}_s$ accompanying a gradual, asymptotic decrease of rotor field $\mathbf{\Psi}_r$ (Fig 3.97). It is very specific to remark an increase of field angle φ_T between field vectors $\mathbf{\Psi}_s$ and $\mathbf{\Psi}_r$ in a manner that ensures the maintenance of a constant torque T_e despite decreasing values of the two vectors.

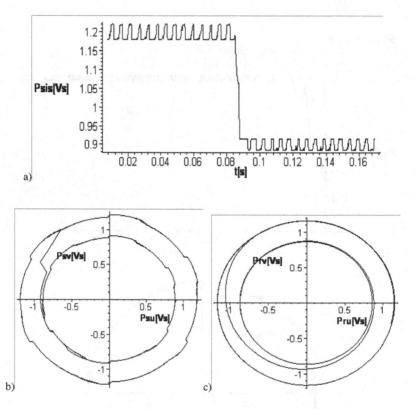

Fig. 3.97 Stepwise reduction (25%) of a stator magnetic flux in steady state of induction 5.5 [kW] motor with DTC while torque maintained on the constant level ($T_{eter} = 0.75\ T_n = 41.5$ [Nm]): a) stator flux $|\mathbf{\Psi}_s|$ b) stator flux trajectory c) rotor flux $|\mathbf{\Psi}_r|$ u,v trajectory

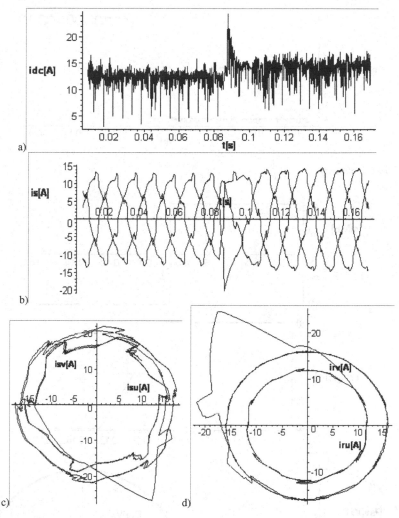

Fig. 3.98 a) DC source current b) stator currents c) stator current u,v trajectory d) rotor current u,v trajectory - after the 25% reduction of the stator flux $|\Psi_s|$

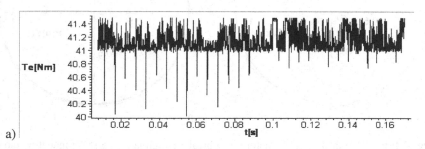

Fig. 3.99 a) electromagnetic torque b) field angle φ_T - after the 25% reduction of the stator flux Ψ_s. Conditions like **Fig. 3.97**

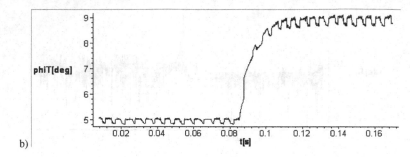

b)

Fig. 3.99 (*continued*)

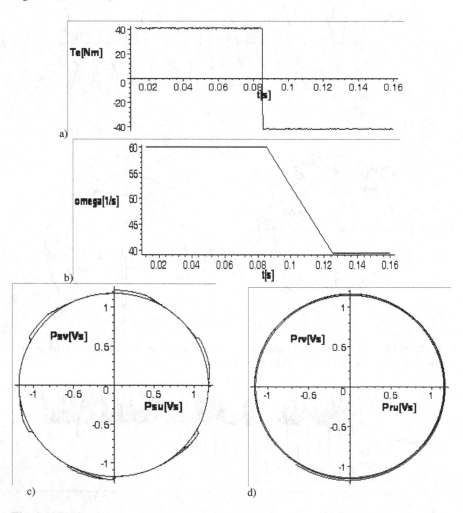

Fig. 3.100 DTC transition from motoring to breaking regime of 5.5 [kW] induction motor (S1): a) electromagnetic torque b) angular speed c) stator and d) rotor magnetic flux trajectories

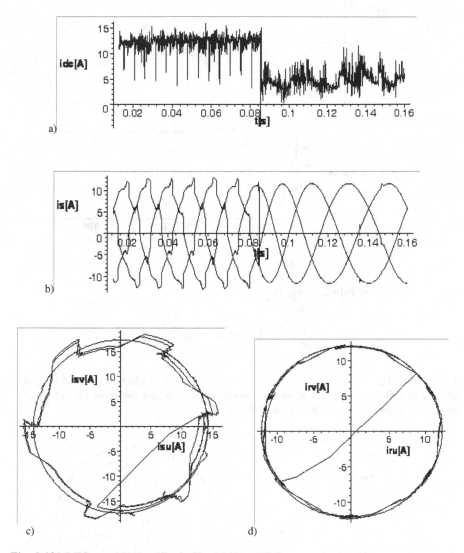

Fig. 3.101 DTC transition... (like in Fig 3.100): a) DC source current b) stator currents c) stator current u,v trajectory d) rotor current u,v trajectory

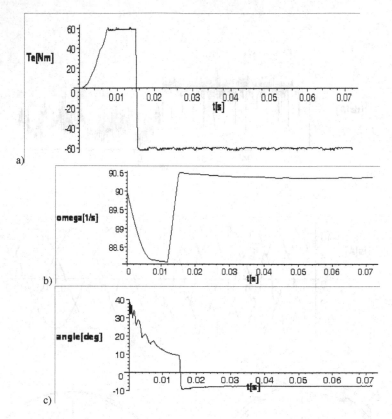

Fig. 3.102 DTC transition to generating regime of 5.5 [kW] induction motor (S1) by the change of switching sequence from the counterclockwise to the clockwise one: a) electromagnetic torque b) angular speed c) field angle φ_T

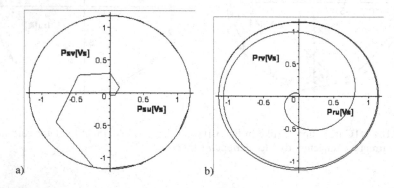

Fig. 3.103 DTC transition to generating regime of 5.5 [kW] induction motor (S1) by the change of switching sequence from the counterclockwise to the clockwise one: a) stator flux $|\Psi_s|$ flux trajectory b) rotor flux $|\Psi_r|$ trajectory

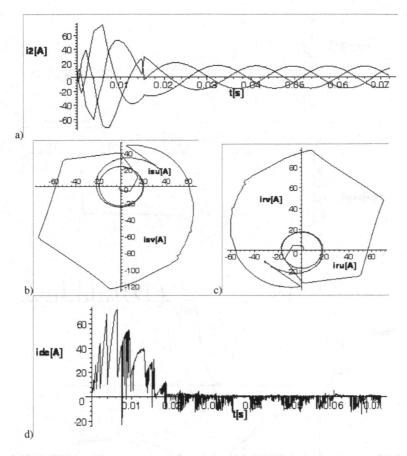

Fig. 3.104 DTC transition to generating regime of 5.5 [kW] induction motor (S1) by the change of switching sequence from the counterclockwise to the clockwise one: a) stator current time curves b) stator current trajectory c) rotor current trajectory d) DC source current

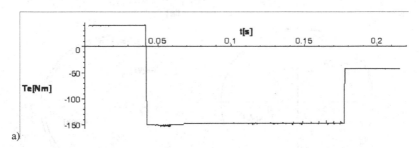

Fig. 3.105 DTC reversing of a of 5.5 [kW] motor drive: a) electromagnetic torque b) angular speed c) field angle

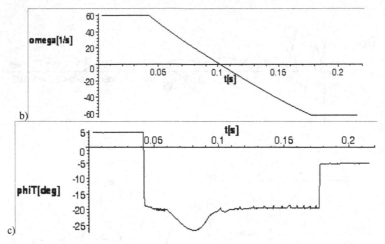

Fig. 3.105 (*continued*)

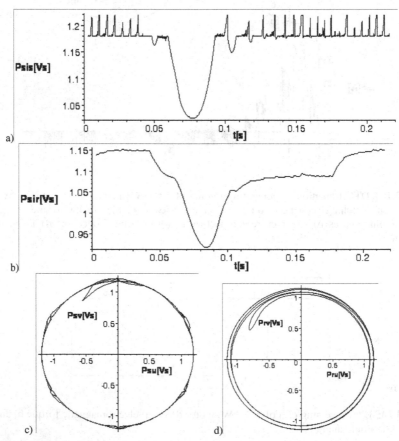

Fig. 3.106 DTC reversing of a of 5.5 [kW] motor drive: a) stator flux $|\Psi_s|$ time curve b) rotor flux $|\Psi_r|$ time curve c) stator flux trajectory d) rotor field trajectory

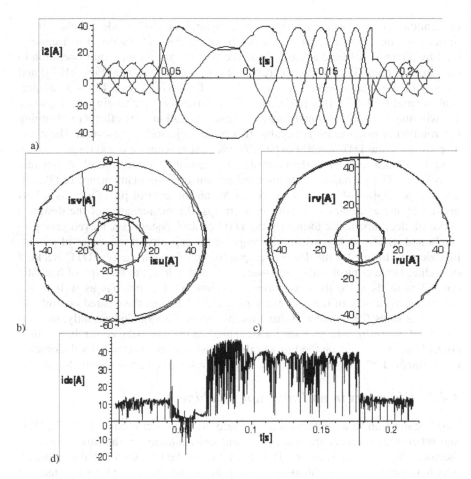

Fig. 3.107 DTC reversing of a of 5.5 [kW] motor drive: a) stator current time curves b) stator current trajectory c) rotor current trajectory d) DC source current

Figs. 3.100 – 3.101 illustrate the transfer of the drive from motor regime of operation to braking as a result of a change of electromagnetic torque, as presented in Fig. 3.100a. The corresponding change of angular speed is presented in Fig. 3.100b. The stabilization of the speed for $t > 0.1$ [s] comes as a consequence of changing load torque T_l and its adaptation to the state of balance. However, this cannot be concluded from the referring figures. The transfer to the braking regime is not accompanied by an increase of stator and rotor current, which is ideally illustrated in Fig. 3.101 c,d. One can note very good characteristics of DTC control in the sense of the maintenance of the designed waveform of electromagnetic torque and flux $|\mathbf{\Psi}_s|$ as well as a considerable overlapping between the curves for the torque and field angle φ_T, which can be made more comprehensible by close examination of relations (3.227) for small values of angle φ_T. Subsequently, Figs. 3.102 – 3.104 present the transfer of the drive controlled by DTC method to the generating regime of operation. This occurs as a result of changing the sequence of semiconductors switching from the 'counterclockwise' mode (Table 3.9) to 'clockwise' one (Table 3.10) for the positive direction of rotation. It is accompanied by a transfer of field angle φ_T from the positive values for motor operation to the negative ones (Fig. 3.102c), negative electromagnetic torque (Fig. 3.102a) and return of energy into the source (Fig. 3.104d). Following Figs 3.105 – 3.107 present reversal of the 5.5 [kW] DTC controlled drive also by changing the sequence of switching of a bridge transistors for a reverse rotation. The effects of changing the rotation sense is quite noticeable in several trajectories presented. These examples prove the DTC control to be effective, stable within a broad range of operating conditions as well as demonstrate its practicality in the control of dynamic waveforms. The prerequisite for this method concerns the maintenance of $|\Psi_s| > 0$ and the possibility of determining vector Ψ_s on the control plane u,v. As it was proved by the examples of conducted start-ups, the maintenance of the designed curve of electromagnetic torque by the DTC control system is achieved very precisely and without delays within the range of the capacity of the motor of providing adequate torque for the determined conditions of system supply. DTC is a kind of sliding mode control realization (see 5.6.1) and belongs to the type of heuristic control methods since the waveforms, the number of switchings as well as the switching frequency of the flux and torque are relative to the desired control and hysteresis loops for torque and flux limitations (Fig. 3.89). Concurrently, the system itself determines the instants for switching on the basis of a currently executed control task as well as on the basis of the precision of the mapping by the observers of the field $|\Psi_s|$, its position and the current value of electromagnetic torque T_e.

3.4.3 Observers in an Induction Machine

The linearization of a model of a drive, field oriented vector control (FOC), DTC and other control procedures require on-line determination of the flux linkage for the stator Ψ_s or for the rotor Ψ_r. The orientation of the flux vector is also required, which involves the determination of its position on the u,v plane for instance by means of orientation angle ρ. These quantities are virtually inaccessible on basis of measurements and need to be calculated in the control system. A mathematical model with data linkage to the measurements of values that are derived very easily and precisely can form the basis for calculations of this type. Such quantities include the current of stator windings, supply voltage, angle of rotor position or rotational speed.

State observers are mathematical objects resulting in calculation algorithms, which perform calculations of inaccessible quantities that are needed to conduct the control process continuously, on the basis of a mathematical model and measurements of certain variables in the drive which are easy to obtain. Such calculations need to be conducted on-line during the control process. The term stability of an observer used here means the decay of estimation error during the determination of the sought quantities caused by disturbance or incorrect initial value along with time passing. Such stability is either examined with the aid of a method based on Liapunov theorem [53], or estimation error decay is demonstrated on the basis of an adequate selection of the observer's parameters. The stability of the observer understood in this manner does not secure its complete accuracy during calculations but confirms the tendency of estimation error to decay in time. The inaccuracy in the estimation of waveforms of the examined variables results from the reasons that are completely beyond the structure of the observer. Such reasons include:

- inaccuracies resulting from simplifications made during the mathematical modeling of the drive serving to establish an observer,
- variability of parameters considered as constant in the algorithm, for example commonly applied quantities α_r, L_m associated with the induction motor that tend to be directly relative to magnetic saturation and temperature,
- inaccuracies of measurements of curves involved in the observer's algorithm from the operating drive,
- inaccuracies of the numerical integration of the observer's equations.

Despite these drawbacks observers have proved their applicability in control even in their more simple forms. The studies devoted to observers, their stability and methods of error reduction are widely discussed in references [22,40,43, 66,77,78]. Several simple observers applied in induction motors are presented below along with some issues pertaining to the maintenance of their stability.

3.4.3.1 Rotor Flux Observer in Coordinates α, β

The most simple flux observer results directly from the model in (3.55). For $\omega_c = 0$ it is a model with a stationary system of axes u,v in relation to the stator, i.e. α, β model.

$$\dot{\hat{\Psi}}_{r\alpha} = -\alpha_r\hat{\Psi}_{r\alpha} - p\Omega_r\hat{\Psi}_{r\beta} + \alpha_r L_m i_{s\alpha}$$
$$\dot{\hat{\Psi}}_{r\beta} = -\alpha_r\hat{\Psi}_{r\beta} + p\Omega_r\hat{\Psi}_{r\alpha} + \alpha_r L_m i_{s\beta}$$

$$(2.234)$$

In the observer (3.234) the estimated values are: $\hat{\Psi}_{r\alpha}, \hat{\Psi}_{r\beta}$ and the values input from measurements include: $\Omega_r, i_{s\alpha}, i_{s\beta}$. The mathematical model for the variables $\Psi_{r\alpha}, \Psi_{r\beta}$ has an identical structure as the one in (2.234), with a note that these variables occur in it in the place of estimated quantities $\hat{\Psi}_{r\alpha}, \hat{\Psi}_{r\beta}$. The errors of the estimated variables are defined as:

$$e_\alpha = \Psi_{r\alpha} - \hat{\Psi}_{r\alpha} \qquad e_\beta = \Psi_{r\beta} - \hat{\Psi}_{r\beta}$$

$$(3.235)$$

The equations of dynamics of the estimation errors are derived by subtracting equations for the observer (3.234) from the equations for the model:

$$\dot{e}_\alpha = -\alpha_r e_\alpha - p\Omega_r e_\beta$$
$$\dot{e}_\beta = -\alpha_r e_\beta + p\Omega_r e_\alpha$$

$$(3.236)$$

The study of the behavior of the estimation errors (3.236) during the disturbances involves the analysis of the dynamics of the error for example with the aid of Liapunov's function. It is a positively determined function based on estimation errors and in the examined case can take the following form:

$$V = e_\alpha^2 + e_\beta^2$$

By calculation of: $\dot{V} = 2(e_\alpha \dot{e}_\alpha + \dot{e}_\beta \dot{e}_\beta)$, and subsequent substitutions (3.236), we obtain:

$$\dot{V} = -2\alpha_r (e_\alpha^2 + e_\beta^2) = -2\alpha_r V \qquad (3.237)$$

From the result (3.237) one can conclude about the stability of the decay of flux estimation error $V(t) = V(0)e^{-2\alpha_r t}$.

3.4.3.2 Rotor Field Observer in Ψ_r, ρ Coordinates

This analysis is based on the model of motor (3.199) in Ψ_r, ρ coordinates:

$$\dot{\rho} = \alpha_r L_m \frac{i_{y\rho}}{\Psi_r} - p\Omega_0 s$$

$$\dot{\Psi}_r = -\alpha_r \Psi_r + L_m \alpha_r i_{x\rho} \qquad (3.238)$$

By planning the observer in the form:

$$\dot{\hat{\rho}} = \alpha_r L_m \frac{\hat{i}_{y\rho}}{\hat{\Psi}_r} - p\Omega_0 s$$

$$\dot{\hat{\Psi}}_r = -\alpha_r \hat{\Psi}_r + L_m \alpha_r \hat{i}_{x\rho} \qquad (3.239)$$

and defining the errors as:

$$e_\Psi = \Psi_r - \hat{\Psi}_r \qquad e_\rho = \rho - \hat{\rho}$$

$$e_x = i_{x\rho} - \hat{i}_{x\rho} \qquad e_y = i_{y\rho} - \hat{i}_{y\rho} \qquad (3.240)$$

we obtain the equations for the dynamics of observer errors:

$$\dot{e}_\rho = \dot{\rho} - \dot{\hat{\rho}} = \alpha_r L_m \left(\frac{i_{y\rho}}{\Psi_r} - \frac{\hat{i}_{y\rho}}{\hat{\Psi}_r} \right) = \alpha_r L_m \frac{-i_{y\rho} e_\Psi + e_y \Psi_r}{\Psi_r \hat{\Psi}_r} \qquad (3.241)$$

$$\dot{e}_\Psi = \dot{\Psi}_r - \dot{\hat{\Psi}}_r = -\alpha_r e_\Psi + \alpha_r L_m e_x$$

In equations (3.241) the error e_ρ is not directly included, but there are errors denoted as e_x, e_y, which result from e_ρ and are transferred to currents $i_{x\rho}, i_{y\rho}$ as a result of transformations. Fig. 3.108 presents geometrical relations resulting from Ψ_r, ρ transformation.

From Fig. 3.108 it stems that: $\rho = \hat{\rho} + e_\rho$, which also represents the relation in (3.240). Hence,

$$
\begin{aligned}
\hat{i}_{x\rho} &= i\cos(\vartheta + e_\rho) = i\cos\vartheta\cos e_\rho - i\sin\vartheta\sin e_\rho = \\
&= i_{x\rho}\cos e_\rho - i_{y\rho}\sin e_\rho \\
\hat{i}_{y\rho} &= i\sin(\vartheta + e_\rho) = i\sin\vartheta\cos e_\rho + i\cos\vartheta\sin e_\rho = \\
&= i_{y\rho}\cos e_\rho + i_{x\rho}\sin e_\rho
\end{aligned}
\tag{3.242}
$$

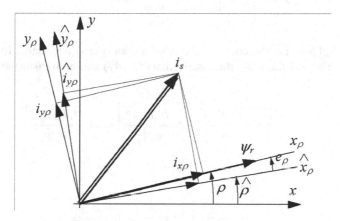

Fig. 3.108 The influence of the orientation error e_ρ on the field oriented stator currents $i_{x\rho}$, $i_{y\rho}$

For the small values of e_ρ in (3.242), it results that:

$$
e_x = e_\rho i_{y\rho} \qquad e_y = -e_\rho i_{x\rho}
\tag{3.243}
$$

As a consequence, on the basis of (3.241, 3.243) the equations for the dynamics of the error can be established in the form:

$$
\begin{bmatrix} \dot{e}_\rho \\ \dot{e}_\Psi \end{bmatrix} = \underbrace{\begin{bmatrix} -\dfrac{\alpha_r L_m i_{x\rho}}{\hat{\Psi}_r} & -\dfrac{\alpha_r L_m i_{y\rho}}{\Psi_r \hat{\Psi}_r} \\ \alpha_r L_m i_{y\rho} & -\alpha_r \end{bmatrix}}_{E} \begin{bmatrix} e_\rho \\ e_\Psi \end{bmatrix}
\tag{3.244}
$$

From the following we can derive the characteristic polynomial:

$$
W(s) = s\begin{bmatrix} 1 \\ & 1 \end{bmatrix} - E = \left(s + \frac{\alpha_r L_m i_{x\rho}}{\hat{\Psi}_r}\right)(s + \alpha_r) + \frac{(\alpha_r L_m i_{y\rho})^2}{\Psi_r \hat{\Psi}_r}
\tag{3.245}
$$

The polynomial in (3.245) can be transformed to take the form:

$$s^2 + s\alpha_r \left(1 + \frac{\Psi_r}{\hat{\Psi}_r}\right) + \alpha_r^2 \frac{(L_m i_s)^2}{\Psi_r \hat{\Psi}_r}$$

(3.246)

with the aid of the relation $\Psi_r = L_m i_{xp}$.

The discriminant of the quadratic equation (3.246) is

$$\Delta = \alpha_r^2 \left[\left(1 + \frac{\Psi_r}{\hat{\Psi}_r}\right)^2 - 4 \frac{(L_m i_s)^2}{\Psi_r \hat{\Psi}_r}\right] < 0$$

(3.247)

always for $|i_{yp}| > 0$, i.e. the curve of the error has an oscillatory shape. In order to assess the values of the roots the discriminant (3.247) can be transformed to take the form:

$$\Delta = \alpha_r^2 \left[\left(1 - \frac{\Psi_r}{\hat{\Psi}_r}\right)^2 - 4 \frac{(L_m i_{yp})^2}{\Psi_r \hat{\Psi}_r}\right] \approx -4 \frac{(\alpha_r L_m i_{yp})^2}{\Psi_r \hat{\Psi}_r}$$

Hence,

$$s_{1,2} = -\frac{1}{2}\alpha_r \left(1 + \frac{\Psi_r}{\hat{\Psi}_r}\right) \pm j \frac{|L_m i_{yp}|}{\sqrt{\Psi_r \hat{\Psi}_r}}$$

and, approximating for $\Psi_r = \hat{\Psi}_r$:

(3.248)

$$s_{1,2} = -\alpha_r \pm j \frac{|L_m i_{yp}|}{\Psi_r} \approx -\alpha_r \pm j \frac{i_{yp}}{i_{xp}}$$

The damping of estimation error curves is proportional to α_r, while the pulsation is relative to the relation i_{yp}/i_{xp}, i.e. the ratio of the electromagnetic torque to the flux.

3.4.3.3 Rotor Field Observer in x, y Coordinates with Speed Measurement

This observer is based on the familiarity of the transformed currents i_{sx}, i_{sy}, which results from the measurement of stator currents and transformation of 0,x,y that involves the need of input of the angle of the rotation of the rotor θ_r. Since speed Ω_r is measured in the system, the determination of the angle of the rotation of the rotor $\dot{\hat{\theta}}_r = \Omega_r$ occurs with the precision range to the constant, i.e. the value of the initial angle $\Omega_r(0)$. The mathematical model (3.56) , for $\omega_c = p\Omega_r$ in the range that is interesting to us takes the form:

$$
\begin{aligned}
\dot{\Psi}_{rx} &= -\alpha_r \Psi_{rx} + \alpha_r L_m i_{sx} \\
\dot{\Psi}_{ry} &= -\alpha_r \Psi_{ry} + \alpha_r L_m i_{sy} \\
\dot{i}_{sx} &= \beta(\alpha_r \Psi_{rx} + p\Omega_r \Psi_{ry}) - \gamma i_{sx} + p\Omega_r i_{sy} + U_{sx}\beta/k_r \\
\dot{i}_{sy} &= \beta(\alpha_r \Psi_{ry} - p\Omega_r \Psi_{rx}) - \gamma i_{sy} - p\Omega_r i_{sx} + U_{sy}\beta/k_r
\end{aligned}
\tag{3.249}
$$

The field observer based on the measured stator winding currents i_s, after the transformation of i_{sx}, i_{sy} that applies the measured speed Ω_r and angle of rotation $\hat{\theta}_r$, is constructed [53] in the following way:

$$
\begin{aligned}
\dot{\hat{\Psi}}_{rx} &= -\alpha_r \hat{\Psi}_{rx} + \alpha_r L_m i_{sx} + \mu\alpha_r e_{ix} - \mu p\Omega_r e_{iy} \\
\dot{\hat{\Psi}}_{ry} &= -\alpha_r \hat{\Psi}_{ry} + \alpha_r L_m i_{sy} + \mu\alpha_r e_{iy} + \mu p\Omega_r e_{ix} \\
\dot{\hat{i}}_{sx} &= \beta(\alpha_r \hat{\Psi}_{rx} + p\Omega_r \hat{\Psi}_{ry}) - \gamma i_{sx} + p\Omega_r i_{sy} + \beta/k_r U_{sx} + k_1 e_{ix} + \\
&\quad + k_2 e_{ix}(\hat{i}_{sx}^2 + \hat{i}_{sy}^2) \\
\dot{\hat{i}}_{sy} &= \beta(\alpha_r \hat{\Psi}_{ry} - p\Omega_r \hat{\Psi}_{rx}) - \gamma i_{sy} - p\Omega_r i_{sx} + \beta/k_r U_{sy} + k_1 e_{iy} + \\
&\quad + k_2 e_{iy}(\hat{i}_{sx}^2 + \hat{i}_{sy}^2)
\end{aligned}
\tag{3.250}
$$

The equations of dynamics of the errors take the following form:

$$
\begin{aligned}
\dot{e}_{\Psi x} &= \dot{\Psi}_{rx} - \dot{\hat{\Psi}}_{rx} = -\alpha_r e_{\Psi x} - \mu\alpha_r e_{ix} + \mu p\Omega_r e_{iy} \\
\dot{e}_{\Psi y} &= \dot{\Psi}_{ry} - \dot{\hat{\Psi}}_{ry} = -\alpha_r e_{\Psi y} - \mu\alpha_r e_{iy} + \mu p\Omega_r e_{ix} \\
\dot{e}_{ix} &= \dot{i}_{sx} - \dot{\hat{i}}_{sx} = -k_1 e_{ix} - k_2 e_{ix}(\hat{i}_{sx}^2 + \hat{i}_{sy}^2) + \beta(\alpha_r e_{\Psi x} + p\Omega_r e_{\Psi y}) \\
\dot{e}_{iy} &= \dot{i}_{sy} - \dot{\hat{i}}_{sy} = -k_1 e_{iy} - k_2 e_{iy}(\hat{i}_{sx}^2 + \hat{i}_{sy}^2) + \beta(\alpha_r e_{\Psi y} - p\Omega_r e_{\Psi x})
\end{aligned}
\tag{3.251}
$$

Verification of the decay of the error involves Liapunov's method with positively determined error function in the form:

$$
V_0 = \frac{1}{2}e_{\Psi x}^2 + \frac{1}{2}e_{\Psi y}^2 + \frac{\mu}{2\beta}e_{ix}^2 + \frac{\mu}{2\beta}e_{iy}^2
\tag{3.252}
$$

After differentiation of (3.252), we obtain:

$$
\dot{V}_0 = \dot{e}_{\Psi x}e_{\Psi x} + \dot{e}_{\Psi y}e_{\Psi y} + \frac{\mu}{\beta}\dot{e}_{ix}e_{ix} + \frac{\mu}{\beta}\dot{e}_{iy}e_{ix}
\tag{3.253}
$$

After the introduction of derivatives (3.251) and ordering things, the result is:

$$\dot{V}_0 = -\alpha_r(e_{\Psi x}^2 + e_{\Psi y}^2) - \frac{\mu}{\beta}(e_{ix}^2 + e_{iy}^2)(k_1 + k_2(\hat{i}_{sx}^2 + \hat{i}_{sy}^2)) \tag{3.254}$$

The result (3.254) indicates a progressive decay of the error function (2.252). The error associated with the estimation of the flux decays with the time constant of $1/\alpha_r$, while the speed of the current estimation error decay e_{ix}, e_{iy} may be controlled by an appropriate selection of the values of k_1, $k_2 > 0$.

3.4.3.4 Observer of Induction Motor Speed Based on the Measurement of Rotor's Position Angle

Induction motors containing the sensors of rotor position, for instance encoders, do not apply speed sensors. The angular speed Ω_r may be calculated on the basis of differentiation of the position signal, which for a discreet determination of the position angle results in considerable noise with high frequency associated with the differentiation. It is, however, possible to avoid it and gain a smooth curve of the estimated speed as a result of application of a simple observer [22]. For standard notations the mathematical model for variables θ_r and Ω_r, i.e., rotor position and angular speed are the following:

$$\dot{\theta}_r = \Omega_r$$
$$\dot{\Omega}_r = \mu \Psi_r i_{y\rho} - \frac{D}{J}\Omega_r - \frac{T_l}{J} \tag{3.255}$$

The proposed observer takes the form:

$$\dot{\hat{\theta}}_r = \hat{\Omega}_r + k_1(\theta_r - \hat{\theta}_r)$$
$$\dot{\hat{\Omega}}_r = \mu \hat{\Psi}_r \hat{i}_{y\rho} - \frac{D}{J}\hat{\Omega}_r - \frac{T_l}{J} + k_2(\theta_r - \hat{\theta}_r) \tag{3.256}$$

The error equations are derived as a result of deducing the sides of (3.255) and (3.256), respectively:

$$\dot{e}_\theta = \dot{\theta}_r - \dot{\hat{\theta}}_r = e_\Omega - k_1 e_\theta$$
$$\dot{e}_\Omega = \dot{\Omega}_r - \dot{\hat{\Omega}}_r = \mu\underbrace{(\Psi_r i_{y\rho} - \mu\hat{\Psi}_r \hat{i}_{y\rho})}_{\approx 0} - \frac{D}{J}e_\Omega - k_2 e_\theta \tag{3.257}$$

Under the assumption that the error in the determination of electromagnetic torque $\mu(\Psi_r i_{y\rho} - \mu \hat{\Psi}_r \hat{i}_{y\rho})$ decays much faster than for the curves of mechanical variables, the equations of the dynamics of errors take the form:

$$\begin{bmatrix} \dot{e}_\theta \\ \dot{e}_\Omega \end{bmatrix} = \begin{bmatrix} -k_1 & 1 \\ -k_2 & -D/J \end{bmatrix} \begin{bmatrix} e_\theta \\ e_\Omega \end{bmatrix} \tag{3.258}$$

The characteristic equation of the error dynamic takes the form:

$$W(s) = \begin{vmatrix} s+k_1 & -1 \\ k_2 & s+D/J \end{vmatrix} = (s+k_1)(s+\frac{D}{J}) + k_2$$

$$W(s) = s^2 + s\left(k_1 + \frac{D}{J}\right) + k_1\frac{D}{J} + k_2$$

(3.259)

The roots of this equation (3.259)

$$r_{1,2} = -\frac{1}{2}\left(k_1 + \frac{D}{J}\right) \pm \sqrt{\left(k_1 + \frac{D}{J}\right)^2 - 4\left(k_2 + k_1\frac{D}{J}\right)}$$

(3.260)

may be formed arbitrarily by selecting k_1, k_2. The damping of the errors $-\frac{1}{2}(k_1+D/J)$ may be adequately large while the curve of the error decay may be oscillatory or exponential depending on the selection of k_2.

3.4.3.5 Flux, Torque and Load Torque Observer in x, y Coordinates

This observer bases on the measurements of i_s, $\theta_r \rightarrow i_{sx}\, i_{sy}$, and the estimated quantities include: $\hat{\Psi}_{rx}, \hat{\Psi}_{ry}, \hat{\Omega}_r, \hat{T}_l$. For simplification purposes we assume that T_l is constant. The corresponding model of the system takes the form:

$$\dot{\Psi}_{rx} = -\alpha_r \Psi_{rx} + \alpha_r L_m i_{sx}$$

$$\dot{\Psi}_{ry} = -\alpha_r \Psi_{ry} + \alpha_r L_m i_{sy}$$

$$\dot{\theta}_r = \Omega_r$$

(3.261)

$$\dot{\Omega}_r = \mu(\Psi_{rx}i_{sy} - \Psi_{ry}i_{sx}) - \frac{D}{J}\Omega_r - \frac{T_l}{J}$$

$$\dot{T}_l / J = 0$$

The observer is designed in the following form [22]

$$\dot{\hat{\Psi}}_{rx} = -\alpha_r \hat{\Psi}_{rx} + \alpha_r L_m i_{sx}$$

$$\dot{\hat{\Psi}}_{ry} = -\alpha_r \hat{\Psi}_{ry} + \alpha_r L_m i_{sy}$$

$$\dot{\hat{\theta}}_r = \hat{\Omega}_r + k_1 e_\theta$$

(3.262)

$$\dot{\hat{\Omega}}_r = \mu(\hat{\Psi}_{rx}i_{sy} - \hat{\Psi}_{ry}i_{sx}) - \frac{D}{J}\hat{\Omega}_r - \frac{\hat{T}_l}{J} + k_2 e_\theta$$

$$\dot{\hat{T}}_l / J = k_3 e_\theta$$

where:

$$e_\theta = \theta_r - \hat{\theta}_r$$

The equations of the dynamics of the error estimation take the form:

$$\dot{e}_x = \dot{\Psi}_{rx} - \dot{\hat{\Psi}}_{rx} = -\alpha_r e_x$$

$$\dot{e}_y = \dot{\Psi}_{ry} - \dot{\hat{\Psi}}_{ry} = -\alpha_r e_y$$

$$\dot{e}_\theta = \dot{\theta}_r - \dot{\hat{\theta}}_r = e_\Omega - k_1 e_\theta$$

$$\dot{e}_\Omega = \dot{\Omega}_r - \dot{\hat{\Omega}}_r = \mu(i_{sy} e_x - i_{sx} e_y) - \frac{D}{J} e_\Omega - e_T - k_2 e_\theta$$

$$(3.263)$$

where:

$$e_\Omega = \Omega_r - \hat{\Omega}_r \qquad e_T = (T_l - \hat{T}_l)/J$$

In accordance with the first two equations for the system (3.263), we assume fast error damping

$$e_x = e_x(0)e^{-\alpha_r t} \qquad e_y = e_y(0)e^{-\alpha_r t} \qquad (3.264)$$

which makes it possible to independently deal with the three remaining equations (3.363), for a decay in the error of torque estimation $\mu(i_{sy} e_x - i_{sx} e_y) \to 0$.

As a result, the dynamics of the errors e_θ, e_Ω, e_T can be restated as:

$$\begin{bmatrix} \dot{e}_\theta \\ \dot{e}_\Omega \\ \dot{e}_T \end{bmatrix} = \begin{bmatrix} -k_1 & 1 & 0 \\ -k_2 & -D/J & -1 \\ -k_3 & 0 & 0 \end{bmatrix} \begin{bmatrix} e_\theta \\ e_\Omega \\ e_T \end{bmatrix} \qquad (3.365)$$

The characteristic polynomial for this system takes the form:

$$W(s) = \begin{vmatrix} s+k_1 & -1 & 0 \\ k_2 & s+D/J & 1 \\ k_3 & 0 & s \end{vmatrix} =$$

$$= s^3 + s^2\left(k_1 + \frac{D}{J}\right) + s\left(k_1\frac{D}{J} + k_2\right) - k_3$$

$$(3.266)$$

The polynomial (3.266) can have three real roots r_1, r_2, r_3 for which case in order to ensure error damping we require that either r_1, r_2, $r_3 < 0$ or a single real number root and two complex ones that are mutually conjugated. In this case it is required that all three real parts are negative. For the solution with three real number roots r_1, r_2, r_3 the gain factors of k_1, k_2, k_3 are calculated on the basis of the general form of the characteristic polynomial

$$W(s) = (s-r_1)(s-r_2)(s-r_3) =$$

$$= s^3 - s^2(r_1 + r_2 + r_3) + s(r_1 r_2 + r_1 r_3 + r_2 r_3) - r_1 r_2 r_3$$

Hence, by comparison with (3.266) we obtain:

$$k_1 = -(r_1 + r_2 + r_3) - D/J$$

$$k_2 = (r_1 r_2 + r_1 r_3 + r_2 r_3) - k_1 \frac{D}{J} \qquad (3.267)$$

$$k_3 = (r_1 r_2 r_3)$$

The relation (3.267) makes it possible to derive the gain factors of k_1, k_2, k_3 for arbitrarily selected values of damping r_1, r_2, r_3. For the complex roots (s-($d\pm j\omega$)) the equation takes the form:

$$W(s) = (s - r_1)(s^2 - 2ds + d^2 + \omega^2) \qquad (3.268)$$

where: - r_1 - is a real root

d, ω - components of complex roots the first of which denotes the damping coefficient of the oscillatory curve while the other the pulsation of the curve. Following an adequate extension:

$$W(s) = s^3 - s^2(r_1 + 2d) + s(2dr_1 + d^2 + \omega^2) - r_1(d^2 + \omega^2)$$

As a result of the comparison of the result with polynomial in (3.266) it is possible to determine the gain of k_1, k_2, k_3 for the desired values of r_1, d, ω:

$$k_1 = -(r_1 + 2d) - D/J$$

$$k_2 = 2dr_1 + d^2 + \omega^2 - k_1 \frac{D}{J} \qquad (3.269)$$

$$k_3 = r_1(d^2 + \omega^2)$$

The presented observer (3.262) has been solved in an effective manner involving the input the numerical values from gauges in order to obtain the desired curve of error decay function. This, unfortunately, does not involve the case for the decay of the errors of flux estimation, which is relative to the time constant $1/\alpha_r$, which was the case in the preceding examples.

3.4.3.6 Stator Flux Observer Ψ_r with Given Rate of Error Damping

The knowledge of stator flux linkage Ψ_s is indispensable for the direct torque control. It also enables one to calculate rotor's flux Ψ_r , which is necessary for vector control (3.52):

$$\Psi_r = \frac{1}{k_r} \Psi_s - \frac{1}{\beta} \Psi_s \qquad (3.270)$$

Concurrently, the mathematical model serving for the calculation of stator flux is the following (3.53):

$$\Psi_s = \mathbf{U}_s - R_s \mathbf{i}_s + \omega_c \mathbf{A}_2 \Psi_s$$

$$\mathbf{i}_s = \frac{\beta}{k_r}(\alpha_r + p\Omega_r \mathbf{A}_2)\Psi_s - \frac{1}{\sigma}(\alpha_s + \alpha_r) + (\omega_c - p\Omega_r)\mathbf{A}_2 \mathbf{i}_s + \frac{\beta}{k_r}\mathbf{U}_s \qquad (3.271)$$

On the basis of this model, while the measurement quantities include $\mathbf{U}_s$, i_s, Ω_r, an observer of stator flux has been designed [66]. The advantage of this observer involves the fact that it is possible to set an arbitrary speed of estimation error damping, which distinguishes it from the preceding rotor flux $\boldsymbol{\Psi}_r$ observers (where error damping was affected by the damping of α_r). The designed observer takes the form:

$$\dot{\hat{\boldsymbol{\Psi}}}_s = \mathbf{U}_s - R_s \mathbf{i}_s + \omega_c \mathbf{A}_2 \hat{\boldsymbol{\Psi}}_s + c_1\left(\mathbf{i}_s - \hat{\mathbf{i}}_s\right)\overline{Z}$$

$$\dot{\hat{\mathbf{i}}}_s = \frac{\beta}{k_r}(\alpha_r + p\Omega_r \mathbf{A}_2)\hat{\boldsymbol{\Psi}}_s - \frac{1}{\sigma}(\alpha_s + \alpha_r) + (\omega_c - p\Omega_r)\mathbf{A}_2 \mathbf{i}_s + \qquad (3.272)$$

$$+ \frac{\beta}{k_r}\mathbf{U}_s + c_2\left(\mathbf{i}_s - \hat{\mathbf{i}}_s\right)\frac{\beta\overline{Z}}{k_r}$$

where: c_1, c_2 - are complex multipliers.

The equations of the dynamics of the error in the complex variables take the form:

$$\dot{\mathbf{e}}_\Psi = \dot{\boldsymbol{\Psi}}_s - \dot{\hat{\boldsymbol{\Psi}}}_s = \omega_c \mathbf{A}_2 \mathbf{e}_\Psi - c_1 \mathbf{e}_i \overline{Z}$$

$$\dot{\mathbf{e}}_i = \dot{\mathbf{i}}_s - \dot{\hat{\mathbf{i}}}_s = \frac{\beta}{k_r}(\alpha_r + p\Omega_r \mathbf{A}_2)\mathbf{e}_\Psi - c_2 \mathbf{e}_i \frac{\beta\overline{Z}}{k_r} \qquad (3.273)$$

In the matrix form the equations of the errors are the following:

$$\begin{bmatrix} \dot{\mathbf{e}}_\Psi \\ \dot{\mathbf{e}}_i \end{bmatrix} = \begin{bmatrix} -j\omega_c & -c_1\overline{Z} \\ \dfrac{\beta}{k_r}(\alpha_r - jp\Omega_r) & -c_2\dfrac{\beta\overline{Z}}{k_r} \end{bmatrix}\begin{bmatrix} \mathbf{e}_\Psi \\ \mathbf{e}_i \end{bmatrix} \qquad (3.274)$$

Hence, the characteristic polynomial follows:

$$W(s) = \begin{vmatrix} s + j\omega_c & c_1\overline{Z} \\ -\dfrac{\beta}{k_r}(\alpha_r - jp\Omega_r) & s + c_2\dfrac{\beta\overline{Z}}{k_r} \end{vmatrix} =$$

$$= s^2 + s\left(c_2\frac{\beta\overline{Z}}{k_r} + j\omega_c\right) + c_1\frac{\beta\overline{Z}}{k_r}\alpha_r + j\frac{\beta\overline{Z}}{k_r}(c_2\omega_c - c_1 p\Omega_r) \qquad (3.275)$$

This is a polynomial with complex parameters. Similarly, $c_1, c_2, \overline{Z}$ can take complex values. In the physical sense it denotes two-dimensional control of flux $\boldsymbol{\Psi}_s$ and current $\mathbf{i}_s$ in u,v coordinates. Having the possibility of setting the real constant ω_c and complex constants

$$c_1 = c_{1R} + c_{1Y} \qquad c_2 = c_{2R} + c_{2Y} \qquad (3.276)$$

and impedance $\overline{Z}$, there are considerable opportunities for the formation of the curves of the estimation error. Assuming the general form of the characteristic polynomial:

$$W(s) = (s - r_1 - j\omega_1)(s - r_2 - j\omega_2) =$$
$$= s^2 + s(-(r_1 + r_2) - j(\omega_1 + \omega_2)) + r_1 r_2 - \omega_1\omega_2 + j(r_1\omega_2 + r_2\omega_1) \qquad (3.277)$$

it is possible to state the equations for the coefficients:

$$c_2 \frac{\beta\overline{Z}}{k_r} + j\omega_c = -(r_1 + r_2) - j(\omega_1 + \omega_2)$$

$$c_1 \frac{\beta\overline{Z}}{k_r}\alpha_r + j\frac{\beta\overline{Z}}{k_r}(c_2\omega_c - c_1 p\Omega_r) = r_1 r_2 - \omega_1\omega_2 + j(r_1\omega_2 + r_2\omega_1) \qquad (3.278)$$

As an example we will present the solution for:

$$\frac{\beta\overline{Z}}{k_r} = 1 \qquad \omega_c = 0 \qquad c_{2Y} = 0 \qquad \omega_2 = -\omega_1 = -\omega \qquad (3.279)$$

In this case, on the basis of (3.278), we obtain:

$$\begin{bmatrix} c_{1R} \\ c_{1Y} \end{bmatrix} = \frac{1}{\alpha_r^2 + (p\Omega_r)^2} \begin{bmatrix} \alpha_r(r_1 r_2 + \omega^2) - p\Omega_r(r_1 - r_2) \\ p\Omega_r(r_1 r_2 + \omega^2) + \omega\alpha_r(r_1 - r_2) \end{bmatrix} \qquad (3.280)$$

$$c_{2R} = -(r_1 + r_2)$$

From this solution (3.280) by appropriate selection of r_1, r_2 it is possible to achieve arbitrarily the decay of the flux estimation error Ψ_s. However, this result is considerably relative to the angular speed Ω_r. The result can serve for the determination of flux Ψ_s for the purposes of DTC control or help one in the determination of rotor flux Ψ_r by using (3.270).

References

[1] Abdin, E.S., Xu, W.: Control design and dynamic performance analysis of a wind turbine-induction generator unit. IEEE Trans. En. Conv. 15, 91–96 (2000)

[2] Agrawal, P.: Power Electronic Systems. In: Agrawal, P. (ed.) Theory and Design, Prentice Hall, Upper Saddle River (2001)

[3] Ai, T.H., Chen, J.F., Liang, T.J.: A random switching method for HPWM full-bridge inverters. IEEE Trans. Ind. Elect. 49, 595–597 (2002)

[4] Amler, G.: A PWM current-source inverter for high quality drives. Eu. P. Electr. J. 1, 21–32 (1991)

[5] Baader, U., Depenbrock, M., Gierse, G.: Direct self control of inverter-fed induction machine: a basis for speed control without speed measurement. IEEE Trans. Ind. Appl. 28, 581–588 (1992)

[6] Bech, M., Blaabjerg, F., Pederson, J.K.: Random modulation techniques with fixed switching frequency for three-phase power converters. IEEE Trans. P. Elect. 15, 753–761 (2000)

[7] Bellini, A., Figalli, G.: On the selection of the commuting instants for induction motor drives. IEEE Trans. Ind. Appl. 5, 501–506 (1979)

[8] Ben Abdeighani, A., Martins, C.A., Roboam, X., et al.: Use of extra degrees of freedom in multilevel drives. IEEE Trans. Industrial Electr. 49, 965–977 (2002)

[9] Bernet, S.: Recent Developments of High Power Converters for Industry and Traction Applications. IEEE Trans. P. Elect. 15, 1102–1117 (2000)

[10] Wu, B.: High Power Converters and AC Drives. John Wiley & Sons, Hoboken (2006)

[11] Bodson, J., Chiasson, J., Novotnak, R.: A systematic approach to selecting flux references for torque maximization in induction motors. IEEE Trans. Contr. Sys. Tech. 3, 388–397 (1995)

[12] Bodson, J., Chiasson, J., Novotnak, R.: Nonlinear speed observer for high-performance induction motor control. IEEE Trans. Ind. Electr. 42, 337–343 (1995)

[13] Boldea, I., Nasar, A.: Vector Control of AC Drives. CRC Press, Boca Raton (1992)

[14] Bose, B.K.: Modern Power Electronics and AC Drives. Prentice-Hall, Englewood Cliffs (2001)

[15] Bowes, S.R., Grewal, S.: Novel space-vector-based harmonic elimination inverter control. IEEE Trans. Ind. Appl. 36, 549–557 (2000)

[16] Buja, G.S.: Optimum output waveforms in PWM inverters. IEEE Trans. Ind. Appl. 1A -16, 38–44 (1980)

[17] Bumby, J.R., Spooner, E., Jagieła, M.: Equivalent circuit analysis of solid-rotor induction machines with reference to turbocharger accelerator applications. IEE Proc. El. P. Appl. 153, 31–39 (2006)

[18] Cassadei, D., Profumo, N., Serra, G.F., et al.: FOC and DTC: Two Viable Schemes for Induction Motor Torque Control. IEEE Trans. P. Electr. 17, 779–787 (2002)

[19] Cassadei, D., Profumo, N., Serra, G.F., et al.: Performance analysis of a speed sensorless induction motor drive based on constant-switching frequency DTC scheme. IEEE Trans. P. Electr. 17, 779–787 (2003)

[20] Chakraborty, C., Hori, Y.: Fast Efficiency Optimization Techniques for the Indirect Vector-Controlled Induction Motor Drives. IEEE Trans. Ind. Appl. 39, 1070–1076 (2003)

[21] Chiasson, J.: A new approach to dynamic feedback linearization control of an induction motor. IEEE Trans. Aut. Contr. 43, 391–397 (1998)

[22] Chiasson, J.: Modeling and High-Performance Control of Electric Machines. John Wiley & Sons Inc., Hoboken (2005)

[23] Corzine, K., Sudhoff, S.D., Whitcomb, C.K.: Performance characteristics of a cascaded two-level converter. IEEE Trans. En. Conv. 14, 433–439 (1999)

[24] Demenko, A.: Time stepping FE analysis of electric motor drives with semiconductor converters. IEEE Trans. Magn. 30, 3264–3267 (1994)

[25] Ekanayak, J.B., Holdsworth, L., Wu, X.G.: Dynamic modeling of doubly fed induction generator wind turbines. IEEE Trans. P. Sys. 18, 803–809 (2003)

[26] Escalante, M.F., Vannier, J.C., Arzande, A.: Flying capacitor multilevel inverters and DTC motor drive applications. IEEE Trans. Ind. Electr. 49, 809–815 (2002)

[27] Espinosa, J.R., Joos, G.: State Variable Decoupling and Power Flow Control in PWM Current-Source rectifiers. IEEE Trans. Ind. Electr. 45, 80–87 (1998)

[28] Gorti, V., Alexander, G.C., Spee, R.: Power balance considerations for brushless doubly-fed machines. IEEE Trans. En. Conv. 11, 687–692 (1996)

[29] Griva, G., Habetler, T.G., Profumo, F., et al.: Performance evaluation of a direct torque controlled drive in the continuous PWM-square wave transition region. IEEE Trans. Power Electr. 10, 464–471 (1995)

[30] Habetler, T.G., Divan, D.M.: Acoustic noise reduction in sinusoidal PWM drives using a randomly modulated carrier. IEEE Trans. P. Electr. 6, 356–363 (1991)

[31] Hammond, P.W.: A new Approach to Enhance Power Quality for Medium Voltage AC Drives. IEEE Trans. Ind. Appl. 33, 202–208 (1997)

[32] Hammond, P.W.: Enhancing reliability of modular medium voltage drives. IEEE Trans. Ind. Electr. 49, 948–954 (2002)

[33] Hashad, M., Iwaszkiewicz, J.: A novel orthogonal-vector-based topology of multilevel inverters. IEEE Trans. Ind. Eletr. 49, 868–874 (2002)

[34] Hava, A.M., Kekerman, R.J., Lipo, T.A.: Carrier-based PWM-VSI overmodulation strategies: analysis, comparison and design. IEEE Trans. P. Electr. 3, 674–689 (1998)

[35] Henriksen, S.J., Betz, R.E., Cook, B.J.: Digital hardware implementation of a current controller for IM variable-speed drives. IEEE Trans. Ind. Appl. 35, 1021–1029 (1999)

[36] Hickiewicz, J., Macek-Kaminska, K., Wach, P.: A simulation of common-bus drives in power plants. Arch. El. 75, 293–302 (1992)

[37] Hickiewicz, J., Macek–Kamińska, K., Wach, P.: Algorithmic methods of induction machines parameters estimation from measured slip–curves. Arch. f. El. 72, 239–249 (1989)

[38] Holmes, D.G.: The significance of zero space vector placement for carrier based PWM schemes. IEEE Trans. Ind. Appl. 32, 1122–1129 (1996)

[39] Holmes, D.G., Lipo, T.A.: Pulse Width Modulation for Power Converters. John Wiley & Sons, New York (2003)

[40] Holtz, J.: Sensorless control of induction machines- with or without signal injection? IEEE Trans. Ind. Electr. 30, 7–30 (2006)

[41] Holtz, J., Beyer, B.: Optimal pulse width modulation for ac servos and low-cost industrial drive. IEEE Trans. Ind. Appl. 30, 1039–1047 (1994)

[42] Holtz, J., Bube, E.: Field oriented asynchronous pulse-width modulation for high performance ac machine drives operating at low switching frequency. IEEE Transactions on Industry Applications 27, 574–581 (1991)

[43] Holtz, J., Juliet, J.: Sensorless acquisition of the rotor position angle of induction motors with arbitrary stator windings. IEEE Trans. Ind. Appl. 40, 591–598 (2004)

[44] Holtz, J., Springob, L.: Reduced harmonics PWM controlled line-side converter for electric drives. IEEE Trans. Ind. Appl. 29, 814–819 (1993)

[45] Hu, J., Dawson, D., Qu, Z.: Robust tracking control of an induction motor. Int. J. Rob. Nonl. Contr., 201–219 (1996)

[46] Hur, H., Jung, J., Nam, K.: A Fast Dynamics DC-link Power- Balancing Scheme for a PWM Converter-Inverter System. IEEE Trans. Ind. Electr. 48, 794–803 (2001)

[47] Ishida, T., Matsuse, K., Sugita, K., et al.: DC control strategy for five-level converter. IEEE Trans. P. Electr. 15, 508–515 (2000)

[48] Jagieła, M.: Analysis of a steady state performance of an induction machine by means of coupling circuit equations to FE. Arch. El. Eng. 52, 317–328 (2003)

[49] Jagieła, M., Bumby, J.R., Spooner, E.: Time-stepping FE analysis of high speed solid rotor induction motors. In: Proc. 41 Symp. El. Mach., vol. 1, pp. 64–67 (2005)

[50] Kawabata, Y., Kawaguchi, K., Nomoto, T., et al.: High efficiency and low acoustic noise drive system using open-winding AC motor and two space-vector modulated inverters. IEEE Trans. Ind. Electr. 49, 783–789 (2002)

[51] Kazimierkowski, M.P., Tunia, H.: Automatic Control of Converter Fed Drives. Elsevier, Amsterdam (1994)

[52] Kazimierkowski, M.P., Krishnan, R., Blaabjerg, R.: Control in Power Electronics. Academic Press, Elsevier (2002)

[53] Khorrami, F., Krishnamurthy, P., Melkote, H.: Modeling and Adaptive Nonlinear Control of Electric Motors. Springer, Heidelberg (2003)

[54] Kim, S.H., Sul, S.K.: Voltage Control Strategy for Maximum Torque Operation of Induction Machine in the Field-Weakening Region. IEEE Trans. Ind. Electr. 44, 512–518 (1997)

[55] Kirilin, R.L., Bech, M.M., Trzynadlowski, A.M.: Analysis of Power and Power Spectral Density in PWM Inverters with Randomized Switching Frequency. IEEE Trans. Ind. Electr. 49, 486–499 (2002)

[56] Kouro, S., Bernal, R., Miranda, H., et al.: High Performance Torque and Flux Control for Multilevel Inverter Fed Induction Motors. IEEE Trans. P. Electr. 22, 2116–2123 (2007)

[57] Krzemiński, Z.: Nonlinear Control of Induction Motors. In: Proc 10-th IFAC Congress, pp. 349–354 (1987)

[58] Kubota, H., Matsuse, K.: Speed Sensorless Field Oriented Control of Induction Machines Using Flux Observer. IEEE Trans. Ind. Appl. 30, 1219–1224 (1994)

[59] Lai, Y.S.: New random inverter control technique for common mode voltage mitigation of motor drives. IEE Proc. El. P. Appl. 146, 289–296 (1999)

[60] Lascu, C., Boldea, I., Blaabjerg, F.: Direct Torque Control of Sensorless Induction Motor Drives: A Sliding Mode Approach. IEEE Trans. Ind. Appl. 40, 582–590 (2004)

[61] Lee, D.C., Lee, G.M.: A novel overmodulation technique for space-vector PWM drives. IEEE Trans. P. Electr. 13, 1144–1151 (1998)

[62] Li, R., Wallace, A., Spee, R.: Determination of converter control algorithms for brushless doubly-fed motor drives using Floquet and Lyapunov techniques. IEEE Trans. En. Conv. 10, 78–85 (1995)

[63] Lyra, R.O.C., Lipo, T.A.: Torque Density Improvements in a Six-Phase Induction Motor with Third Harmonic Current Injection. IEEE Trans. Ind. Appl. 38, 1351–1360 (2001)

[64] Marchesoni, M., Tenca, P.: Diode-clamped multilevel converters: A practical way to balance DC link voltages. IEEE Trans. Ind. Electr. 49, 752–765 (2002)

[65] Marino, R., Peresada, S., Tomei, P.: Global adaptive output feedback control of induction motors with uncertain rotor resistance. IEEE Trans. Aut. Contr. 44, 967–983 (1999)

[66] Martin, P., Rouchon, P.: Two simple flux observers for induction motors. Int. J. Ad. Contr. Sig. Proc. 14, 171–175 (2000)

[67] Mcmahon, R.A., Roberts, P.C., Wang, X., et al.: Performance of the BDFM as a generator or motor. IEEE Proc. El. P. Appl. 153, 289–299 (2006)

[68] Mengoni, M., Zarri, L., Tani, A., et al.: Stator Flux Vector Control of Induction Motor Drive in the Field Weakening Region. IEEE Trans. P. Electr. 23, 941–948 (2008)

[69] Mohan, N., Undeland, T.M., Robbins, W.P.: Power Electronics-Converters. Applications and Design, vol. 3. John Wiley & Sons, New York (2003)

[70] Mondal, M., Bose, B.K., Oleschuk, V., et al.: Space vector pulse width modulation of three-level inverter extending operation into overmodulation region. IEEE Trans. P. Electr. 18, 604–611 (2003)

[71] Na, S.H., Young, Y.G., Lim, Y., et al.: Reduction of audible switching noise in induction motor drives using random position space vector PWM. IEEE Proc. El. P. Appl. 149, 195–200 (2002)

[72] Narayanan, G., Ranganathan, V.T.: Synchronized PWM strategies based on space vector approach. Part 1: Principles of wave form generation. IEEE Proc. El. P. Appl. 146, 267–275 (1999)

[73] Narayanan, G., Ranganathan, V.T.: Synchronized PWM strategies based on space vector approach. Part 2: Performance assessment and application to V/f drives. IEEE Proc. El. P. Appl. 146, 276–281 (1999)

[74] Nash, J.N.: Direct Torque Control, Induction Motor Vector Control without Encoder. IEEE Trans. on Ind. Appl. 33, 333–341 (1997)

[75] Novotnak, R.T., Chiasson, J., Bodson, M.: High-Performance Motion Control of an Induction Motor with Magnetic Saturation. IEEE Trans. Contr. Sys. Techn. 7, 315–327 (1999)

[76] Novotny, D.W., Lipo, T.A.: Vector Control and Dynamics of AC Machines. Clarendon Press, Oxford (1996)

[77] Orlowska-Kowalska, T., Migas, P.: Analysis of Induction Motor Speed Estimation Quality using Neural Networks of Different Structure. Arch. El. Eng. 50, 411–425 (2001)

[78] Orlowska-Kowalska, T., Pawlak, M.: Induction Motor Speed Estimation Based on Neural Modeling Method. Arch. El. Eng. 49, 35–48 (2000)

[79] Ortega, R., Barabanov, N., Valderrama, G.E.: Direct Torque Control of Induction Motors: Stability Analysis and Performance Improvement. IEEE Trans. Aut. Contr. 46, 1209–1222 (2001)

[80] Paap, G.C.: The Influence of Space Harmonics on Electromechanical Behavior of Asynchronous Machines. Dissertation, Politechnika Łódzka, Lodz (1988)

[81] Peng, F.Z.: A Generalized Multilevel Inverter Topology with Self Voltage Balancing. IEEE Trans. Ind. Appl. 37, 611–618 (2001)

[82] Rams, W.: Synthesis of Dynamic Properties of Asynchronous Machines. Dissertation, Academy of Mining and Metallurgy, Krakow (1973)

[83] Rashid, M.H.: Power Electronics Handbook. Academic Press, New York (2001)

[84] Raviraj, V.S.C., Sen, P.C.: Comparative study of Proportional-Integral, Sliding Mode, and Fuzzy Logic Controllers for Power Converters. IEEE Trans. Ind. Appl. 33, 518–524 (1997)

[85] Richardson, J., Kukrer, O.: Implementation of a PWM Regular Sampling Strategy for AC Drives. IEEE Trans. P. Electr. 6, 649–655 (1989)

[86] Rodriguez, J., Lai, J.S., Peng, F.Z.: Multilevel inverters: A Survey of Topologies, Controls, and Applications. IEEE Trans. Ind. Electr. 49, 724–738 (2002)

[87] Ryuh, M., Kim, J.H., Sul, S.K.: Analysis of multiphase space Vector PWM based o multiple d-q spacer concept. IEEE Trans. P. Electr. 20, 1364–1371 (2005)

[88] Schibili, N.P., Nguyen, T., Rufer, A.C.: A Three-Phase Multilevel Converter for High Power Induction Motors. IEEE Trans. P. Electr. 13, 985–987 (1998)

[89] Seok, J.K., Kim, J.S., Sul, S.K.: Overmodulation Strategy for High Performance Torque Control. IEEE Trans. P. Electr. 13, 784–792 (1998)

[90] Sobczyk, T.J., Warzecha, A.: Alternative Approach to Modeling Electrical Machines with Non-linear Magnetic Circuit. Arch. El. Eng. 46, 421–434 (1997)

[91] Stephan, J., Bodson, M., Chiasson, J.: Real Time Estimation of Induction Motor Parameters. IEEE Tran. Ind. Appl. 30, 746–759 (1994)

[92] Sudhoff, S.D., Aliprantis, C.D., Kuhn, B.T., et al.: Experimental Characterization Procedure for Use with an Advanced Induction Machine Model. IEEE Trans. En. Conv. 18, 48–56 (2003)

[93] Tolbert, L.M., Peng, F.Z., Habetler, T.G.: Multilevel Converters for large Electric Drives. IEEE Trans. Ind. Appl. 35, 36–43 (1999)

[94] Toliyat, H.A.: Analysis and Simulation of Five-Phase Variable-Speed Induction Motor Drives under Asymmetrical Connections. IEEE Trans. P. Electr. 13, 748–756 (1998)

[95] Trzynadlowski, A.M.: Introduction to Modern Power Electronics. John Wiley & Sons Inc., New York (1992)

[96] Trzynadlowski, A.M.: The Field Orientation Principle in Control of Induction Motors. Kluwer Academic Publishers, Hingham (1994)

[97] Trzynadlowski, A.M., Bech, M.M., Blaabjerg, F.: Optimization of Switching Frequencies in limited-Pool Random Space Vector PWM Strategy for Inverter-Fed Drives. IEEE Trans. P. Electr. 16, 852–857 (2002)

[98] Uddin, M.N., Radwan, T.S., Rahman, M.: Performances of fuzzy-logic based indirect vector control for induction motor drive. IEEE Trans. Ind. App. 138, 1219–1225 (2002)

[99] Vas, P.: Vector Control of AC Machines. Clarendon Press, Oxford (1990)

[100] Vas, P.: Sensorless Vector and Direct Torque Control. Clarendon Press, Oxford (1998)

[101] Wach, P.: Algorithmic Method of Design and Analysis of Fractional-Slot Windings of AC Machines. El. Eng. 81, 163–170 (1998)

[102] Wach, P.: Multi-Phase Systems of Fractional-Slot Windings of AC Electrical Machines. Arch. El. Eng. 46, 471–486 (1997)

[103] Wach, P.: Optimization of Fractional – Slot Windings of AC Electrical Machines. Arch. El. Eng. 47, 69–80 (1998)

[104] Wach, P.: Asymmetric Fractional – Slot Windings. Arch. of El. Eng. 48, 365–374 (2000)

[105] Wai, R.J., Lee, J.D., Lin, M.: Robust Decoupled Control of Direct Field-Oriented Induction Motor Drive. IEEE Trans. Ind. Electr. 52, 837–854 (2005)

[106] Wu, B., Dewan, S., Slemon, G.: PWM-CSI Inverter Induction Motor Drives. IEEE Trans. Ind. Appl. 28, 64–71 (1992)

[107] Xu, H., Toliyat, H.A., Peterson, L.J.: Five-Phase Induction Motor Drives with DSP-Based Control System. IEEE Trans. P. Electr. 17, 524–533 (2002)

[108] Xu, X., Novotny, D.W.: Selection of the Flux Reference for Induction Machine Drives in the Field Eeakening Region. IEEE Trans. Ind. Appl. 28, 1353–1358 (1992)

Chapter 4
Brushless DC Motor Drives (BLDC)

Abstract. This chapter deals with properties and control of brushless DC motors drives with a permanent magnet excitation (BLDC, PMDC). In the first part characteristics of contemporary permanent magnets (PM) that are used in electric motors are presented along with the simplified ways of their modeling. As an example a pendulum is given, which consists of swinging coil over a stationary PM, and the influence of PM modeling simplifications upon the dynamic trajectories of movement is discussed. Further on, a model of PMDC drive is derived on a transformed *d-q* and also a contra model in which no transformation of variables is used, with the commutation taking place according to the state of physical (natural) variables. However, the problem of nonholonomic constraints is not undertaken while dealing with PMDC modeling. In a classical DC motor with mechanical commutator, the existence of such constraints is evident, because the connection of each armature's coil to the external circuit depends on the rotor position. In case of electronic commutation, a switching of windings' supply is controlled also by the rotor position angle, but topographic structure of circuits remains fixed and the switching is carried out by a abrupt changes of impedance values of power electronic switches. In this chapter various characteristics and transient curves for BLDC drives are presented and a comparison is made between results obtained from both types of models: *d-q* transformed and untransformed ones. It gives the possibility of justifying the choice regarding the kind of the model to be used in particular applications, depending on the dimension of a whole system and required rigorousness of results. The results presented cover the operation of DC drives with and without control system intervention. The PID control is discussed in its application to a given profile of speed and rotor position movement and also inverse dynamics method is introduced. Numerous examples of DC drive problems are included, employing two typical BLDC motors with given data.

4.1 Introduction

For 150 years DC direct current machines have played an important role in electric drives. The basic advantages associated with their application in drives include: easy adjustment of rotational speed, uncomplicated start-up and reversal, stable operation for small speeds as well as good dynamic properties ensuring fast reaction to changing parameters of power supply. DC machines with classical design consist of a stator with electromagnets which are powered by the excitation current and whose role is to generate a magnetic field of excitation. The windings of the rotating armature are connected to the mechanical commutator with the graphite-metal brushes slipping over it. They constitute the electrical connection between the movable armature windings and external circuits. The mechanical commutator, which is a device mostly of historical significance, plays the role of a mechanical rectifier, which converts AC current with the frequency corresponding to the rotor's rotational speed into the DC current outside the armature. At its time, the mechanical commutator was an outstanding device, which was however troublesome during exploitation and costly in terms of cost of investment. It was also the weakest link in the system in the sense of reliability of operation as well as required frequent service and regular overhaul. A modern brushless DC machine (BLDC) displays two fundamental differences in contrast to the mechanically commutated DC machines [13,15,17,28,42]. First of all, it does not have a mechanical commutator over which the brushes forming the electrical node used to slip. A static electronic commutator is used in its place, whose role is the commutation of the current in armature windings in the function of the angle of rotor position θ_r. Hence, the principle governing DC machines is preserved, i.e. the machine is self-commutating. In the characteristics of the machine the basic effect involves the fact that along with the increase of the load the machine tends to slow down unless it is supplied with external speed control for the stabilization of the speed. As a result of this slowing of rotational speed the armature current tends to increase and this leads to a new equilibrium point of the operation. The second relevant difference between a classical mechanically commutated machine and up-to-date motor involves the replacement of electromagnets exciting the main magnetic field with adequately selected permanent magnet assembly [12,22]. This solution is rendered possible as a result of magnetic parameters and other utility parameters and ultimately the commercial value of permanent magnets. They contain rare earth elements, such as neodymium (Nd), samarium (Sr) among others. The application of permanent magnets improves the efficiency of a machine since there are no power losses in the excitation windings and leads to the decrease of machine mass. However, in terms of the construction and thermal requirements of machine operation there is no advancement since the permanent magnets installed in the machine and providing the excitation flux require adequate operating conditions which do not permit the deterioration or a decay of the magnetic field from the magnets. These requirements basically involve the limitation of the temperature inside the permanent magnet motor, limitation of the influence of armature reaction in a way that ensures that irreversible demagnetization of magnets does not occur and not extending air gaps in order to prevent overloading of permanent magnets. One has to bear in mind that the permanent magnet DC machines (PMDC) have to be designed in manner that ensures their operation over a number of years without deterioration of the exploitation parameters. Another important difference between the classical mechanically commutated DC machine and a brushless one concerns the number of windings of the armature and, subsequently, the current waveform on the DC side. In a commutator machine the usual number of windings varies around a couple dozen, as a consequence of which there is an adequately large number of commutator segments. In connection with this, DC current contains very small pulsations since the commutation occurs every couple of degrees of rotor's angle of rotation. Consequently, the electromagnetic torque generated by this machine tends to demonstrate small pulsations. In BLDC machines of the most common engineering design there are three phase windings of the armature, which is reflected by three branches of an electronic commutator (rectifier controlled by the angle of rotor position). This results in considerable current and torque pulsations generated by the machine since the commutation occurs every 60° of the angle of rotor rotation, alternatively in the anode and cathode group of the electronic commutator. It is obviously possible to increase the number of the armature windings and number of commutator branches thus leading to the reduction of current pulsation; however, two negative effects follow. One of them is associated with the need to use a more extensive and expensive electronic commutator, while the other one involves an increase of commutation losses and decrease of the efficiency of the drive. A final remark that can be made at the beginning of this introduction is that brushless DC machines with permanent magnets can vary considerably in terms of their engineering structure. First of all, there can be minute machine serving as servodrives in technology, household appliances and vehicles. Besides, there are larger machines, which are applied in electric drives of automatically controlled devices, including drives in manipulator joints. Finally, there are high power machines with the parameters of the drives used in industrial machinery, for example in steel mills or vessels. BLDC machines may have a various number of phases, have cylindrical construction and in some applications they can have a form of a disk with immobile armature and rotating magnets. The final solution can serve for use in low revolution gearless drives. BLDC machines need not have low revolution ranges, as ones discussed before, but also can operate under rotational speeds exceeding 10,000 [rev/min]. The number of the available versions is large and still growing.

4.2 Permanent Magnet – Basic Description in the Mathematical Model

Permanent magnets have been in use for a long time. They have been applied as components of technical devices for nearly 200 years. The acquaintance with the physics of magnetic materials and principles governing magnetization on the micro level has occupied the attention of scientists for the last 50 years while the technicalities of the process of production of up-to-date composite magnets has dated since 1980s. At present we are familiar with permanent magnets with stable magnetic properties on condition of not exceeding admissible temperatures with high value of unitary internal energy, magnetic induction under the magnet in excess of 1[T] and a broad magnetization loop. The basic characteristics of the permanent magnet is presented in Fig 4.1.

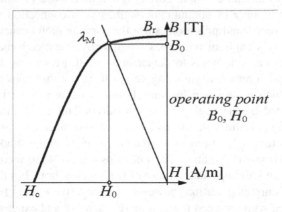

Fig. 4.1 General characteristic of a permanent magnet with an air gap load curve given by λ_M

For the description of the operation of permanent magnet we normally present the characteristics of magnetization merely in the II quadrant of the coordinate system since this is the operating range of a magnet. In a simple magnetic circuit consisting of a permanent magnet, air gaps and a small ferromagnetic core ($\mu = \infty$) that is used for closing the magnetic circuit, one can state that

$$\oint \mathbf{H}\,d\mathbf{l} = H_0\,l_M + H_\delta\delta = 0 \tag{4.1}$$

Under the assumption that leakage flux is neglected in such a circuit the following relation is fulfilled

$$B_0\,S_M = B_\delta S_\delta \quad \rightarrow \quad B_0 = B_\delta \frac{S_\delta}{S_M} \tag{4.2}$$

and since: $H_\delta = B_\delta/\mu\mu_0$ as a result, on the basis of (4.1) and (4.2) one can state that:

$$\lambda_M = \frac{B_0}{H_0} = -\frac{l_M}{\delta}\frac{S_\delta}{S_M}\mu\mu_0 \tag{4.3}$$

In the above formulae:

- H_0, B_0, l_M, S_M - are: magnetic field strength and induction in the magnet and, consecutively, the length and internal cross section of the permanent magnet
- H_δ, B_δ. δ, S_δ - are: magnetic field strength and induction in the air gap and consecutively, the length and cross section of the air gap

- μ, μ_0 - is magnetic permeability, relative one and that of the air
- λ_M - denotes, in accordance with (4.3) the unit magnetic conductance of the magnetic circuit also known as the inclination of the straight line of magnet load.

The inclination of the straight with the directional coefficient λ_M corresponds to the conductance of the air gap in a simple magnetic circuit and the cross section between the straight line and characteristics of the magnet determines the operating point H_0,B_0 of the permanent magnet in a given magnetic circuit. Concurrently, the product of H_0,B_0 determines the unitary energy of the magnet (per unit of volume) at a given operating point H_0,B_0. For a certain inclination of the straight line λ_M the rectangle with the sides marked as H_0,B_0 has the largest area for a given characteristic of operation and this specific operating point determines the maximum operating energy $(H\,B)_{max}$ for the magnetic material from which the magnet is formed. A given material is optimal in terms of magnetic properties when it has concurrently a large value of induction of the magnetic remnant B_r and intensity of the coercion of the magnetic field $|\text{-}H_c|$, as well as the large value of the maximum operating energy $(H\,B)_{max}$. An ideal would involve a magnet with a nearly rectangular magnetization loop for large values of B_r and $|\text{-}H_c|$ since it ensures a large and nearly constant induction under a magnet with a wide range of loads. As a result of the wide application of rare earth elements in magnets, they are able to come closer to this specific requirements to a much larger degree (Fig 4.2).

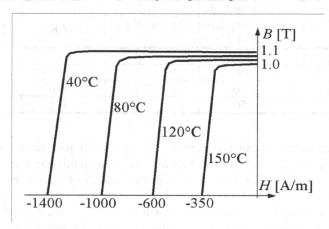

Fig. 4.2 A family of magnetizing curves of a rare earth permanent magnet, for different temperature values of operation

As one can conclude from z Fig. 4.2, the increase of temperature has a considerable effect on the magnetization characteristics of up-to-date permanent magnets based on rare earth elements. There is a certain, small reduction of the value of the remnant induction B_r and very large decrease of the absolute value of $|\text{-}H_c|$ that is the intensity of the magnetic coercion. Too high an ambience temperature of a magnet results in the deterioration of the range of adequate conditions for the magnet to operate. This comes as a consequence of the fact that following the change of $|-H_c|$ the inclination of λ_M, i.e. the characteristic of magnet loading is limited, which means that the admissible air gap in the magnetic circuit is considerably smaller. The fundamental parameters of major families of permanent magnets, i.e. ferrite magnets, alloy based ones with aluminum, nickel and cobalt (AlNiCo) and two major groups of magnets with rare earth elements, i.e. ones with samarium (Sr) and neodymium (Nd), are presented in Tables 4.1 and 4.2.

Table 4.1 Basic magnetic properties of the main PM materials

Family of a PM materials	B_r [T]	$\|-H_c\|$ [kA/m]	$(HB)_{max}$ [kJ/m^3]
Ferrites	0.4	250	30
Al Ni Co	1.1	150	80
Sm Co	1.1	750	200
Nd Fe B	1.2	850	300

Table 4.2 Basic temperature parameters of the main PM materials

Family of a PM materials	Maximum operating temperature [$^{\circ}$C]	Currie's point [$^{\circ}$C]	κ_{Br} [%/$^{\circ}$K]	κ_{Hc} [%/$^{\circ}$K]
Ferrites	300	440	-0.2	0.4
Al Ni Co	500	820	-0.03	0.0
Sm Co	300	750	-0.05	-0.25
Nd Fe B	150	300	-0.15	-0.6

The tables contain mean and approximated values of parameters taken from various references in a manner that does not reflect any particular magnetic material available in the market. One can note that the details of the materials summarized in the tables are offered commercially in various alloy combination, as composites or sinters, as it is the case for ferrites. The particular materials described in manufacturers' catalogues display various properties despite belonging to a single family. From the data in Tables 4.1 and 4.2 one can conclude that neodymium magnets are suitable for operation with lower operating temperatures while the ones with samarium display much better properties in higher temperature ranges. There are couple of methods of modeling on the macroscopic scale of PMs applied in electromechanical devices. We mean here simplified modeling, such that makes it possible to present the operation of electromechanical transducers and enable their modeling and simulation of operation in drive systems. One of the methods involves the replacement of the magnet with a compact turn with zero resistance and an adequately adapted self-inductance and circulating current i_{f0} in this turn. The reverse effect of the armature on the magnet occurs as a result of the armature current i_a via the mutual inductance M. The product $M\,i_{f0}$ corresponds to the magnetic flux Ψ_{f0} by means of which the permanent magnet affects the armature circuit. Another quite simple way involves the presentation of permanent magnet flux in the mathematical model in the form

$$\Psi_f = M i_f$$
$$i_f = i_{f0} - i_a M / L_f \tag{4.4}$$

The effect of the armature is modeled using the term $-i_a M/L_f$, which reduces the conventional magnetizing current i_{f0} originating from the permanent magnet. The simplest way of modeling the current originating from PM coupled with a given circuit is the adoption of its value Ψ_f as a constant. This involves disregarding armature currents during the operation of a machine for a small air gap in the magnetic circuit. It also corresponds to the operation of the magnet in the initial section of magnetization characteristic of a magnet produced from alloys of rare earth elements (Fig. 4.2). None of the presented here PM modeling methods accounts for the magnetization characteristics under the effect of the temperature rise. In order to present the discussed PM modeling methods, below is found an example of a servomechanism with a movable coil swinging above the magnet.

Example 4.1 Pendulum coil over PM.

A simplified model of the electromechanical system in which a pendulum coil moves in the field of a immobile PM is presented in Fig. 4.3.

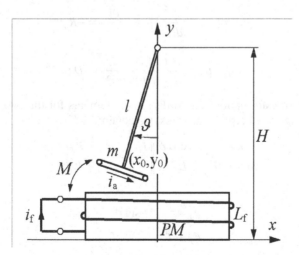

Fig. 4.3 Model of a pendulum coil over PM

The kinetic energy of the system is:

$$T = \frac{1}{2}L_f i_f^2 + \frac{1}{2}L_a i_a^2 + M_{af} i_a i_f + \frac{1}{2}m(\dot{x}_0^2 + \dot{y}_0^2) \tag{4.5}$$

while the potential energy:

$$U = mgy_0$$

This system has three degrees of freedom and after the introduction of generalized coordinates:

$$\mathbf{q} = (Q_f, Q_a, \vartheta)$$
$$i_f = \dot{Q}_f \quad i_a = \dot{Q}_a$$

(4.6)

With $J = ml^2$, Lagrange's function for this system takes the form

$$L = T - U = \frac{1}{2} L_f i_f^2 + \frac{1}{2} L_a i_a^2 + M_{af} i_a i_f + \frac{1}{2} J \dot{\vartheta}^2 -$$
$$- mg(H - l\cos\vartheta)$$

(4.7)

Assuming in a simplified form that

$$M_{af} = M \cos\vartheta$$

(4.8)

it is possible to determine the equations of motion for this system as Lagrange's equation:

1^0 $q = Q_f$ $\dfrac{d}{dt}\left(\dfrac{\partial L}{\partial i_f}\right) - \dfrac{\partial L}{\partial Q_f} = -R_f i_f$ (4.9)

2^0 $q = Q_a$ $\dfrac{d}{dt}\left(\dfrac{\partial L}{\partial i_a}\right) - \dfrac{\partial L}{\partial Q_a} = -R_a i_a$ (4.10)

3^0 $q = \vartheta$ $\dfrac{d}{dt}\left(\dfrac{\partial L}{\partial \dot{\vartheta}}\right) - \dfrac{\partial L}{\partial \vartheta} = -D\dot{\vartheta}$ (4.11)

After adequate transformations and ordering of elements for the two equations of motion regarding the electrical variables, we obtain:

$$\begin{bmatrix} L_f & M\cos\vartheta \\ M\cos\vartheta & L_a \end{bmatrix} \begin{bmatrix} i_f \\ i_a \end{bmatrix} = \begin{bmatrix} e_f - R_f i_f \\ e_a - R_a i_a \end{bmatrix}$$

(4.12)

$$e_a = \dot{\vartheta} i_a M \sin\vartheta$$
$$e_f = \dot{\vartheta} i_f M \sin\vartheta$$

(4.13)

After the transformation to the normal form, the system (4.12) in the matrix notation is:

$$\begin{bmatrix} i_f \\ i_a \end{bmatrix} = \frac{1}{L_f L_a - M^2 \cos^2\vartheta} \begin{bmatrix} L_a & M\cos\vartheta \\ M\cos\vartheta & L_f \end{bmatrix} \begin{bmatrix} e_f - R_f i_f \\ e_a - R_a i_a \end{bmatrix}$$

(4.14)

The maximum value of the magnetic flux coupled with the coil is:

$$\Psi_f = M i_f \tag{4.15}$$

Concurrently, the equation for the mechanical motion of the pendulum for variable ϑ, takes the form

$$J\ddot{\vartheta} = -i_a \Psi_f \sin\vartheta - mgl\sin\vartheta - D\dot{\vartheta} \tag{4.16}$$

where the first right hand side term denotes electromechanical torque braking the motion of the pendulum:

$$T_e = -i_a \Psi_f \sin\vartheta \tag{4.17}$$

The mathematical model presented in (4.14), (4.16) is applicable with regard to the first, least simplified way of modeling PMs. A more simplified magnet model involves disregarding of the modeling of the magnet by means of a separate differential equation and the presentation of the effect of the armature in the form resulting from (4.4). In this case, Lagrange's equation takes the form:

$$L = \frac{1}{2}\left(L_a - M^2\cos^2\vartheta/L_a\right)i_a^2 + M\cos\vartheta i_a i_{f0} +$$
$$+ \frac{1}{2}J\dot{\vartheta}^2 - mg(H - l\cos\vartheta) \tag{4.18}$$

This leads to two equations of motion for variables $q = (Q_a, \vartheta)$ in the form:

$$\left(L_a - 2M^2\cos^2\vartheta/L_a\right)i_a = M\dot{\vartheta}\sin\vartheta(i_{f0} - 4i_aM/L_a\cos\vartheta) - R_a i_a$$
$$J\ddot{\vartheta} = -Mi_a(i_{f0}\sin\vartheta + \frac{M}{L_a}i_a\sin2\vartheta) - mgl\sin\vartheta - D\dot{\vartheta} \tag{4.19}$$

The final and most simplified model which disregards the effect of the armature is gained for $i_f = i_{f0}$, $M_f i_{f0} = \Psi_{f0}$. In this case:

$$L_a\dot{i}_a = \dot{\vartheta}\Psi_f\sin\vartheta - R_a i_a$$
$$J\ddot{\vartheta} = -Mi_a\Psi_f\sin\vartheta - mgl\sin\vartheta - D\dot{\vartheta} \tag{4.20}$$

The operation of the models for the parameters of the system with the values of

$$L_f = 10.5[H] \quad L_a = 0.01[H] \quad M = 0.08[H] \quad R_a = 0.3[\Omega]$$
$$m = 0.05[kg] \quad i_{f0} = 5.0[A] \quad \Psi_f = 0.4[Wb]$$
$$D = 10^{-5}[N/s] \quad l = 0.15[m] \quad J = 0,022[Nms^2] \tag{4.21}$$

is presented in a series of figures (Figs. 4.4 – 4.7), which illustrate the motion of this system for the initial position of the pendulum $\vartheta_0 = 36°$.

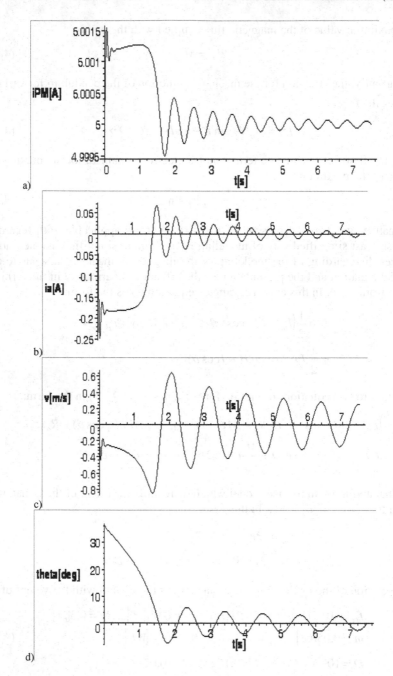

Fig. 4.4 Swinging motion of the pendulum coil, for $\vartheta_0 = 36°$, computed by the model (4.14 – 4.16): a) i_f current of PM b) i_a coil current c) $\dot{\vartheta}$ angular velocity of the pendulum d) sway angle ϑ e) electromagnetic torque T_e

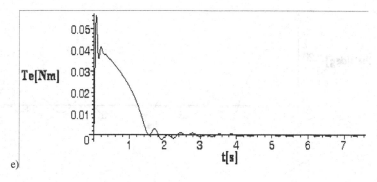

e)

Fig. 4.4 (*continued*)

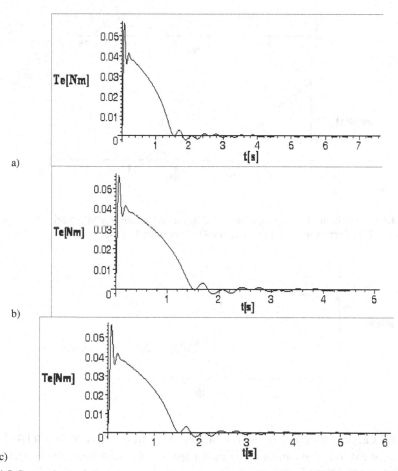

a)

b)

c)

Fig. 4.5 Comparison of electromagnetic torque computed with different PM model simplifications: a) full model (4.14 – 4.16) b) armature reaction model (4.4) c) constant Ψ_f model

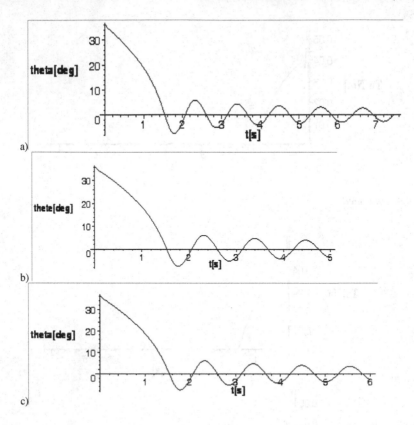

Fig. 4.6 Comparison of swinging motion - ϑ angle, for different PM models: a) full model (4.14 – 4.16) b) armature reaction model (4.4) c) constant Ψ_f model

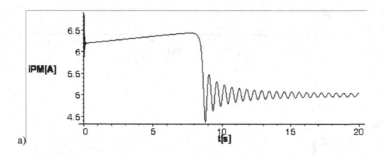

Fig. 4.7 Characteristics of movement for a very strong field linkage $\Psi_f = 1.0$ [Wb]: a) i_f current of PM b) i_a coil current c) $\dot\vartheta$ angular speed of the pendulum d) sway angle ϑ e) electromagnetic torque T_e

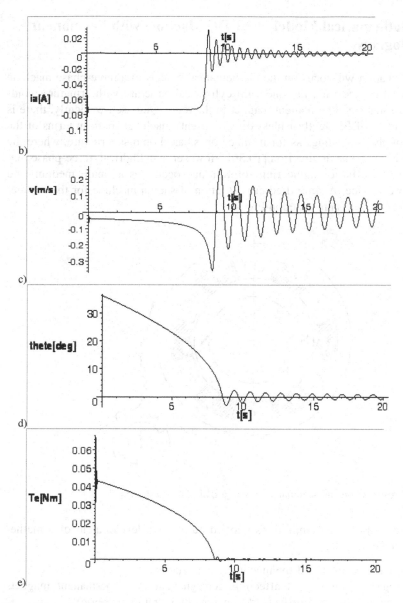

Fig. 4.7 (*continued*)

On the basis of the results presented in Fig. 4.5 and Fig. 4.6 one can conclude that the differences in terms of the curves for the variables characterizing the pendulum motion gained for various versions of the PM model simplifications are inconsiderable and the magnet model for $\Psi_f = const$ is acceptable for the modeling of motion parameters of the drive.

4.3 Mathematical Model of BLDC Machine with Permanent Magnets

The presentation will focus on the mathematical models of a three-phase machine with typical engineering, i.e. one with cylindrical structure with armature windings in the stator and permanent magnet in the rotor. Quite self-evidently, there is a wide variety of BLDC (brushless direct current) machines, both in terms of the number of phase windings as for instance ones based on disk structure, where the major field has an axial direction [10,36]. However, a cylindrical, three-phase machine forms the basic engineering solution and occurs as a small, medium and large power device. A simplified cross-section of such a machine for the number of pole pairs $p = 1$ is presented in Fig. 4.8.

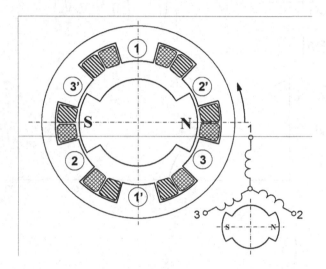

Fig. 4.8 Cross-section and schematic view of a BLDC motor, for $p = 1$

The basic simplifying assumptions applied during the development of a mathematical model include:

- complete symmetry of the machine's construction,
- disregarding of factors affecting demagnetization of permanent magnets during the operation (effects of armature, temperature increase)
- disregarding high order MMF harmonics of armature windings.

The remaining, more detailed assumptions associated with the development of the mathematical model will be presented during the course of its derivation. For such an electromechanical transducer and lack of elements serving for the accumulation of potential energy, the Lagrange's function is equal to kinetic co-energy:

$$L = T = \frac{1}{2}L_{11}(\varphi)i_1^2 + \frac{1}{2}L_{22}(\varphi)i_2^2 + \frac{1}{2}L_{33}(\varphi)i_3^2 +$$
$$+ M_{12}(\varphi)i_1 i_2 + M_{13}(\varphi)i_1 i_3 + M_{23}(\varphi)i_2 i_3 + \quad\quad (4.22)$$
$$+ \Psi_{1r}(\varphi)i_1 + \Psi_{2r}(\varphi)i_2 + \Psi_{3r}(\varphi)i_3 + \frac{1}{2}J\dot{\theta}_r^2$$

which can also take the form of a matrix notation:

$$L = \frac{1}{2}\mathbf{i}_a^T \mathbf{L}_a(\varphi)\mathbf{i}_a + \Psi_{af}^T(\varphi)\mathbf{i}_a + \frac{1}{2}J\dot{\theta}_r^2 \quad\quad (4.23)$$

where:

$$\mathbf{L}_a(\varphi) = \begin{bmatrix} L_{11}(\varphi) & M_{12}(\varphi) & M_{13}(\varphi) \\ M_{21}(\varphi) & L_{22}(\varphi) & M_{23}(\varphi) \\ M_{31}(\varphi) & M_{32}(\varphi) & L_{33}(\varphi) \end{bmatrix} \quad\quad (4.24)$$

- matrix of armature inductance

$$\Psi_{af}(\varphi) = \begin{bmatrix} \Psi_{1f}(\varphi) \\ \Psi_{2f}(\varphi) \\ \Psi_{3f}(\varphi) \end{bmatrix}$$

- vector of the coupling between permanent magnet flux and armature windings,

$$\mathbf{i}_a = \begin{bmatrix} i_1 & i_2 & i_3 \end{bmatrix}^T \quad\quad (4.25)$$

- vector of armature current.

The particular components of the inductance matrix of the armature windings account for the variable reluctance of the rotor and dissipation flux of the armature windings and for the purposes of simplification can be presented as follows:

$$L_{11}(\varphi) = L_a - M_s \cos 2\varphi$$
$$L_{22}(\varphi) = L_a - M_s \cos(2\varphi + a) \quad\quad (4.26)$$
$$L_{33}(\varphi) = L_a - M_s \cos(2\varphi - a)$$

$$M_{12}(\varphi) = M_{21}(\varphi) = -\frac{1}{2}L_m - M_s \cos(2\varphi - a)$$

$$M_{13}(\varphi) = M_{31}(\varphi) = -\frac{1}{2}L_m - M_s \cos(2\varphi + a) \quad\quad (4.27)$$

$$M_{23}(\varphi) = M_{32}(\varphi) = -\frac{1}{2}L_m - M_s \cos 2\varphi$$

$$\Psi_{af}(\varphi) = \Psi_f \begin{bmatrix} \sin\varphi \\ \sin(\varphi - a) \\ \sin(\varphi + a) \end{bmatrix} \tag{4.28}$$

where: $a = 2\pi/3$.

The angle φ accounts for the number of pole pairs in the machine

$$\varphi = p\theta_r \tag{4.29}$$

while the self-inductance of the armature windings is assumed in the form which identifies the leakage inductance:

$$L_a = L_{\sigma s} + L_m \tag{4.30}$$

After accounting for these remarks, the inductance of the armature windings (4.24) can be restated in the following form:

$$\mathbf{L}_a(\varphi) = L_{\sigma s} \begin{bmatrix} 1 & & \\ & 1 & \\ & & 1 \end{bmatrix} + L_m \begin{bmatrix} 1 & -\frac{1}{2} & -\frac{1}{2} \\ -\frac{1}{2} & 1 & -\frac{1}{2} \\ -\frac{1}{2} & -\frac{1}{2} & 1 \end{bmatrix} -$$
$$- M_s \begin{bmatrix} \cos 2\varphi & \cos(2\varphi - a) & \cos(2\varphi + a) \\ \cos(2\varphi - a) & \cos 2\varphi & \cos 2\varphi \\ \cos(2\varphi + a) & \cos 2\varphi & \cos(2\varphi - a) \end{bmatrix} \tag{4.31}$$

The equation of the mechanical motion of the machine can be derived from Lagrange's equation for a variable denoting rotation angle θ_r:

$$\frac{d}{dt}\frac{\partial L}{\partial \dot{\theta}_r} - \frac{\partial L}{\partial \theta_r} = -T_l - D\dot{\theta}_r \tag{4.32}$$

or:

$$J\dot{\Omega}_r = T_e - T_l - D\Omega_r \tag{4.33}$$

where:

$$T_e = \frac{\partial L}{\partial \theta_r} = \frac{1}{2}\mathbf{i}_a^T\left(\frac{\partial}{\partial \theta_r}\mathbf{L}_a(\varphi)\right)\mathbf{i}_a + \mathbf{i}_a^T\frac{\partial}{\partial \theta_r}\Psi_{af}(\varphi) \tag{4.34}$$

$$\varphi = p\theta_r$$

The expression in (4.34) defines the electromagnetic torque of the machine and involves two terms. The first of them denotes the reluctance torque of the machine, which comes as a consequence of the reactions of armature current with the salient poles of the rotor with magnets:

$$T_{er} = pM_s \mathbf{i}_a^T \begin{bmatrix} \sin 2\varphi & \sin(2\varphi - a) & \sin(2\varphi + a) \\ \sin(2\varphi - a) & \sin 2\varphi & \sin 2\varphi \\ \sin(2\varphi + a) & \sin 2\varphi & \sin(2\varphi - a) \end{bmatrix} \mathbf{i}_a \qquad (4.35)$$

Concurrently, the other term of the expression (4.34) denotes the principal torque of the machine resulting from the interaction between armature currents with permanent magnets' excitation flux.

$$T_{ef} = \mathbf{i}_a^T p \Psi_f \begin{bmatrix} \cos\varphi \\ \cos(\varphi - a) \\ \cos(\varphi + a) \end{bmatrix} \qquad (3.36)$$

In addition, BLDC machines have another component of the torque, i.e. cogging torque beside the reluctance related one (4.35). It is present as a result of the reaction of the principal flux with the armature teeth. In the presented model (4.34) it is, however, not encountered since the harmonics associated with the stator slots are disregarded. This omission is admissible since the designers throughout their engineering efforts [3,8,9,18,21,27,31,35], tend to effectively aim at the minimization of this component of the torque.

4.3.1 Transformed Model Type d-q

The structure of the inductance matrix of the armature (4.31) suggests the application of the orthogonal transformation, similar as in the case concerning a three-phase induction machine. In this case we will apply transformation $\mathbf{T}_r$ (3.34) for $\omega_c = 0$.

Thus,

$$\mathbf{T}_{ra} = \sqrt{\frac{2}{3}} \begin{bmatrix} \dfrac{1}{\sqrt{2}} & \dfrac{1}{\sqrt{2}} & \dfrac{1}{\sqrt{2}} \\ \cos\varphi & \cos(\varphi + a) & \cos(\varphi - a) \\ -\sin\varphi & -\sin(\varphi + a) & -\sin(\varphi - a) \end{bmatrix} \qquad (4.37)$$

$$\dot{\mathbf{T}}_{ra} = \dot{\varphi} \frac{\partial \mathbf{T}_{ra}}{\partial \varphi} = p\dot{\theta}_r \mathbf{A}_3 \mathbf{T}_{ra}$$

The transformation of equations for the electric circuits of the armature will be conducted in the general form derived from of Lagrange's equations for electric variables:

$$\frac{d}{dt} \frac{\partial L}{\partial \mathbf{i}_a} - \frac{\partial L}{\partial \mathbf{Q}_a} = \mathbf{U}_a - R_a \mathbf{i}_a \qquad (4.38)$$

which in the consideration of $\dfrac{\partial L}{\partial \mathbf{Q}_a} = \mathbf{0}$ offers the following matrix equation:

$$\frac{d}{dt}\left(\mathbf{L}_a(\varphi)\mathbf{i}_a\right)+\frac{d}{dt}\Psi_{af}(\varphi)=\mathbf{U}_a-R_a\mathbf{i}_a \tag{4.39}$$

and, subsequently,

$$\mathbf{L}_a(\varphi)\frac{d\mathbf{i}_a}{dt}+\dot{\varphi}\frac{\partial\mathbf{L}_a(\varphi)}{\partial\varphi}\mathbf{i}_a+\dot{\varphi}\Psi_f\mathbf{C}(\varphi)=\mathbf{U}_a-R_a\mathbf{i}_a \tag{4.40}$$

By multiplication of the above by $\mathbf{T}_{ra}$ and by application of the orthogonality condition of the transformation matrix (4.37) we can note that:

$$\underbrace{\mathbf{T}_{ra}\mathbf{L}_a(\varphi)\mathbf{T}_{ra}^T}_{\mathbf{L}^*}\mathbf{T}_{ra}\frac{d\mathbf{i}_a}{dt}+\dot{\varphi}\underbrace{\mathbf{T}_{ra}\frac{\partial\mathbf{L}_a(\varphi)}{\partial\varphi}\mathbf{T}_{ra}^T}_{\mathbf{L}_\varphi^*}\underbrace{\mathbf{T}_{ra}\mathbf{i}_a}_{\mathbf{i}_a^*}+$$

$$+\dot{\varphi}\Psi_f\,\mathbf{T}_{ra}\mathbf{C}(\varphi)=\underbrace{\mathbf{T}_{ra}\mathbf{U}_a}_{\mathbf{U}_a^*}-\underbrace{\mathbf{T}_{ra}\mathbf{i}_a}_{\mathbf{i}_a^*} \tag{4.41}$$

The expression $\mathbf{T}_{ra}\dfrac{d\mathbf{i}_a}{dt}$ is transformed in the following manner:

$$\mathbf{T}_{ra}\frac{d\mathbf{i}_a}{dt}=\frac{d}{dt}\left(\mathbf{T}_{ra}\mathbf{i}_a\right)-\left(\frac{d}{dt}\mathbf{T}_{ra}\right)\mathbf{i}_a=\frac{d}{dt}\mathbf{i}_a^*-\mathbf{A}_3\mathbf{i}_a^* \tag{4.42}$$

where $\mathbf{A}_3$ is a skew-symmetric matrix in the form:

$$\mathbf{A}_3=\begin{bmatrix}0 & 0 & 0\\ 0 & 0 & 1\\ 0 & -1 & 0\end{bmatrix} \tag{4.43}$$

Following the transformation, the matrix equation (4.41) takes the form:

$$\mathbf{L}^*\frac{d\mathbf{i}_a^*}{dt}+\dot{\varphi}\left(\mathbf{L}_\varphi^*-\mathbf{L}^*\mathbf{A}_3\right)\mathbf{i}_a^*+\dot{\varphi}\Psi_f\mathbf{C}^*=\mathbf{U}_a^*-R_a\mathbf{i}_a^* \tag{4.44}$$

The components of the equation (4.44) are transformed as follows:

$$\mathbf{i}_a^*=\mathbf{T}_{ra}\mathbf{i}_a=\begin{bmatrix}0\\ i_q\\ i_d\end{bmatrix}\qquad\qquad\mathbf{U}_a^*=\mathbf{T}_{ra}\mathbf{U}_a=\begin{bmatrix}0\\ u_q\\ u_d\end{bmatrix} \tag{4.45}$$

$$\mathbf{L}^* = \mathbf{T}_{ra}\mathbf{L}_a(\varphi)\mathbf{T}_{ra}^T = \begin{bmatrix} L_{\sigma s} & & \\ & L_{\sigma s} + \frac{3}{2}L_m & \\ & & L_{\sigma s} + \frac{3}{2}L_m \end{bmatrix} -$$

$$- \begin{bmatrix} 0 & & \\ & \frac{3}{2}M_s & \\ & & -\frac{3}{2}M_s \end{bmatrix} = \begin{bmatrix} L_{\sigma s} & & \\ & L_q & \\ & & L_d \end{bmatrix} \tag{4.46}$$

where:

$$L_q = L_{\sigma s} + \frac{3}{2}(L_m - M_s) \qquad L_d = L_{\sigma s} + \frac{3}{2}(L_m + M_s) \tag{4.47}$$

$$\left(\mathbf{L}_\varphi^* - \mathbf{L}^*\mathbf{A}_3\right) = \begin{bmatrix} 0 & 0 & 0 \\ 0 & 0 & -L_d \\ 0 & L_q & 0 \end{bmatrix} \tag{4.48}$$

$$\mathbf{C}^* = \mathbf{T}_{ra}\mathbf{C}(\varphi) = \begin{bmatrix} 0 & \sqrt{\dfrac{3}{2}} & 0 \end{bmatrix}^T \tag{4.49}$$

As a result, the transformed equations for the armature circuits (4.44) take the form:

$$\begin{bmatrix} L_q & 0 \\ 0 & L_d \end{bmatrix}\begin{bmatrix} \dot{i}_q \\ \dot{i}_d \end{bmatrix} + p\dot{\theta}_r \begin{bmatrix} 0 & -L_d \\ L_q & 0 \end{bmatrix}\begin{bmatrix} i_q \\ i_d \end{bmatrix} + p\dot{\theta}_r\Psi_f\begin{bmatrix} \sqrt{\frac{3}{2}} \\ 0 \end{bmatrix} = \begin{bmatrix} u_q - R_a i_q \\ u_d - R_a i_d \end{bmatrix}$$

or, alternatively,

$$L_q \frac{di_q}{dt} = u_q - R_a i_q + p\dot{\theta}_r(L_d i_d - \sqrt{\tfrac{3}{2}}\Psi_f)$$

$$L_d \frac{di_d}{dt} = u_d - R_a i_d - p\dot{\theta}_r L_q i_q \tag{4.50}$$

The determination of the particular expressions of the transformed voltages u_q, u_d (4.45) is associated with the need to consider the problem of the commutation of armature currents, which occurs in the function of the angle of the rotor position θ_r. This issue will be discussed later. Concurrently, the quadratic form (4.34) which, determines the electromagnetic torque can be transformed in the following manner:

$$T_e = \frac{1}{2}p\mathbf{i}_a^* \begin{bmatrix} 0 & 0 & 0 \\ 0 & 0 & -3M_s \\ 0 & -3M_s & 0 \end{bmatrix}\mathbf{i}_a^* + p\Psi_f\mathbf{i}_a^*\mathbf{C}^*$$

which offers the following result:

$$T_e = pi_q(-3M_s i_d + \sqrt{\tfrac{3}{2}}\Psi_f) \tag{4.51}$$

The expression (4.51) which determines the electromagnetic torque of the motor, after the transformation, takes an uncomplicated form: the first term denoting the reluctance torque is relative to the product of axial currents i_d, i_q and is proportional to inductance M_s associated with the basic harmonic of the reluctance of the air gap, while the other term denotes the principal torque proportional to the product of the magnetic flux Ψ_f and current i_q in the transverse axis of the machine. There is a complete analogy here to the commutator DC machine.

4.3.2 Untransformed Model of BLDC Machine with Electronic Commutation

The application of the model that does not involve the transformation of the coordinate system has a number of advantages. For the case of a motor with electronic commutation there is a possibility of a more realistic modeling of commutation and, thus, gaining results more precisely, including the electric variables over time. The commutation as well as the parameters of the switching transistors can be taken into consideration more precisely in a manner that is required for a specific problem of drive control. Secondly, for the lack of transformation, the modeling of the machine and drive itself can account for a number of asymmetries and differences in terms of parameters, which renders it possible to simulate the emergency states of the drive. In an untransformed model we consider that the armature windings are connected in a star (Fig. 4.9), which take the form of adequate constraint equations. Here we will apply the matrices of constraints $\mathbf{W}_{ir}$ and $\mathbf{W}_{ur}$ (3.78) and (3.79) for the respective currents and phase voltages of the motor

$$\begin{bmatrix} i_1 \\ i_2 \\ i_3 \end{bmatrix} = \underbrace{\begin{bmatrix} 1 & 0 \\ -1 & -1 \\ 0 & 1 \end{bmatrix}}_{\mathbf{W}_{ir}} \begin{bmatrix} i_1 \\ i_3 \end{bmatrix}$$

i.e. $\mathbf{i}_a = \mathbf{W}_{ir}\mathbf{i}_{a13}$ \hfill (4.52)

$$\begin{bmatrix} u_{12} \\ u_{32} \end{bmatrix} = \underbrace{\begin{bmatrix} 1 & -1 & 0 \\ 0 & -1 & 1 \end{bmatrix}}_{\mathbf{W}_{ur}} \begin{bmatrix} u_1 \\ u_2 \\ u_3 \end{bmatrix}$$

i.e. $\mathbf{U}_{13} = \mathbf{W}_{ur}\mathbf{U}_a$ \hfill (4.53)

As a result of the multiplication of the left-hand side of the equation in (4.40) by the matrix of constraints $\mathbf{W}_{ur}$ and introducing the vector of armature currents

$$\mathbf{i}_{a13} = \begin{bmatrix} i_1 & i_3 \end{bmatrix}^T \tag{4.54}$$

we obtain:

$$\underbrace{\mathbf{W}_{ur}\mathbf{L}_a(\varphi)\mathbf{W}_{ir}}_{\mathbf{L}_{13}(\varphi)}\frac{d\mathbf{i}_{a13}}{dt} + \dot{\varphi}\underbrace{\mathbf{W}_{ur}\frac{\partial\mathbf{L}_a}{\partial\varphi}(\varphi)\mathbf{W}_{ir}\,\mathbf{i}_{a13}}_{\mathbf{L}_{13}^\varphi(\varphi)} +$$

$$+\dot{\varphi}\Psi_f\underbrace{\mathbf{W}_{ur}\mathbf{C}(\varphi)}_{\mathbf{C}_{13}(\varphi)} = \underbrace{\mathbf{W}_{ur}\mathbf{U}_a}_{\mathbf{U}_{13}} - \underbrace{\mathbf{W}_{ur}\mathbf{R}_a\mathbf{W}_{ir}\,\mathbf{i}_{a13}}_{\mathbf{R}_{13}} \tag{4.55}$$

$$\mathbf{L}_{13}(\varphi) = \begin{bmatrix} 2L_s - 3M_s\cos(2\varphi + \pi/3) & L_s + 3M_s\cos(2\varphi - \pi/3) \\ L_s + 3M_s\cos(2\varphi - \pi/3) & 2L_s + 3M_s\cos 2\varphi \end{bmatrix} \tag{4.56}$$

$$\mathbf{L}_{13}^\varphi = 6M_s\begin{bmatrix} \cos(2\varphi + \pi/6) & \cos(2\varphi + \pi/6) \\ \cos(2\varphi + \pi/6) & -\sin 2\varphi \end{bmatrix} \tag{4.57}$$

$$\mathbf{C}_{13}(\varphi) = \sqrt{3}\begin{bmatrix} \cos(\varphi + \pi/6) \\ -\sin\varphi \end{bmatrix} \tag{4.58}$$

$$\mathbf{R}_{13} = R_a\begin{bmatrix} 2 & 1 \\ 1 & 2 \end{bmatrix} \tag{4.59}$$

After the transformation, the equations for the variables of the armature currents i_1, i_3 (4.54) take the following form:

$$\mathbf{L}_{13}(\varphi)\frac{d\mathbf{i}_{a13}}{dt} + p\dot{\theta}_r\left(\mathbf{L}_{13}^\varphi\mathbf{i}_{a13} + \Psi_f\mathbf{C}_{13}(\varphi)\right) = \mathbf{U}_{13} - \mathbf{R}_{13}\mathbf{i}_{a13} \tag{4.60}$$

At this point it is possible to pass on to the transformation of the quadratic form of electromagnetic torque (4.34) by introduction of constraints (4.52) and (4.53) for the connection of the armature windings into a star:

$$T_e = \frac{1}{2}\mathbf{i}_{a13}^T\mathbf{W}_{ir}^T\left(\frac{\partial}{\partial\theta_r}\mathbf{L}_a(\varphi)\right)\mathbf{W}_{ir}\mathbf{i}_{a13} + \mathbf{i}_{a13}^T\mathbf{W}_{ir}^T\frac{\partial}{\partial\theta_r}\Psi_{af}(\varphi) \tag{4.61}$$

Since $\mathbf{W}_{ir}^T = \mathbf{W}_{ur}$, the expression takes the form

$$\mathbf{W}_{ir}^T\frac{\partial\mathbf{L}_a}{\partial\theta_r}(\varphi)\mathbf{W}_{ir} = \mathbf{W}_{ur}\frac{\partial\mathbf{L}_a}{\partial\theta_r}(\varphi)\mathbf{W}_{ir} = p\mathbf{L}_{13}^\varphi(\varphi) \tag{4.62}$$

and the other term is transformed into:

$$\mathbf{i}_{a13}^T\mathbf{W}_{ir}^T\frac{\partial}{\partial\theta_r}\Psi_{af}(\varphi) = \mathbf{i}_{a13}^T\mathbf{W}_{ur}\Psi_f\mathbf{C}(\varphi) = \Psi_f\mathbf{i}_{a13}^T\mathbf{C}_{13}(\varphi) \tag{4.63}$$

Hence,

$$T_e = p\mathbf{i}_{a13}^T\left(\tfrac{1}{2}\mathbf{L}_{13}^\varphi(\varphi)\mathbf{i}_{a13} + \Psi_f \mathbf{C}_{13}(\varphi)\right) \tag{4.64}$$

After adequate algebraic operations we obtain:

$$T_e = 3pM_s\left(i_1^2\cos(2\varphi-\pi/6) - i_3^2\sin 2\varphi + i_1 i_3\cos(2\varphi+\pi/6)\right) + \\ + \sqrt{3}p\Psi_f\left(i_1\cos(\varphi+\pi/6) - i_3\sin\varphi\right) \tag{4.65}$$

The expression (4.65) presents electromagnetic torque of BLDC machine in the natural coordinates i_1, i_3 without transformation, for the connection of three-phase armature windings in a star.

4.3.3 Electronic Commutation of BLDC Motors

Commutation in a brushless DC machine involves the switching of the armature current to particular phase windings depending on the position of the rotor angle θ_r. In a traditional brushless DC machine this occurred as a result of application of a mechanical commutator consisting of isolated copper segments with armature windings connected to them. Over this commutator the graphite brushes would slip thus receiving the current while the position of the brushes was fixed in space. In such a manner the commutation occurred naturally depending on the position of the rotor. The electronic commutation is ensured by the converting bridge while the switching of the current between the windings also occurs in the function of the rotor's position angle, and the signal responsible for the control of the switchings is obtained from the position sensor measuring the angle of rotation θ_r. As a principle, such sensors are optical, including encoders and induction based ones, i.e., resolvers.

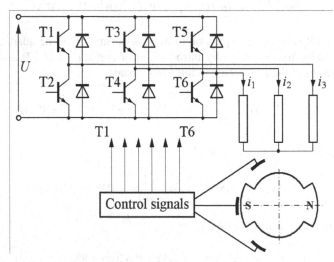

Fig. 4.9 Basic scheme of a bipolar 3-phase BLDC motor supply

In any case, however, the commutation angle needs to be set at an appropriate value, which in a traditional DC machine was the role of the correct positioning of the brushes in a commutator. Fig. 4.9 presents the standard transistor bridge fulfilling the role of an electronic commutator for a motor with three phase windings in the armature and bipolar supply of the windings connected in a star. The bipolar supply means that in the armature windings the current flows in both directions, i.e. the current flowing through windings is AC.

The angular scheme of the commutation of BLDC motor for a positive direction of rotor motion is presented in Fig. 4.10.

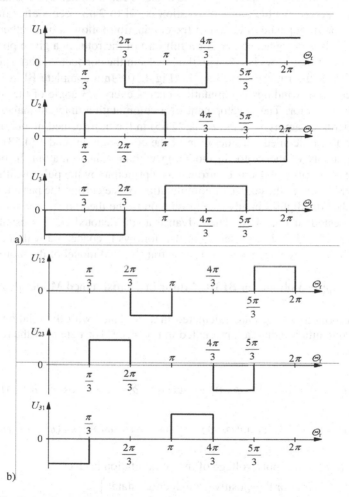

Fig. 4.10 Typical scheme of current commutation in BLDC motor's 3- phase armature in relation to rotor position angle: a) conducting of individual phase windings b) conducting of pairs of star-connected windings

A typical commutation diagram for three-phase windings connected in a star involves simultaneous conduction of two phase belts while the third one remains in OFF state. The simultaneous conduction of all three belts occurs only in very short commutation periods when we have to do with the transfer of the conduction from the belt that is about to terminate the operation to another phase belt, which in accordance with the commutation diagram takes the turn in starting commutation. A singular phase belt in a three-phase BLDC machine conducts over the period corresponding to the angle of rotation, i.e. $2\pi/3$ and subsequently takes a break over the time corresponding to the angle of rotation, i.e. $\pi/3$. The subsequent conduction period for the angle of rotation equal to $2\pi/3$ occurs after this break; however, for an opposite direction of the conduction followed by another break in conduction. It is designed so that for a full turn of the rotor in a given phase winding the current that flows is AC with the breaks in the conduction corresponding to the rotation of the rotor over $\pi/3$ angle (Fig 4.10). In a complete BLDC machine with three phase windings commutation occurs every $\pi/3$ angle of the rotation of the machine's rotor. The development of a commutation diagram makes it possible to determine supply voltages u_d, u_q (4.45) in the transformed model of the motor and perform detailed consideration of the commutation model (4.53) to be applied for supply of the motor in modeling without transformation. In both cases the value of supply voltage is controlled as a principle by the pulse width modulation (PWM). Due to the course of commutation of the current the particular phases are switched on slightly in advance in relation to the theoretical commutation diagram presented in Fig. 4.10. This advance angle denoted as δ is usually in the range from 25° - 35°. The presentation that follows is concerned with the determination of the supply voltages u_d, u_q for the transformed model of the motor.

4.3.3.1 Supply Voltages of BLDC Motor in Transformed Model u_q, u_d

The transformed voltages are calculated in accordance with the relation in (4.45) and the commutation diagram presented in Fig. 4.10. The details of the relation are as follows:

$$u_q = \frac{\sqrt{2}}{3} U k_u \big(c_1 \cos(\varphi - \delta) + c_2 \cos(\varphi - \delta - a) + c_3 \cos(\varphi - \delta + a) \big)$$

$$u_d = -\frac{\sqrt{2}}{3} U k_u \big(c_1 \sin(\varphi - \delta) + c_2 \sin(\varphi - \delta - a) + c_3 \sin(\varphi - \delta + a) \big)$$

(4.66)

where: U - supply voltage of the commutation bridge

$$c_1, c_2, c_3 = \begin{cases} 1 & \text{for the positive conduction state} \\ -1 & \text{for the reverse conduction state} \\ 0 & \text{for the OFF state} \end{cases} \text{- conduction factors}$$

$$k_u = \frac{t_p}{T_{PWM}} \quad \text{- pulse width factor (PWM control)}$$

$\varphi = p\theta_r$ - electrical rotation angle (4.29)

t_p - conduction time within a pulse period T_{PWM}

$t_z = T_{PWM} - t_p$ - recuperation time within a pulse period T_{PWM}

δ - advance angle

$a = 2\pi/3$

Over the period of T_{PWM} for the duration of the supply t_p respective switches are in the ON state and the voltage $U_s = U$ is fed to the windings. Concurrently, when we have to do with control without energy recuperation, the closure of the phase takes place and the current flows through the return diode and one of the transistors of the bridge, and $U_s = 0$. For the control with energy recuperation all transistors are in the OFF state and the energy is returned to the source through the two of the return diodes for the voltage of the motor $U_s = -U$. This occurs in the section of the control period t_z. The above description of a single pulse with the period of T_{PWM} offers an explanation to the issue of calculation of output voltage of the commutation bridge (4.67) for both types of bridge control. The coefficient k_u makes it possible to calculate the mean values of the voltages u_d, u_q. These means are determined on the basis of formulae in (4.66) while the functions of the conduction factors c_1, c_2, c_3 are determined according to commutation scheme in Fig. 4.10:

$$u_{q\,av} = \frac{3\sqrt{2}}{\pi} U_{ph} k_u \sin(\delta + \pi/2)$$

$$u_{d\,av} = \frac{3\sqrt{2}}{\pi} U_{ph} k_u \cos(\delta + \pi/2) \tag{4.67}$$

The examples of the waveforms for u_d, u_q are presented in Figs. 4.11 and 4.12. In both cases we apply an angle depending PWM coefficient k_u calculated from the relation:

$$k_u = k_{u1}(1 - \exp(-\varphi/T_u \pi)) \tag{4.68}$$

where:

     T_u - is the angular constant of voltage increment.

The exponential character of voltage increase u_d, u_q offers the possibility of the smooth motor start-up.

4.3.3.2 Modeling of Commutation in an Untransformed Model of BLDC

The modeling of commutation in an untransformed system for a three-phase windings of the armature can have a various degree of detail. In this chapter we will present a method that is considerably simplified and, subsequently, apply it in

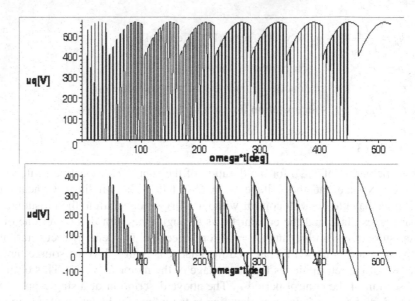

Fig. 4.11 Voltages u_q and u_d as a function of rotation angle for: $\delta = 0$, $Tp = 7.2°$, $U = 400$ [V], $k_{ul} = 1$, $T_u = 0.9$ [s]

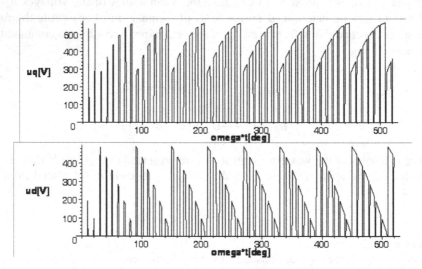

Fig. 4.12 Voltages u_q and u_d as a function of rotation angle for $\delta = 30°$, $T_p = 10°$, $u = 400$ [V], $k_{ul} = 1$, $Tu = 1.8$ [s]

examples. It takes into consideration the fact that adequate supply voltage is connected to the particular pairs of phase windings connected in a star, i.e. winding no. 1-2, 2-3, 3-1 via a commutation bridge. During the commutation we will distinguish two states: first, when the commutation begins during the connection of the source voltage to the windings, i.e. during the active part t_p of the supply pulse

T_p and the other state, when the commutation begins during the passive part of the pulse while the energy is returned from the windings to the source or during the closure of the winding. In both these states it is possible to adequately model resistances and voltages occurring in the particular electric circuits in the given state, i.e. the resistances of the motor windings, electronic switches as well as the blocking resistance R_b in the circuit of unsupplied phase winding. It will be illustrated by appropriate examples.

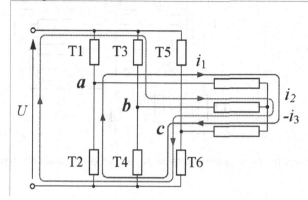

Fig. 4.13 Currents flow during commutation $+i_1 \rightarrow +i_2$ for $\varphi = \pi / 3$, and an active part t_p of the period T_p

We will consider the commutation occurring in the active part of the pulse t_p, which takes place for the angle of rotation $\varphi = \pi / 3$, Fig. 4.10, where we have to do with the switching of the current from $+i_1$ to current $+i_2$, i.e. the termination of the conduction in the positive direction in the winding in phase 1 and commencement of the conduction in the positive direction by the winding in phase 2. During that time in the remaining winding of phase 3 the current flows continuously in the conventional negative direction. This situation is illustrated in Fig. 4.13. The target circuit after the commutation supplied with voltage U is marked through transistors T3, T6 ($+i_2$, $-i_3$), while the decaying current in the winding of phase 1 is closed in the circuit with the return diode D2 that is antiparallel to transistor T2, since the transistor T1 has just been closed, and conducting transistor T6 connected to the phase winding 3. In the state presented in the figure the potential of point a amounts to 0, potential of point b is U and the potential of point c is 0. Hence, the voltages $U_{12} = -U$, $U_{32} = -U$. After the commutation, i.e. after the current $+i_1 \rightarrow 0$ has decayed, the potential of the point a will change to ½ U and the respective voltages will be $U_{12} = -½ U$, $U_{32} = -U$, while the resistance $R_1 = R_b$, which means that it will assume the value of the blocking resistance. For a better illustration of the considerations we will additionally examine the commutation for the angle of rotation $\varphi = \pi$. In this case (see Fig. 4.14) the commutation involves a change $+i_2 \rightarrow +i_3$, while $-i_1$ is continuing its flow. Decaying current $+i_2$ is closed across the return diode D4 of the transistor T4 and transistor T2. The potentials of the particular points *a,b,c* associated with the beginning of the phase windings in this state amount to: 0, 0, *U*. As a consequence, line-to-line voltages supplying the windings are as follows: $U_{12} = 0$, $U_{32} = U$.

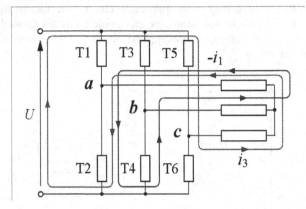

Fig. 4.14 Currents flow during commutation $+i_2 \rightarrow i_3$ for $\varphi = \pi$, and an active part t_p of the period T_p

Concurrently, after the termination of the commutation the potential of the point *b* changes to ½*U*, which results in the following values of line-to-line voltages supplying the windings $U_{12} = -\frac{1}{2}U$ and $U_{32} = \frac{1}{2}U$. The situation during the commutation over the passive section of pulse T_p can be presented for two various alternatives of control of the commutation bridge, i.e. for the case when during the commutation the energy returns to the source and the opposite one when the energy is not returned to the voltage source *U*. At the beginning, we will consider the first of the cases, when in the passive section of the pulse the energy is recuperated. For this case we will consider commutation $+i_1 \rightarrow i_2$, ($\varphi = \pi/3$), i.e. the same as in Fig. 4.13, but for the passive section of the PWM pulse. During this commutation transistor T1 is just switched off terminating supply to the phase winding 1, transistor 3 is not switched on because of the passive period and transistor T6 is switched off to facilitate recuperation of energy. The decaying current flows through diodes D2 and D5 against the voltage of the source. In this state the potentials of points *a,b,c* are respectively equal to 0, ½*U*, *U* and consequently $U_{12} = -\frac{1}{2}U$, $U_{32} = \frac{1}{2}U$. After the commutation is finished potentials of all three *a,b,c* points are the same and equal to ½*U* and inter-phase voltages are $U_{12} = U_{32} = 0$. This state is presented in Fig. 4.15.

Concurrently, for the other version of commutation without energy return to the source during the passive part of the period T_p, the transistor T3 is not switched on, while transistor T6 is switched on - continuing conduction, and the decaying current of windings 1 and 3 flows through T6 and D2 in a shorted circuit. At that state the potentials of *a,b,c* points are respectively 0, ½*U*, 0 and inter-phase voltages are $U_{12} = U_{32} = -\frac{1}{2}U$. After the commutation there is no current and like in

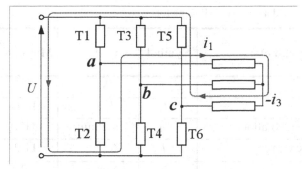

Fig. 4.15 Currents flow during commutation $+i_1 \rightarrow +i_2$ for $\varphi = \pi / 3$ in a passive part of the period T_p, with energy recovery

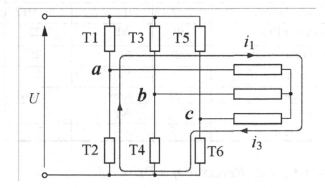

Fig. 4.16 Currents flow during commutation $+i_1 \rightarrow +i_2$ for $\varphi = \pi / 3$, in a passive part of the period T_p, without energy recovery

the previous case all three clam potentials are equal to $\frac{1}{2}U$ and in consequence $U_{12} = U_{32} = 0$. An illustration of this is found in Fig. 4.16.

4.4 Characteristics of BLDC Machine Drives

The presentation in this chapter will be devoted to the computer simulations of characteristics of brushless DC drives. Such issues include: start-up, braking and drive reversal, control of rotational speed and tracking control of the drive as well as its reaction to variable parameters of the supply and loading. Since these detailed issues can be illustrated with the aid of adequately selected results of dynamic calculations, it is important to select motors for the demonstration of the drive in operation beside the presentation of topics devoted to mathematical modeling of the drive. For theses purposes the parameters of two BLDC motors are presented: one with a smaller and the other with the higher power and different supply voltages. A summary of the parameters is found in Table 4.3.

Table 4.3 Rated data of two PMDC motors

Rated Parameters	Symbol	Unit	Motor A	Motor B
Power	P_n	kW	0.95	6.6
Voltage	U_n	V	120	400
Velocity of rotation	n_n	rev/min	3100	2600
Armature current	I_{tn}	A	11.9	23.4
Torque	T_n	Nm	3.0	25.0
Efficiency	η_n	%	90.4	92.6
Windings self-inductance	L_s	H	0.016	0.030
Mutual inductance	M_s	H	0.0012	0.003
PM excitation flux	Ψ_{fn}	Wb	0.22	1.0
Windings resistance	R_s	Ω	0.25	0.50
Moment of inertia	J_s	Nms^2	0.018	0.15
Damping factor	D	Nms	0.0002	0.002
Pulse width	T_{PWM}	deg	2	2
Commut. advance angle	δ	deg	35	30

4.4.1 Start-Up and Reversal of a Drive

4.4.1.1 Drive Start-Up

Start-up forms the basic issue associated with the motion of a drive and, hence, the motor drive and the control system have to fulfill a number of prerequisites in order to ensure the appropriate course of the process. These prerequisites include: possibility of start-up from every initial position, start-up with a required load as well as limitation of the start-up current to the values acceptable by the motor and the supply system. The process of start-up of BLDC motor is further impeded as a result of occurrence of parasitic torques, i.e. reluctance torque and cogging torque. The two effects are reduced during the process of motor design in a manner that they are not manifested too strongly during the start-up. The limitation of the start-up current can be achieved in two ways: by incremental increase of the voltage supplying the armature using PWM method or as a result of controlling the start-up current by means of PWM method as well, relative to the instantaneous value of the current. A smooth increase of the voltage during the start-up can be achieved in numerous manners. In simulation models applied for the demonstration of the start-up curves it is achieved by exponential increase of k_u coefficient (4.68), i.e. the one denoting the active part of the pulse. The figures that follow illustrate the start-up curves for a motor with higher power, i.e. motor marked B in Table 4.3 as well as for a smaller motor A under a rated load. The comparison will involve the start-up curves for the d-q transformed model as well as for the untransformed motor.

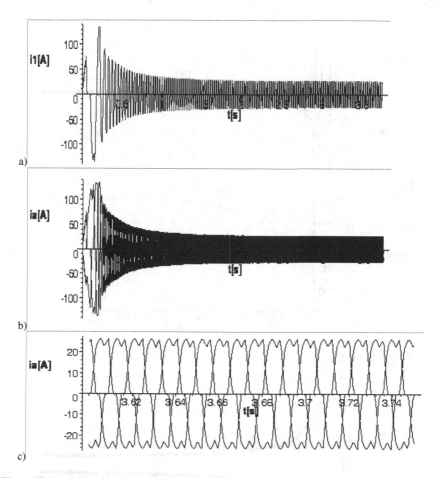

a)

b)

c)

Fig. 4.17 Starting of the B (6.6 [kW]) motor with a voltage regulation: a) armature phase current b) 3-phase currents c) currents in the steady state. The results obtained from the untransformed model of BLDC

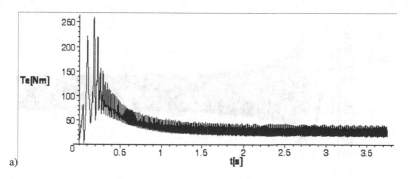

a)

Fig. 4.18 The same course as in Fig 4.17 for: a) electromagnetic torque b) reluctance torque c) rotational speed

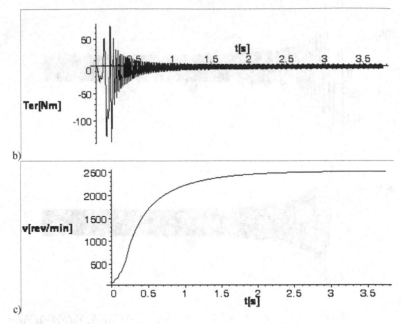

b)

c)

Fig. 4.18 (*continued*)

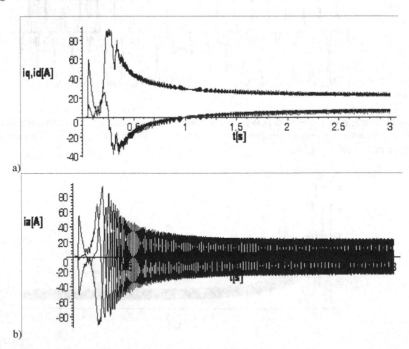

a)

b)

Fig. 4.19 Starting of the B (6.6 [kW]) motor. Results obtained by the transformed model of BLDC: a) transformed i_d, i_q currents, b) transformed back armature currents i_a

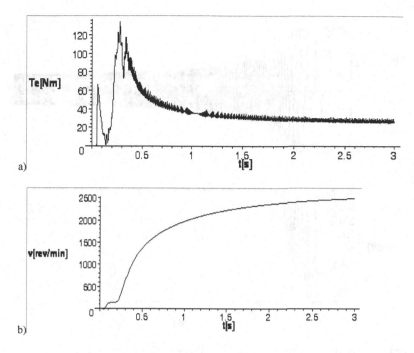

Fig. 4.20 Starting of the B (6.6 [kW]) motor. Results obtained by the transformed model of BLDC: a) electromagnetic torque b) rotor's speed

The two figures that follow, i.e. Figs. 4.21 and 4.22 present the start-up of the same motor, however, for the application of a current delimiter. The operating principle of the device involve the division or multiplication of the pulse width coefficient k_u by a reduction coefficient *red* in the subsequent pulses (4.69) depending on whether the value of the current in any of the phases exceeds or does not reach the value of the imposed limitation I_r.

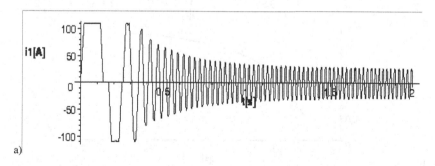

Fig. 4.21 Starting of the B (6.6 [kW]) motor with a current delimiter set for I_r = 110 [A]: a) single phase current b) 3-phase currents c) $U \cdot k_u$ armature voltage

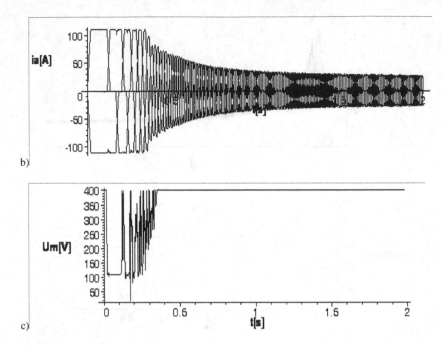

b)

c)

Fig. 4.21 (*continued*)

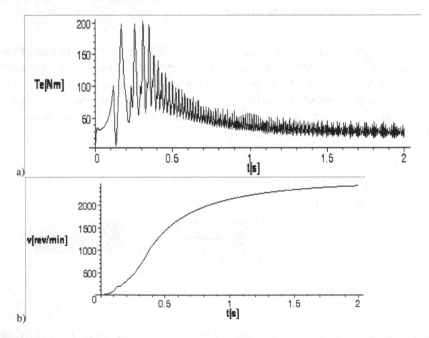

a)

b)

Fig. 4.22 The same course as in Fig 4.21 but for: a) electromagnetic torque b) rotational speed

The illustrations that follow, i.e. Fig. 4.23 and Fig. 4.24 present the curves of the start-up of the smaller motor (motor A, Table 4.3) gained as a result of applying untransformed and transformed models of the BLDC motor. As one can conclude, the two curves are very similar, in particular with regard to the mapping of electromechanical variables. Considerable differences are noted in terms of the current curves since in the transformed model the commutation is not as precisely modeled. In conclusion, in terms of the quality of the modeling untransformed model is a better one, while the basic advantage of the transformed model involves the 10 to 20 times decreased cost of simulations. For these reasons the transformed model of BLDC motor presents more advantages during simulations of large electromechanical systems in which a greater number of drives is present. Concurrently, the reduction of the duration of the calculations forms a considerable premise in favor of the execution of the simulations of the operation of the system.

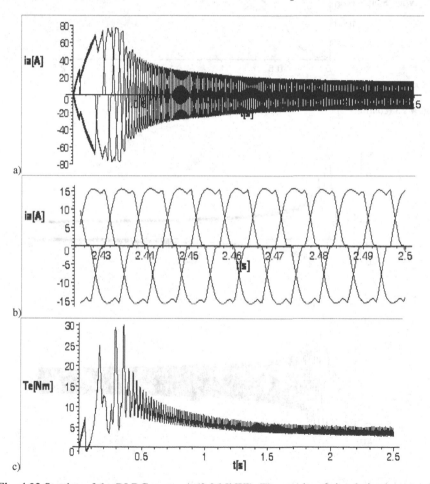

Fig. 4.23 Starting of the BLDC motor A (0.96 [kW]). The results of simulation by untransformed model: a) armature currents b) shape of current curves c) electromagnetic torque d) reluctance torque e) rotational speed

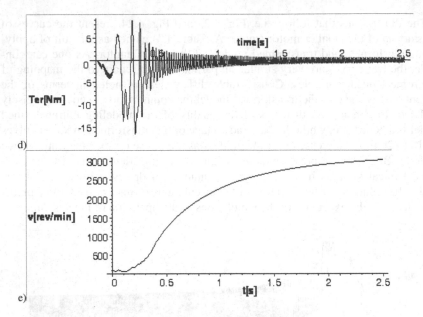

Fig. 4.23 (*continued*)

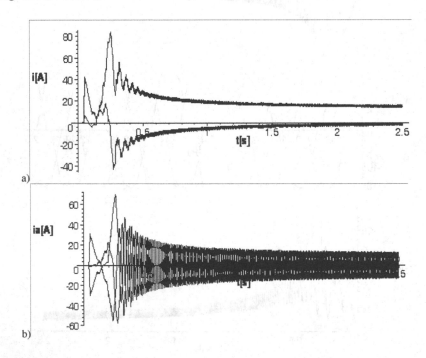

Fig. 4.24 Starting of the BLDC motor A (0.96 [kW]). The results of simulation by transformed model: a) *d-q* currents b) armature currents c) shape of armature currents d) electromagnetic torque e) rotational speed

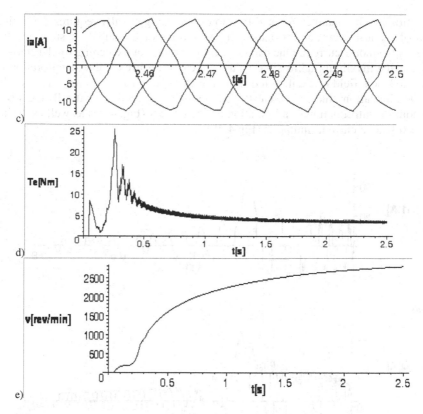

Fig. 4.24 (*continued*)

4.4.1.2 Reversing DC Motor

Brushless DC machine, just as the commutator machine based one can serve very well for operation in both directions of rotation. However, it is necessary that two fundamental conditions are met: the system of the supply and control has to be prepared for such circumstances and a particular construction of the cooling system or an adequate air-flow across the machine has to be provided. A separate issue is associated with the operation of the drive at small rotational speed and ensuring that heat is carried away in such conditions, thus, that the temperature inside the machine does not exceed the permitted limit. This may be associated with an application of a machine with independently driven fans. In order to perform the start-up of a BLDC motor in the reverse direction, it is necessary to change the sequence of the supply of the motor phases and reverse the value of the delay angle δ. A similar course of action is assumed for the case when one intends to perform reversing of a motor during its operation. The switching on of the reverse direction of rotation in accordance with the preceding description first results in a period of deceleration, named counter-current braking. After that, when the drive reaches zero speed of rotation the drive starts the operation in the reverse direction. Such a manner adopted for motor reversal generally requires the limitation of the armature current due to the conditions of the supply system and the admissible motor current. One can note at this point that exceeding by the start-up current the maximum admissible value can cause an irreversible deterioration of the machine's field of excitation originating from permanent magnets. The results of the simulation studies present the curves of the reversing of the BLDC drive without the introduction of a limitation on the current (Fig. 4.25) as well as during a considerable current limitation (Fig 4.26).

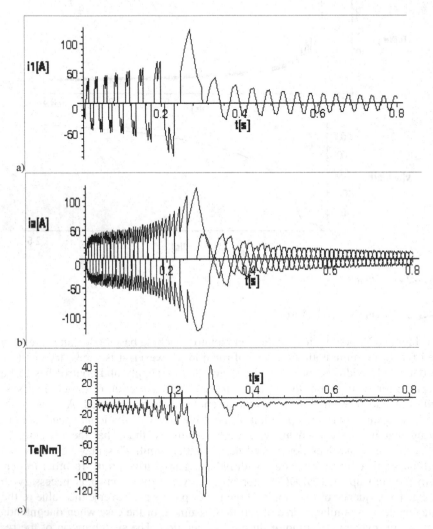

Fig. 4.25 Reversing of PMDC motor (A) without current limitation: a) single phase current b) 3-phase armature currents c) electromagnetic torque d) reluctance component of the torque e) rotational speed

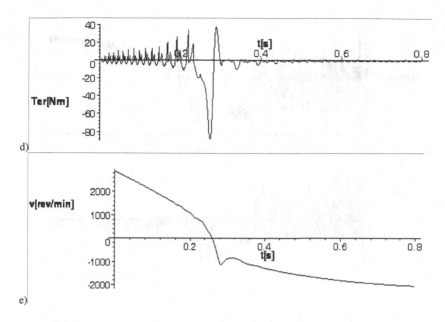

d)

e)

Fig. 4.25 (*continued*)

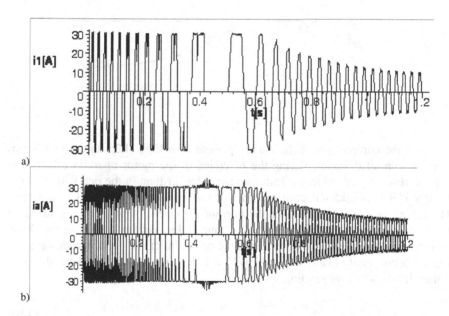

a)

b)

Fig. 4.26 Reversing of PMDC motor (A) with current delimiter set on 30 [A]: a) single phase current b) 3-phase armature currents c) electromagnetic torque d) reluctance component of the torque e) angular speed

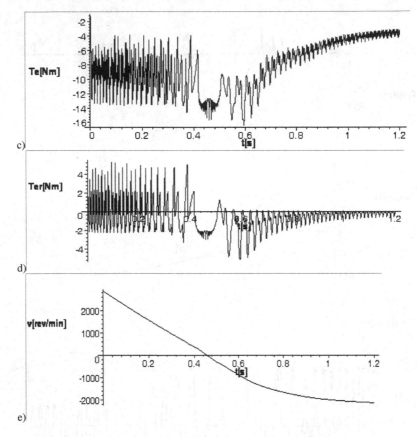

c)

d)

e)

Fig. 4.26 (*continued*)

From the comparison of the curves presented in Figs. 4.25 and 4.26 it stems that the current delimiter during the reversing of the motor operates effectively; but the reversing of the motor lasts two times longer than in the one without a delimiter if it is considered until the time of the transfer of the speed across zero. However, complete control over the current is present, which ensures safety of permanent magnets and the electronic commutation system supplying the drive. Besides, one can observe that the effective operation of the delimiter, designed as the a fraction multiplier *red* of the PWM coefficient, considerably depends on its value. It follows the algorithm:

$$
\begin{aligned}
i_a > I_r &\quad \rightarrow \quad k_u(n+1) = k_u(n) * red \\
i_a \leq I_r &\quad \rightarrow \quad k_u(n+1) = k_u(n) / red
\end{aligned}
\tag{4.69}
$$

For the curves presented in Fig. 4.26, the value of his factor is set at *red* = 0.3.

4.4.2 Characteristics of BLDC Machine Drive

The term drive characteristics denotes the graphical representation of a set of points representing the operation of a drive relative to the selected parameters characterizing its operation. The parameter considered as the independent variable is found on the X-axis, while the Y-axis denotes the values of the examined variable considered as the output variable. Often the same figure contains a family of the characteristics for which the particular curve differ in terms of another parameter that is very important for the presentation of the operation of the drive, i.e. one for whose course its value is constant. If on the X-axis we find a parameter that is not time, then such a characteristic can be termed as the steady-state characteristic (curve). This characteristic forms a set of points for which the dynamic trajectory finds final steady state, if such a stationary state exists at all. The entire static characteristic informs at which point of the operation the drive is currently found after the termination of the dynamic process, i.e. for instance start-up, braking, change of the parameters of the supply or loading for a given parameter on the X-axis. One should note, however, that a change in the state of the operation of the drive (dynamic trajectory) does not overlap with the static characteristic, since if this were the case, the duration of the execution of the designed trajectory would be infinitely long. One can say that the trajectory begins and ends at the static characteristic; however, its curve is different than the one for the characteristic since it occurs in a determined, finite and very often short time. The shorter the time, the further the trajectory is from the static characteristic curve. Another type of characteristic is the one in which X-axis contains time t. In such a case the curve takes the form of a time history for a given variable and generally has a different waveform for other parameters of the drive operation. It is also relative to the initial conditions from which the curve originated. If for such a curve there is a steady state, the steady value of this state forms a component of an corresponding static characteristic. The static characteristics can be derived in a number of ways. For a ready drive we can use a method of measurements for appropriate possibilities of variation of the parameters of the supply and load of the examined drive. If we have a mathematical model of a drive available, static characteristics can be derived by definition by conducting dynamic calculations and performing simulations until the steady state is obtained. Such calculations have to be conducted separately for each point that determines the characteristic. Concurrently, there is a possibility of assuming adequately favorable initial conditions, whose dynamic trajectory leads sufficiently fast to a steady state. Despite that, it is a cost and time-consuming enterprise. Another effective method involves the substitution of hypothetical steady states to the mathematical model in the form of differential equations of motion and converting the model into a system of algebraic equations. This procedure has been followed for instance during the introduction of the equivalent diagram of the induction motor in chapter 3.2.3. If we are capable of effectively gaining such a reduction of the differential model to an algebraic model, as a result we will obtain static characteristics of the drive in the form of functional relations between the parameters and variables. However, one has to bear in mind that not all of the obtained characteristics have to be available as a result of the termination of the dynamic curve or need not be available for arbitrary initial conditions, i.e. from any starting point. From that it stems that, as a principle, the mathematical models of electromechanical systems are non-linear. The static characteristics of the BLDC can be gained from transformed model (4.50) by algebraization after substituting constant functions for its variables:

$$\dot{\theta}_r = \Omega_r = const$$

$$i_q = I_q = const \quad \rightarrow \quad \frac{di_q}{dt} = 0 \qquad\qquad (4.70)$$

$$i_d = I_d = const \quad \rightarrow \quad \frac{di_d}{dt} = 0$$

The transformed voltages $u_{q\,av}$, $u_{d\,av}$ that are present in this model are described by relations (4.67) and account for the relation with commutation advance angle δ. As a result of the substitutions of the fixed variables (4.70) in the mathematical model in the form in (4.50,4.51) we obtain a system of three algebraic equations in the form:

$$u_{qav} = R_a I_q - p\Omega_r \left(L_d I_d - \sqrt{\frac{3}{2}}\Psi_f \right)$$

$$u_{dav} = R_a I_d + p\Omega_r L_q I_q \qquad\qquad (4.71)$$

$$p\left(\sqrt{\frac{3}{2}}\Psi_f I_q - 3M_s I_d I_q \right) - T_l - p\Omega_r D = 0$$

The non-linear system of algebraic equations (4.71) accounts for three variables of the steady state of the drive (I_q, I_d, Ω_r), parameters of the supply $u_{q\,av}$, $u_{d\,av}$ relative to U, δ, parameters of the load T_l, D and engineering parameters of the drive, such as p, L_q, L_d, M_s, R_a, Ψ_f. Such a static model makes it possible to determine the characteristics for selected variables in the subject of the examination. Important examples include mechanical characteristics Ω_r, I_q, I_d, $I_a = f(T_l)$, i.e. characteristics in the function of the load torque for the remaining parameters with constant values, including parameters of the supply. The non-linear system of algebraic equations of the steady state (4.71), that are cubic in relation to variables (I_q, I_d, Ω_r), can be solved effectively using numerical methods, whose applications are widely found in a number of popular mathematical packages. In this case mathematical package MAPLE V was applied in order to gain the further presented characteristics. The voltages $u_{q\,av}$, $u_{d\,av}$ of the transformed model calculated in accordance with (4.67) contain phase voltages U_{ph}, whose value for a typical supply of the BLDC motor are assumed in the form

$$U_{ph} = U / \sqrt{3}, \qquad\qquad (4.72)$$

which forms a simplification by assuming sinusoidal waveforms of the i_1, i_2, i_3 currents in the particular phase windings of the motor. The algebraic model of the motor under of the supply of phase windings in accordance with (4.72) has made it possible to determine the static characteristics in the both researched BLDC motors. For the 6.6 [kW] motor the characteristics in the function of the commutation advance angle δ are presented in Figs. 4.27 and 4.28. These are the respective families of characteristics I_q, I_d, $I_a = f(\delta)$ and Ω_r, $\eta = f(\delta)$, for four values of the load torque $T_l = 15, 25, 35, 45$ [Nm]. More detailed descriptions are found in the captions under the figures.

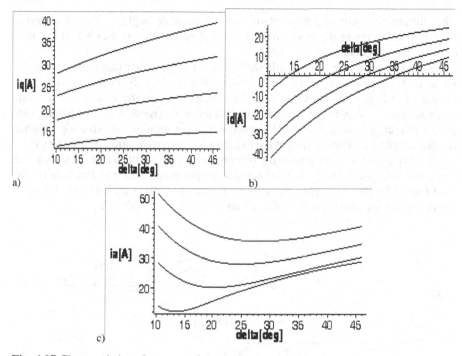

Fig. 4.27 Characteristics of currents of the 6.6 [kW] BLDC motor as a function of advance angle δ: a) I_q transformed current for $T_l = 45, 35, 25, 15$ [Nm], (top –down) b) I_d transformed current for $T_l = 15, 25, 35, 45$ [Nm] (top –down) c) I_a armature current for $T_l = 45, 35, 25, 15$ [Nm], (top –down)

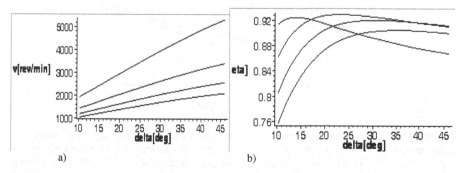

Fig. 4.28 Characteristics of speed and efficiency of the 6.6 [kW] BLDC motor as a function of advance angle δ: a) Ω_r speed [rev/min] for $T_l = 15, 25, 35, 45$ [Nm], (top –down) b) efficiency η for $T_l = 15, 25, 35, 45$ [Nm], (top-down at $\delta = 10°$)

From the presented curves stem a number of conclusions. I_q transformed current does not depend much on advance angle δ, but mainly on load torque T_l, as it decides on electromagnetic torque value at the steady state. According to (4.71), it is equal to:

$$T_e = T_l + Dp\Omega_r \tag{4.73}$$

The current I_d is considerably relative to the advance angle δ since this corresponds to a change in the position of the axis of the brushes in a classical DC machine with a mechanical commutator.

The commutation advance angle should be set in a manner that ensures that the value of I_d is close to zero, i.e. in the range of $\delta = 25°...35°$ in the characteristic presented in Fig. 4.27b. For this case the operation of the drive occurs at a minimum armature current I_a and maximum efficiency η. The selection of higher values in this range makes it possible to ensure the operation of the drive for a higher rotational speed Ω_r (Fig. 4.28a) at the expense of the deterioration of efficiency.

Another group of characteristics presented in Figs. 4.29 and 4.30 shows the same variables as formerly but the results are presented is in the function of the load torque T_l. The other parameter in the figures that offers a distinction between the particular waveforms is the advance angle $\delta = 10°, 20°, 30°, 40°$.

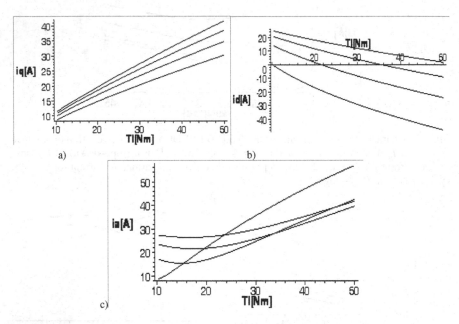

Fig. 4.29 Characteristics of currents for the 6.6 [kW] motor in a function of load torque T_l: a) I_q transformed current for $\delta = 40°, 30°, 20°, 10°$ (top – down) b) I_d transformed current for $\delta = 40°, 30°, 20°, 10°$ (top – down), c) I_a armature current for $\delta = 40°, 30°, 20°, 10°$ (top – down, at $T_l = 10$ [Nm])

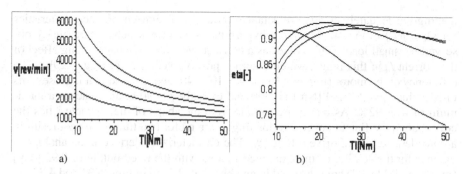

Fig. 4.30 Characteristics of rotor's speed and efficiency factor for the 6.6 [kW] motor in a function of load torque T_l: a) rotational speed for $\delta = 40°, 30°, 20°, 10°$ (top – down) b) efficiency factor for $\delta = 10°, 20°, 30°, 40°$ (top – down, at $T_l = 10$ [Nm])

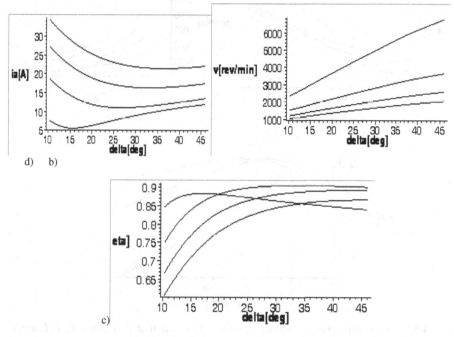

Fig. 4.31 Characteristics for the 0.95 [kW] motor, in a function of advance angle δ: a) I_a armature current for $T_l = 6.0, 4.5, 3.0, 2.5$ [Nm] (top – down) b) rotational speed for $T_l = 1.5, 3.0, 4.5, 6.0$ [Nm] (top – down) c) efficiency factor for $T_l = 1.5, 3.0, 4.5, 6.0$ [Nm] (top – down, at $\delta = 10°$)

In the commentary of the information found in the sets of characteristics in Figs. 4.29 and 4.30 one can conclude that the family of the characteristics $I_q = f(T_l)$ generally presents the involvement of the reluctance torque in the total torque T_e of the motor. The largest share of the reluctance torque T_{er} occurs for $\delta = 10°$ and this characteristic lies the lowest in its family. This observation is confirmed by waveform $I_d = f(T_l)$, where for $\delta = 10°$, I_d has negative and decreasing values, thus leading to adequately positive reluctance torque T_{er}. The mechanical characteristics for $\Omega_r = f(T_l)$ in Fig. 4.30a have a hyperbolic waveform which is particularly observed for small loads. This comes as a consequence of the demagnetizing effect of the current I_d in this range, which assumes positive values there (Fig. 4.29b). The efficiency of the motor (Fig. 4.30b) for $\delta = 10°...40°$ and the load that is close to its rated value $T_l = 20...30$ [Nm] is high and exceeds $\eta > 90\%$, and reaches a maximum of over 92%. As one can conclude from the shape of the characteristics the curves are quite flat and even overloading of the motor two times does not result in a considerable loss of drive efficiency. The characteristics derived in an analogical manner for the smaller of the examined motors with the rated output of 0.95 [kW] (motor A, Table 4.3) are presented in an abbreviated form in Figs. 4.31 and 4.32.

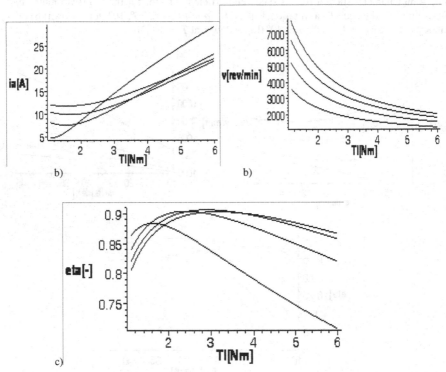

Fig. 4.32 Characteristics for the 0.95 [kW] motor in a function of load torque T_l: a) I_a armature current for $\delta = 45°, 35°, 25°, 15°$ (top – down, at $T_l = 1$ [Nm]) b) rotational speed for $\delta = 45°, 35°, 25°, 15°$ (top – down) c) efficiency factor for $\delta = 15°, 25°, 35°, 45°$ (top – down, at $T_l = 1$ [Nm])

The static characteristics for an untransformed model of the BLDC motor (4.60), (4.65) for two-phase control cannot be gained simply in the algebraic form since they are relative to the rotor's angle. Obviously, the equations with the periodically variable coefficients and solutions for the steady state in accordance with Floquet's theorem are referred to in literature with regard to mathematical models of electric machines. However, for the case of such an abbreviated mathematical model and low cost of calculations, the static characteristics can be derived as a set of stationary points in the dynamic state. This approach has another advantage, namely, that it presents whether a given static state of a drive is possible to achieve from a given initial state. Numerous examples indicate that this is not always the case, in particular with regard to BLDC motors with a higher share of the reluctance torque or cogging torque. The static characteristics of a 6.6 [kW] motor gained as a result of this method are presented in Figs. 4.33 and 4.34.

From the comparison of characteristics derived on the basis of the transformed and untransformed models of the motor one can conclude that both of them look very similar. The only relevant difference regards the waveform marking the current of the armature $I_a = f(T_l)$. In the transformed model (Fig. 3.29c) the values of the current for a small load are considerably higher than the ones gained on the basis of the untransformed model (Fig. 4.33a). This comes as a consequence of the course of the term I_d in the transformed model. For higher loads the relevance of the differences starts to fade.

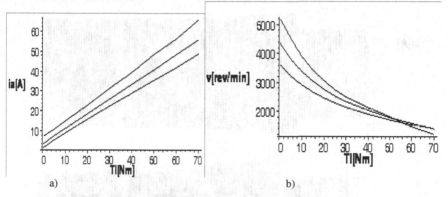

a)   b)

Fig. 4.33 Steady-state characteristics for 6.6 [kW] BLDC motor in a function of load torque, computed by untransformed model: a) I_a armature current for δ = 40°, 30°, 20° (top – down) b) Ω_r rotor velocity for = 40°, 30°, 20° (top – down)

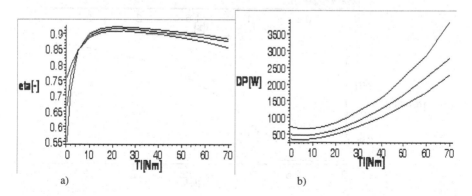

a)   b)

Fig. 4.34 Steady-state characteristics for 6.6 [kW] BLDC motor in a function of load torque, computed by untransformed model: a) efficiency factor for δ = 20°, 30°, 40° (top – down) b) total losses for δ = 40°, 30°, 20°, (top – down)

At this occasion one can note that the two models applied in this case, i.e. the transformed and the untransformed ones are not completely equivalent. By definition in the transformed model an assumption is made that each of the three phases of the armature is independently supplied. In addition, the influence of commutation is disregarded. The untransformed model accounts for the constraints imposed by two-phase supply and involves commutation between the star connected windings while the switchings occur during the rotation of the rotor in the function of its position in accordance with the diagram in Fig. 4.10. The advantage of the transformed model is that it is very simple and does not pose any problems during calculations. This plays an important role in a complex regulation system in which a single BLDC motor forms one of many components of the system as a drive in one of the joints, for instance as an industrial manipulator. The static characteristics comprise the sets of the possible steady states of the drive. However, the

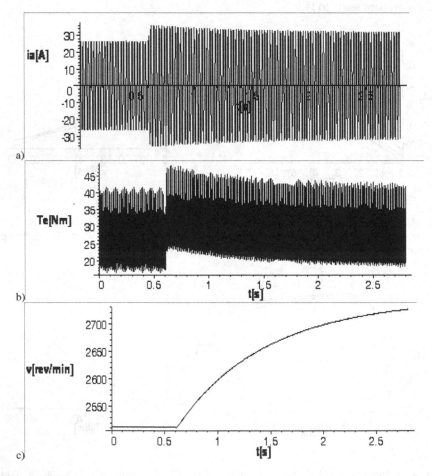

Fig. 4.35 Transients following stepwise change of advance angle δ: 30°→40°, for 6.6 [kW] motor, computed by untransformed model of BLDC: a) I_a currents b) electromagnetic torque c) rotor's speed

transfer between two points on the characteristics occurs as a result of transients and, hence, transfer is not always possible since we have to do with a non-linear dynamic system. For the purposes of illustration the figures that follow present the transients for the dynamic states that occur during the change of the parameters in a system with BLDC motor. Figs. 4.35 and 4.36 present transients resulting from a abrupt change of the advance angle δ: $30° \rightarrow 40°$ for an untransformed and transformed models of the motor, respectively.

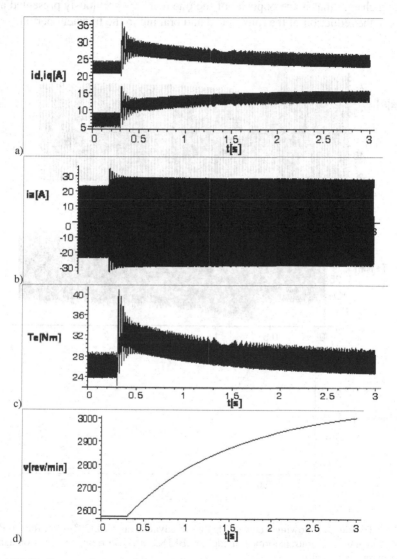

Fig. 4.36 Similar transients as in Fig 4.35, but computed by transformed model: a) I_q, I_d currents b) I_a armature currents c) electromagnetic torque d) rotor's speed

For the case of the results gained from untransformed model the increase of the speed Ω_r is smaller since in a two-phase motor supply the response of the drive to the change in the advance angle δ is limited in comparison to the case of independent supply of three phases, which is additionally confirmed by the static characteristics of the drive presented earlier in this section. The two figures that follow, i.e. Figs. 4.37 and 4. 38 present transients for the respective transformed and untransformed model after a stepwise change in the advance angle δ: 30°→20°. This is a change that is the opposite of the one that was previously presented as it results in the reduction of the rotor speed and braking in the transient period.

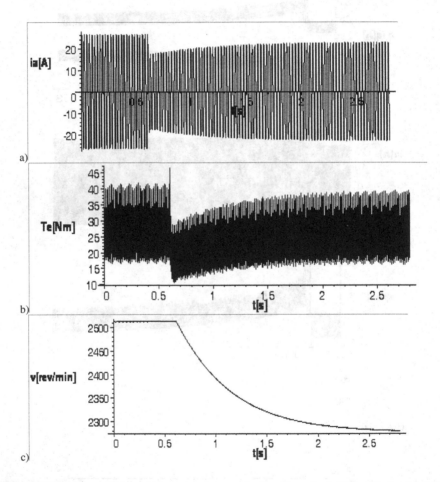

Fig. 4.37 Transients following stepwise change of advance angle δ: 30°→20°, for 6.6 [kW] motor, computed by untransformed model of BLDC: a) I_a currents b) electromagnetic torque c) rotor's velocity

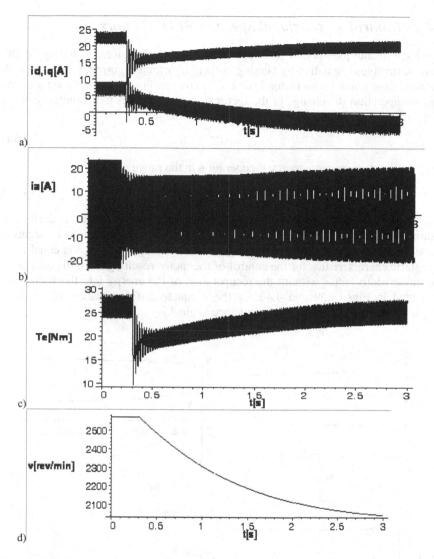

Fig. 4.38 Similar transients as in Fig. 4.37, but computed by transformed model: a) I_q, I_d transformed currents b) I_a armature currents c) electromagnetic torque d) rotor's speed

The comparison between the results of calculations for the untransformed and transformed models indicates a greater decrease of the rotor's speed (Fig. 4.38d versus Fig. 4.37c) accompanied by an adequately higher increase of the speed (Fig. 4.36d versus Fig. 4.35c) for the case of the results gained using the transformed model in which the windings are not connected. The same results are gained on the basis of static characteristics, for instance from the comparison of the results in Fig. 4.30a with the ones in Fig. 4.33b.

4.4.3 Control of Rotational Speed in BLDC Motors

The basic technique applied for the control of the DC motors, including BLDC motors, involves regulation by altering voltage U, which is currently realized with the aid of the pulse width factor k_u of PWM control. For the systems without energy recuperation the change of the factor k_u, which realizes the complete change of the rotor's speed, occurs approximately in the range:

$$k_u = 0.02...1.0 \tag{4.74}$$

while in the systems with energy recuperation in the range

$$k_u = 0.51...1.0 \tag{4.75}$$

The difference for the both types of the control results from the fact that during the return of the energy the motor over this period is fed with a voltage with negative value $-U$, so that for $k_u = 0.5$ the mean value of the supply voltage is equal to 0. The static characteristics for the control of the motor resulting from the change of the pulse width factor without the recuperation of the energy into the source are presented in Figs. 4.39 and 4.40 for the adequate different values of the load torque T_l and various values of the advance angle δ.

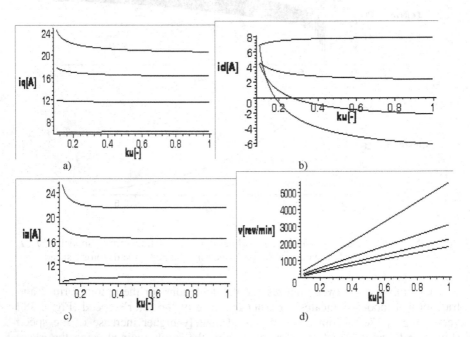

Fig. 4.39 Characteristics for 0.95 [kW] motor in a function of k_u factor , without energy recuperation, for $\delta = 35°$: a) I_q transformed current for $T_l = 6.0, 4.5, 3.0, 2.5$ [Nm] (top – down) b) I_d transformed current for $T_l = 1.5, 3.0, 4.5, 6.0$ [Nm] (top – down) c) I_a armature current for $T_l = 6.0, 4.5, 3.0, 2.5$ [Nm] (top – down) d) rotational speed for $T_l = 1.5, 3.0, 4.5, 6.0$ [Nm] (top – down)

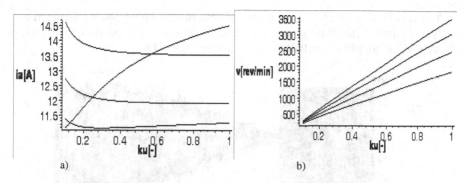

Fig. 4.40 Characteristics for 0.95 [kW] motor in a function of k_u factor, without energy re-cuperation, for $T_l = 3$ [Nm], a) I_a armature current for $\delta = 15°, 45°, 35°, 25°$ (top – down, at $k_u = 1$) b) rotational speed for $\delta = 45°, 35°, 25°, 15°$ (top – down)

Similar characteristics are displayed for the control with energy recuperation; however, the change of the pulse width factor is limited in accordance with (4.75). Selected characteristics for this type of control are presented in Fig. 4.41.

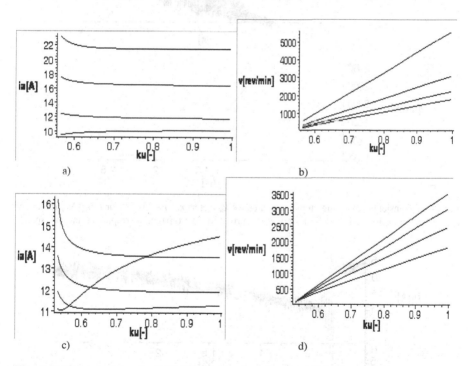

Fig. 4.41 Characteristics for 0.95 [kW] motor in a function of k_u factor, with energy recu-peration: a) I_a armature current for $\delta = 35°, T_l = 6.0, 4.5, 3.0, 2.5$ [Nm] (top – down) b) rotational speed for $\delta = 35°, T_l = 1.5, 3.0, 4.5, 6.0$ [Nm] (top – down) c) I_a armature current for $T_l = 3.0$ [Nm], $\delta = 15°, 45°, 35°, 25°$, (top – down, at $k_u = 1$) d) rotational velocity for $T_l = 3$ [Nm], $\delta = 45°, 35°, 25°, 15°$ (top – down)

The two figures that follow present transient state, which occurs after decreasing pulse width factor from $k_u = 1$ to $k_u = 0.75$, change that is equivalent to the reduction of the supply to the half of the source voltage value.

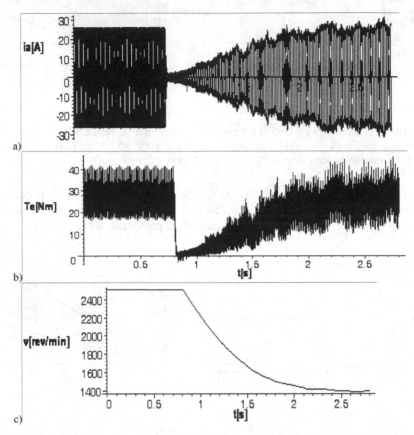

Fig. 4.42 Transients following stepwise change of k_u factor: k_u: $1 \rightarrow 0.75$, for 6.6 [kW] motor by use of the untransformed model: a) i_a armature current b) electromagnetic torque c) rotational speed

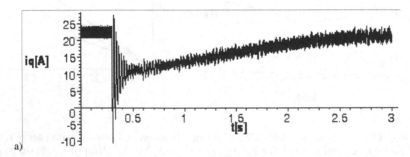

Fig. 4.43 Similar transients as in Fig 4.42, but computed by use of the transformed model of BLDC: a) i_q current b) i_q, i_d currents, c) i_a armature current, d) electromagnetic torque, e) rotor speed

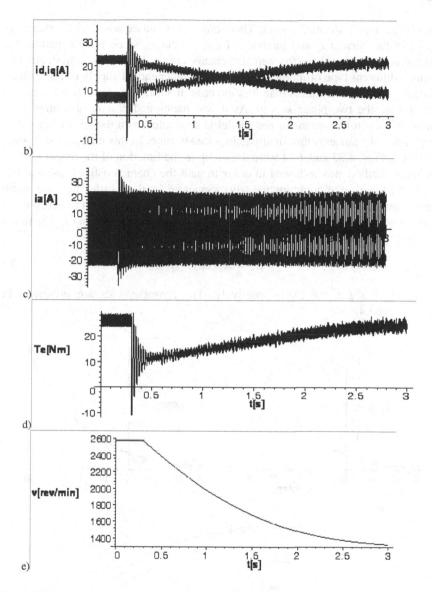

Fig. 4.43 (*continued*)

The characteristics presented in the two figures (Figs. 4.42 and 4.43) are similar with the only exception of the waveform for the armature currents. For the case of the waveform resulting from the application of the transformed model in the transient state the armature current does not decrease during the braking of the drive but shows an increase. This is opposite than in the case for transients gained on the

basis of the untransformed model. This comes as a consequence of the sharp decrease of the current i_q and increase of the current i_d, since in this manner the
transformed model realizes the stepwise changes of the motor load. Physically this
means a different type of motor braking for the independent supply of three phase
windings without constraints than it is the case in a three-phase system connected
in a star for the two-phase supply. As it was mentioned before, obtaining static
characteristics in an untransformed model is associated with the need of calculating a series of transients that finally gain a steady state. In this manner the characteristics in Figs. 4.33 and 4.34 were drawn up in the function of the machine load.
The same method was followed in order to gain the characteristics presented below in the function of commutation advancement δ for three different pulse width
factor values $k_u = 1, 0.9, 0.8$. This was conducted in a system with energy recuperation so that the anticipated values of rotational speed are found in the range
that is in agreement with the formula below

$$\Omega_r = \Omega_n (2k_u - 1) \qquad (4.76)$$

that is Ω_n, 0.8 Ω_n, 0.6 Ω_n, respectively. The characteristics are presented in
Figs. 4.44 and 4.45.

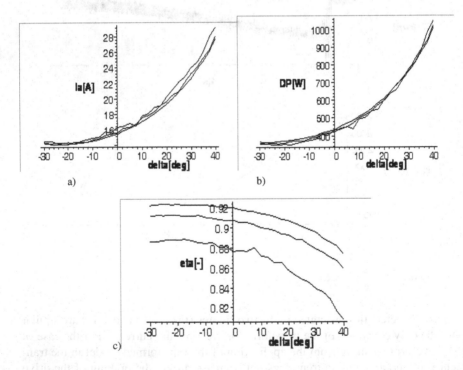

Fig. 4.44 Characteristics for 6.6 [kW] motor in a function of advance angle δ, for $k_u = 1$,
0.9, 0.8 and $T_l = T_n$: a) I_a armature current b) total energy losses $\Sigma\Delta P$ c) efficiency factor η

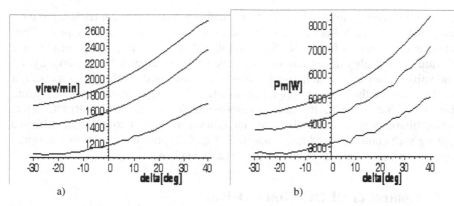

a)   b)

Fig. 4.45 Characteristics for 6.6 [kW] motor in a function of advance angle δ, for $k_u = 1$, 0.9, 0.8 and $T_l = T_n$: a) rotational speed b) mechanical power P_m

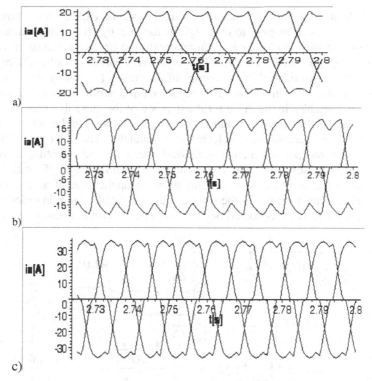

a)

b)

c)

Fig. 4.46 Shapes of the steady-state i_a current for: a) $\delta = -15°$ b) $\delta = 15°$ c) $\delta = 45°$; $k_u = 1$ and $T_l = T_n$

The characteristics gained in this way are not smooth since they are formed on the basis of a limited number of points for the variable δ, i.e. about 30 points and, in addition, the final steady state of the drive is difficult to determine in a comparable way for each final point. The characteristics presented in Fig. 4.44, i.e. for armature current I_a, total losses $\Sigma\Delta P$ and, within a certain range, efficiency η indicate that for a steady load these values are only slightly relative to factor k_u. Concurrently, the curves in Fig. 4.45, i.e. rotational speed Ω_r and mechanical power P_m are considerably dependent on the value of the voltage and, consequently, on the value of factor k_u, which directly affects the value of the voltage. As a result, it is the value of the voltage supplying armature, here represented by pulse width factor k_u, that is the basic variable responsible for the control of BLDC drive, while high energy efficiency is to be maintained and it is not considerably affected during such control procedure. In addition, Fig. 4.46 presents waveforms of armature current i_a for various values of the advance angle δ.

4.5 Control of BLDC Motor Drives

4.5.1 Control Using PID Regulator

As it is indicated by static characteristics and transients, BLDC motor drives are controlled in a similar manner to other DC motors, i.e. by changing armature voltage. In a system with an electronic commutator this occurs as a result of the change of the pulse width factor k_u. This may happen at every particular pulse as a result of modifying the signal controlling the series of the pulses. In most cases we have to do with discreet control in which the value of the pulse width factor k_u is relative to the values of variables in the drive for the duration of the pulse that precedes. Control is relative to the angle of rotation θ_r, just as presented earlier on during the discussion of the operation of the electronic commutator. The other values that are responsible for the control include rotational speed Ω_r and armature current i_a. Since for the purposes of the control it is necessary that the angle of rotation is familiar, one of the common solutions involves the application of an encoder for determination of the position of the rotor and a differential system in order to indirectly obtain speed Ω_r. It is sometimes the case that the armature current is applied

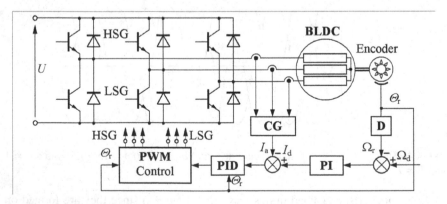

Fig. 4.47 BLDC drive control system with encoder and PID regulator. HSG – high side gates, LSG – low side gates

for regulation as the value that is an auxiliary one and can be used for the purposes of limiting control, just as presented in the example devoted to start-up of a motor. Sometimes the control of BLDC drive does not apply position sensors and the necessary rotor angle is obtained indirectly on the basis of measurements of voltages and currents in armature windings by use of state observers. This method will be presented in the further part of the current section. A typical control system for a BLDC motor drive with a position sensor is presented in Fig. 4.47.

This section will be concerned with the presentation of examples of BLDC drive control systems with speed stabilization for changing loads and stepwise

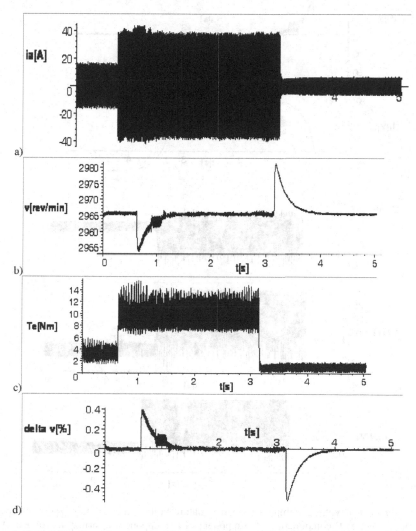

Fig. 4.48 Stabilization of rotor speed by PID regulator ($kP = 60$; $kI = 300$; $kD = 0.5$) after stepwise load torque change T_l: 3 [Nm] → 9 [Nm] and consequently 9 [Nm] → 1 [Nm]: a) armature currents b) rotor speed c) electromagnetic torque d) speed error [%]

change of the input value for a constant motor load. Such a basic control task using PID regulator reveals the ability of BLDC motors to operate in the drive control based systems on changing the value of the supply voltage. Fig. 4.48 presents results of simulations conducted using an untransformed model of the motor for stabilizing rotational speed for a stepwise change of the load torque from the rated value of $T_l = 3$ [Nm] initially to $T_l = 9$ [Nm], i.e. three times overloading the motor followed by a stepwise change to reach the value of $T_l = 1$[Nm], which is one third of the rated load.

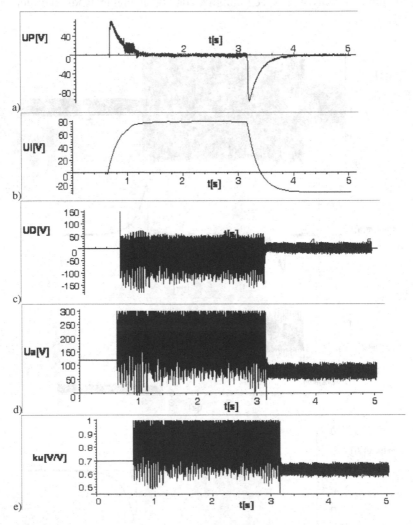

Fig. 4.49 Control voltage components and armature voltage during stabilization of speed (see Fig 4.48): a) P component b) I component c) D component d) armature voltage e) k_u factor

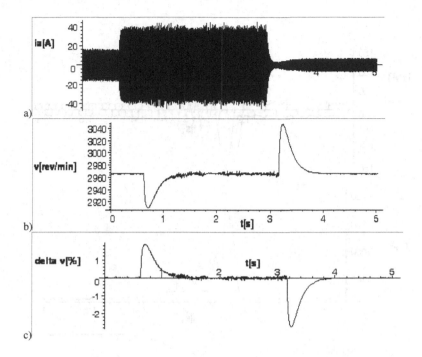

Fig. 4.50 Stabilization of rotor speed by PID regulator (see Fig 4.48) but for different control values: $kP = 10$; $kI = 50$; $kD = 0.1$: a) armature currents b) rotor speed c) speed error

Figs. 4.48 – 4.50 illustrate a correct stabilization of the rotor speed for very dynamic changes of the load. For high gain values of PID regulator (Fig. 4.48) the error of speed regulation is in the range $|\varepsilon_\Omega| = 0.4\%$, while for a lower gains (Fig. 4.50) this error ranges around $|\varepsilon_\Omega| = 2\%$, which denote values that are approximately proportional to the gains applied for kI, kP, kD factors. Stability state is achieved after around 0.5 [s]. Beneficial conditions for regulation are secured by a surplus of the regulation that involves input of the supply voltage $U = 300$ [V] to a motor while the rated value of the voltage is $U = 120$ [V]. As one can see for such an intensive regulation the system of the supply, commutator and motor itself have to be designed to withstand the maximum value of the voltage (300 [V]). The following waveforms, i.e. the ones presented in Figs. 4.51 and 4.52 illustrate PID regulation for the stepwise change of input speed $\Omega_d : 500 \rightarrow 1500$ [rev/min] for both untransformed and transformed models of the motor.

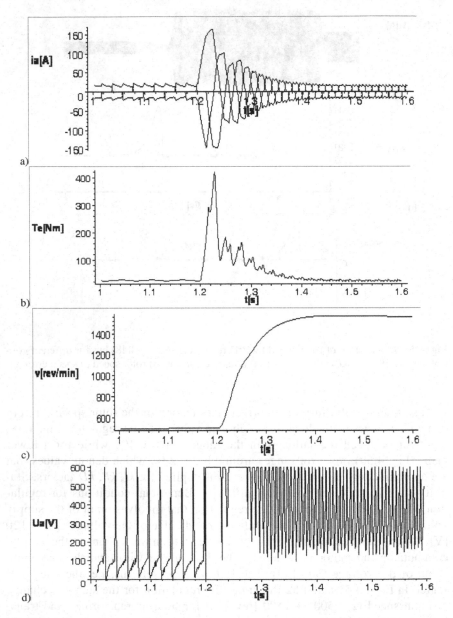

Fig. 4.51 PID regulation of rotor speed of the 6.6 [kW] motor, after stepwise change of required value of speed Ω_d : 500 → 1500 [rev/min]. Results for untransformed model and: kI = 1000, kP = 500, kD = 10: a) stator currents b) electromagnetic torque c) rotor speed d) armature voltage

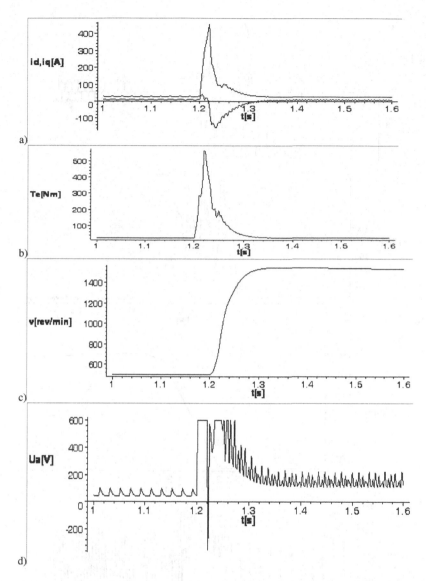

Fig. 4.52 PID regulation of rotor speed – transients like in Fig 4.51, but resulting from transformed d,q model: a) d,q currents b) electromagnetic torque c) rotor speed d) armature voltage

The waveforms in Figs. 4.51 and 4.52 indicate that the transients resulting from the application of transformed and untransformed models are very similar. The regulation is realized quickly and effectively despite large difference in terms of the target speed. However, this happens under the assumption of accessibility of the higher value of supply voltage $U = 1.5U_n$ and an additional condition that the motor is capable of the generation of a surge torque of about $T_e = 500$ [Nm],

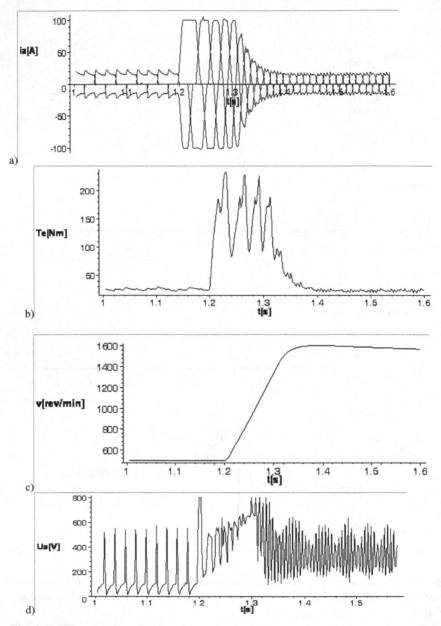

Fig. 4.53 PID regulation of rotor speed of the 6.6 [kW] motor, with armature current limitation to $i_r = 110$ [A], after stepwise change of required value of speed Ω_d: 500 → 1500 [rev/min]. Results for untransformed model and $kI = 1000$, $kP = 500$, $kD = 10$: a) stator currents b) electromagnetic torque c) rotor speed d) armature voltage

i.e., 20 times the value of T_n. Such regulation properties can be gained as a result of using a motor and supply system that is oversized in relation to the rated value during operation. For smaller requirements regarding the regulation speed it is possible to apply a more economical supply system and introduce current delimiter in the motor design. The two figures that come below, i.e. Figs. 4.53 and 4.54 present the regulation of the motor for a stepwise change of the input speed and simultaneous application of a current delimiter which wouldn't allow armature currents surges that exceed set multiple of the rated value, as in the previous examples. Fig. 4.53 presents the results for the stepwise increase of the input value of the rotational speed while the curve in Fig. 4.54 presents the example involving the reduction of the input value. For both cases the armature current is limited to the value of $i_r = 110$ [A], which is about five times the value of the rated current.

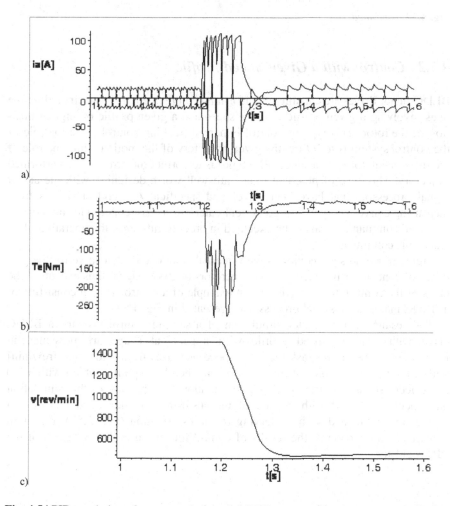

Fig. 4.54 PID regulation of rotor speed of the 6.6 [kW] motor, with armature current limitation to $i_r = 110$ [A], after stepwise change of required value of speed Ω_d: 1500 → 500 [rev/min]. Results for untransformed model and $kI = 1000$, $kP = 500$, $kD = 10$: a) stator currents b) electromagnetic torque c) rotor speed d) armature voltage

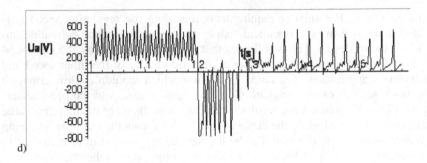

d)

Fig. 4.54 (*continued*)

4.5.2   Control with a Given Speed Profile

BLDC motor drives are capable of performing complex control tasks including the ones involving a given profile of drive speed and a given profile of angular position of the rotor. This type of control is named tracking control since the role of the control system is to follow the given trajectory of the motion while the role of the drive regulator (for instance, PID type) is to correct the error of the performed trajectory. This type of practical issues are well worth designing with the aid of signal processors that have been developed specifically to collect signals from measuring converters, perform numerical calculations associated with the control and send command signals to be executed in order to influence the operation of an electronic commutator.

Signal processors are equipped with internal components that serve for the purposes of control for instance several generators of PWM signal thus executing the tasks of the control of an converter. An example of a control system consisting of a BLDC motor and a signal processor is presented in Fig. 4.55.

The results of a computer simulation of a sample control task for a BLDC drive with a given speed profile with trapezoidal shape are presented in Figs. 4.56 – 4.58. In this task the rotor speed increases from 0 to 2.500 [rev/min] within 0.8 [s] and remains at this level for another 1.2 [s], after which within 1.0 [s] it decreases to 500 [rev/min]. The calculations for the case of this simulation have been conducted with the aid of untransformed mathematical model of a motor (Fig. 4.56) and with preserving *d,q* transformation (Fig. 5.57). Fig. 4.58 contains a comparison of the results of control for various values of gains of the PID regulator.

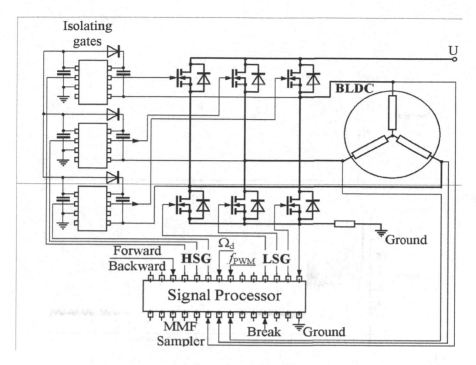

Fig. 4.55 Control system of a BLDC motor, without a rotor position gauge with a signal processor

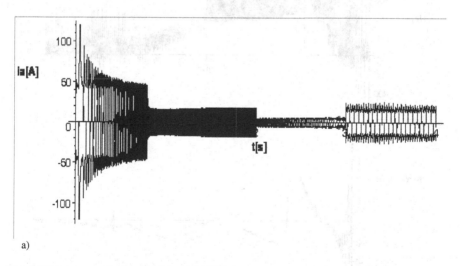

a)

Fig. 4.56 Rotor speed control of the 6.6 [kW] PMDC motor, according to trapezoidal shape of speed trajectory, under the nominal load of $T_l = 25$ [Nm]. PID regulator settings are: $kI = 1000$, $kP = 500$, $kD = 10$. Untransformed model employed: a) armature currents b) rotor speed c) electromagnetic torque d) motor voltage e) speed error [rev/min]

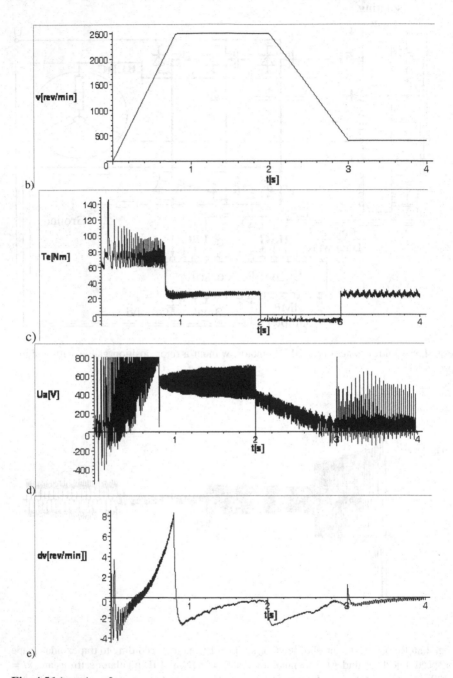

Fig. 4.56 (*continued*)

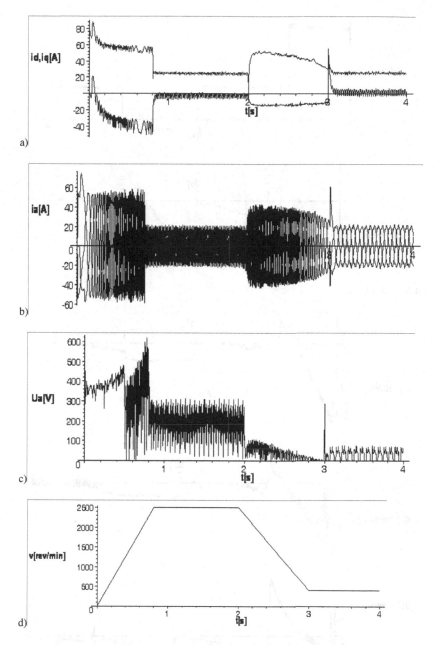

Fig. 4.57 The same control problem as in Fig 4.56, but transformed d,q model of BLDC is employed: a) d,q transformed armature currents b) armature currents c) motor voltage d) rotor speed e) electromagnetic torque f) speed error

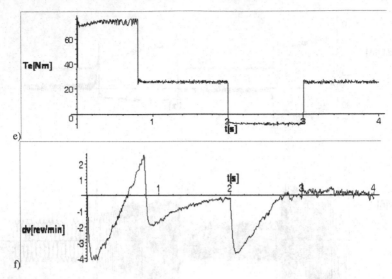

e)

f)

Fig. 4.57 (*continued*)

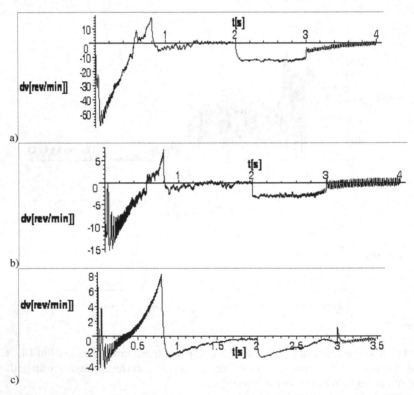

a)

b)

c)

Fig. 4.58 Tracking error for trapezoidal speed control of the PMDC motor, like in Fig 4.56, but for various PID regulator settings: a) *kI* = 100, *kP* = 50, *kD* = 1 b) *kI* = 400, *kP* = 200, *kD* = 2 c) *kI* = 1000, *kP* = 500, *kD* = 10

On the basis of the results presented in Fig. 4.58 one can conclude that the error of the drive control is roughly proportional to the values of regulator's gains.

4.5.3   Control for a Given Position Profile

The application of a signal processor as an electronic system that combines the properties of a converter of measured quantities, a digital calculation processor and a generator of control signals offers the possibility of performing complex issues regarding a BLDC motor drive control with an alternative of eliminating a position sensor [19,24,30,38,40]. Fig. 4.55 presents a system including a signal processor that collects voltage signals from a EMF sampler and with the aid of such a system that is capable of determining the periods during which the transistors are in the ON state without direct measurement of the angle of rotation. The question of the sensorless control forms a complex task since it requires the determination of the position of the rotor even before start-up when the induced EMF is too small in order to determine rotor position on its basis. Subsequently, after the required threshold is exceeded, on the basis of the induced EMF in the unsupplied phase of the armature winding at a desired instant the control signal for the electronic commutator occurs. Fig. 4.59 presents EMF induced between clamps 1,2 during the start-up of the motor.

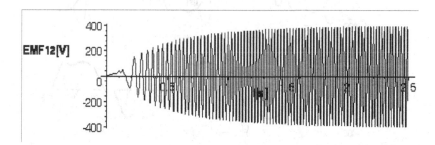

Fig. 4.59 EMF induced in armature winding of a BLDC motor shortly after start-up

For the operation of the motor with higher speed the determination of the instant of the commutation on the basis of the measured currents and EMF in the windings is easier since there is a considerable angular correlation between the two values that are relative to the position of the rotor. Fig. 4.60 presents the correlation between SEM and armature currents and Fig. 4.61 contains the correlation of the voltages between the clamps and currents.

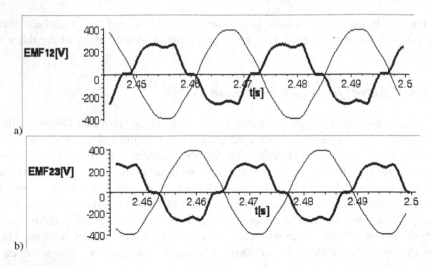

Fig. 4.60 EMF and armature currents for BLDC operation with a normal speed: a) e_{12}, i_1 b) e_{23}, i_2

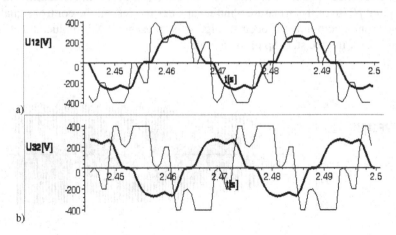

Fig. 4.61 Phase-to phase voltages and armature currents for BLDC operation with a normal speed: a) u_{12}, i_1 b) u_{32}, i_2

As it has already been mentioned, during the start-up the situation is more complex and in order to determine the rotor position it is possible to apply the pulse method [29,30]. The application of this method, however, requires that the rotor has a variable reluctance during the rotation, i.e. a situation which involves the salient pole rotor. Fig. 4.62 presents electromotive forces (EMFs) and currents for the case of the start-up, for which the waveforms confirm the lack of simple correlation between the curves.

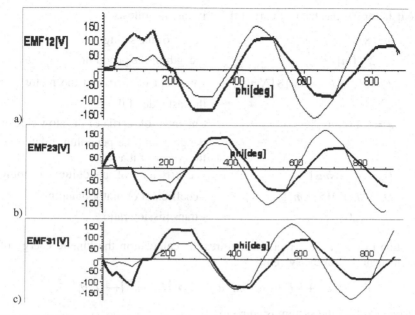

Fig. 4.62 EM Forces and currents during starting of BLDC motor: a) e_{12}, i_3 b) e_{23}, i_1 c) e_{31}, i_2

An example of the control of the reversible BLDC drive including the regulation of the position and the speed is presented in the following example.

Example 4.2 Fig. 4.63 presents a diagram of the drive of a large, massive pendulum (swing) with a controlled amplitude and period of oscillations diverging from a natural period, which anyhow is dependent on damping of the motion. The drive applies a motor with a rated power of 0.95 [kW] (Table 4.3).

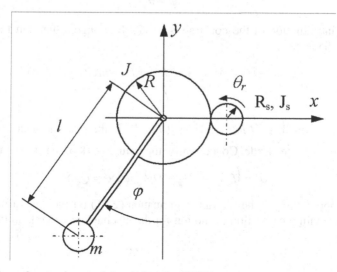

Fig. 4.63 View of a massive pendulum driven by BLDC motor

The data for the mechanical part of the drive are as follows:

$m = 8\,[kg]$ — mass of the swing

$l = 2.8\,[m]$ — length of the pendulum

$J_r = J_s + J_g = 0.018\,[Nms^2]$ — moment of inertia of the motor and the fast side of the gear

$J = 0.1\,[Nms^2]$ — moment of inertia of the slow side of the gear and the pendulum relative to the center of mass

$D_\varphi = 1.2\,[Nms]$ — coefficient of pendulum damping

$D_r = 0.00015\,[Nms]$ — coefficient of motor damping

$k_g = 120\,[-]$ — transmission ratio

Application of Lagrange's method requires the calculation the kinetic energy of the mechanical part of the system:

$$T = \tfrac{1}{2}J\dot{\varphi}^2 + \tfrac{1}{2}J_r\dot{\theta}_r^2 + \tfrac{1}{2}ml^2\dot{\varphi}^2 = \tfrac{1}{2}\dot{\varphi}^2\big(J + ml^2\big) + \tfrac{1}{2}J_r\dot{\theta}_r^{\,2}$$

The virtual work of the system is equal to:

$$\delta A = (-D_\varphi\dot{\varphi})\delta\varphi + (T_e - D_r\dot{\theta}_r)\delta\theta_r$$

where: T_e - denotes the electromagnetic torque of the motor.

In its mechanical part the system has a single degree of freedom. Since the calculations are to be performed from the point of view of the motion of the pendulum, in this case $q_1 = \varphi$ should be adopted as generalized coordinate. The constraints resulting from the transmission take the form:

$$k_g\varphi = \theta_r \tag{4.77}$$

After the introduction of the constraints (4.77), Lagrange's function for this system takes the form:

$$L = T - U = \tfrac{1}{2}\dot{\varphi}^2\underbrace{\big(J + k_g^2 J_r + ml^2\big)}_{J_p} - mgl(1 - \cos(\varphi)) \tag{4.78}$$

where: - $J_p = ml^2\big(1 + J/(ml^2) + k_g^2 J_r/(ml^2)\big)$ - is the total moment of inertia reduced to the low speed side. Concurrently, the virtual work of the system is equal to:

$$\delta A = \big(T_e k_g - (D_\varphi + D_r k_g^2)\dot{\varphi}\big)\delta\varphi = Q_\varphi\delta\varphi \tag{4.79}$$

After the application of the Lagrange's equation (2.51) for the generalized coordinate φ we obtain the equation of motion for the mechanical variable in the form:

$$\frac{d}{dt}\frac{\partial L}{\partial\dot{\varphi}} - \frac{\partial L}{\partial\varphi} = Q_\varphi \tag{4.80}$$

After substitutions in the Lagrange's function (4.78), we have:

$$\ddot{\varphi} = \left(-mgl\sin\varphi + T_e k_g - (D_\varphi + D_r k_g^2)\dot{\varphi}\right)/J_p \qquad (4.81)$$

The assumed pulsation of the motion of the pendulum is equal to the pulsation of the mathematical pendulum with moment of the inertia ml^2, which differs from J_p (4.78), hence:

$$\Omega = \sqrt{\frac{g}{l}} \cong 1.9\,[1/s] \qquad (4.82)$$

The assumed motion of the pendulum (*Example 4.2a*) involves the maintenance of the undamped fluctuations with the amplitude φ_0 and pulsation Ω (4.82).

Hence, the required trajectory is determined as follows:

$$\begin{aligned}
\varphi_1 &= \varphi_0 \sin(\Omega t + \pi/2) \\
\omega_1 &= \dot{\varphi}_1 = \varphi_0 \Omega \cos(\Omega t + \pi/2) \\
a_1 &= \ddot{\varphi}_1 = \dot{\omega}_1 = -\varphi_0 \Omega^2 \sin(\Omega t + \pi/2)
\end{aligned} \qquad (4.83)$$

The other type of the given motion (*Example 4.2b*) is the exponential start-up of the pendulum from the initial angle φ_0 to angle φ_a with the time constant of T_a.

$$\varphi_2 = \left(\varphi_0 + (\varphi_a - \varphi_0)(1 - e^{-\frac{t}{T_a}})\right)\sin(\Omega t + \pi/2)$$

$$\begin{aligned}
\omega_2 = \dot{\varphi}_2 &= \frac{\varphi_a - \varphi_0}{T_a} e^{-\frac{t}{T_a}} \sin(\Omega t + \pi/2) + \\
&+ \Omega\left(\varphi_0 + (\varphi_a - \varphi_0)(1 - e^{-\frac{t}{T_a}})\right)\cos(\Omega t + \pi/2)
\end{aligned} \qquad (4.84)$$

$$a_2 = \ddot{\varphi}_2 = \left(-\frac{\varphi_a - \varphi_0}{T_a^2} e^{-\frac{t}{T_a}} - \Omega^2(\varphi_a - \varphi_0)(1 - e^{-\frac{t}{T_a}})\right)\sin(\Omega t + \pi/2) +$$

$$2\Omega\frac{\varphi_a - \varphi_0}{T_a} e^{-\frac{t}{T_a}} \cos(\Omega t + \pi/2)$$

The mathematical model of the system also in the mechanical part is nonlinear and this is so to the larger degree the greater is amplitude of the motion of the pendulum φ_0, φ_a. This could be explained by the linearizing approximation of the pendulum swing

$$\sin\varphi \cong \varphi \qquad \varphi < 1 \qquad (4.85)$$

that is fulfilled with much increasing error for higher values of φ. The regulation of the drive follows as a result of the application of PIDD regulator, i.e. the one that is responsible for the control of 4 types of error:

$$\varepsilon_I = \int_t (\varphi_{1(2)} - \varphi)d\tau$$

$$\varepsilon_\varphi = \varphi_{1(2)} - \varphi \qquad\qquad (4.86)$$

$$\varepsilon_\omega = \omega_{1(2)} - \omega$$

$$\varepsilon_a = a_{1(2)} - a$$

The figures that follow, i.e. Figs. 4.64 to Fig 4.69 present the results of the regulation for *Example 4.2a*, which denotes swinging of the pendulum with a given amplitude. But the first of the figures (Fig. 4.64) shows the free motion of the pendulum with damping and without drive regulation. This gives the basis for the comparisons for examples analyzed further in which the regulation system and the actuator, i.e. BLDC motor and the transmission are engaged to perform the motion in accordance with the given trajectory.

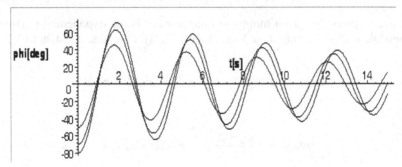

Fig. 4.64 Free swing of the pendulum without any intervention of a regulation system, for φ_0 = -50°, -70°, -80°

The figures that come below present the motion of the pendulum and regulation curves resulting in the achievement of the designed trajectory (4.83).

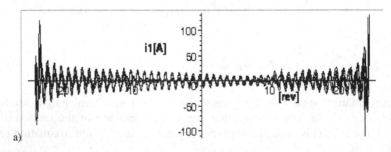

Fig. 4.65 Curves for electromechanical variables for the PIDD controlled pendulum swinging, φ_0 = -70°: a) $i_1(\varphi)$ current b) armature currents c) electromagnetic torque $T_e = f(t)$ d) electromagnetic torque $T_e = f(\varphi)$ e) k_u factor f) U_d/U motor voltage curve

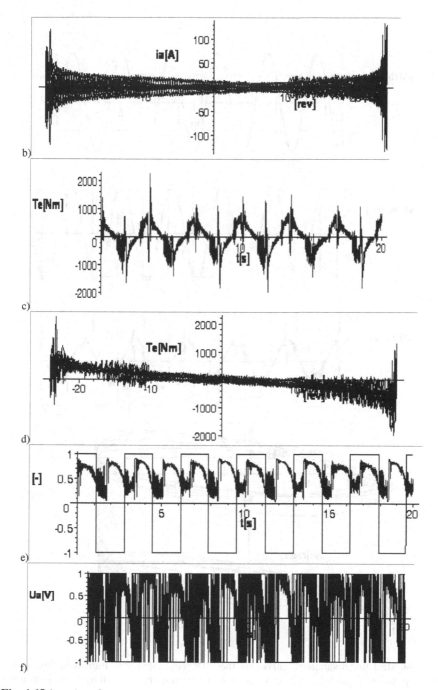

Fig. 4.65 (*continued*)

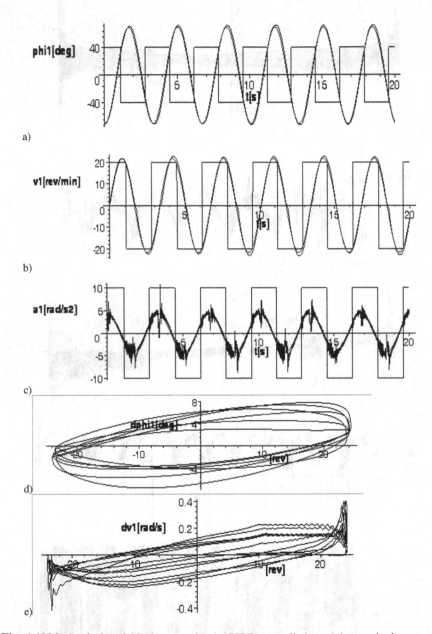

Fig. 4.66 Mechanical variables' curves for the PIDD controlled pendulum swinging, $\varphi_0 = -70°$: a) position angle φ_1 b) angular velocity ω_1 c) angular acceleration a_1 d) position error ε_φ e) speed error ε_ω

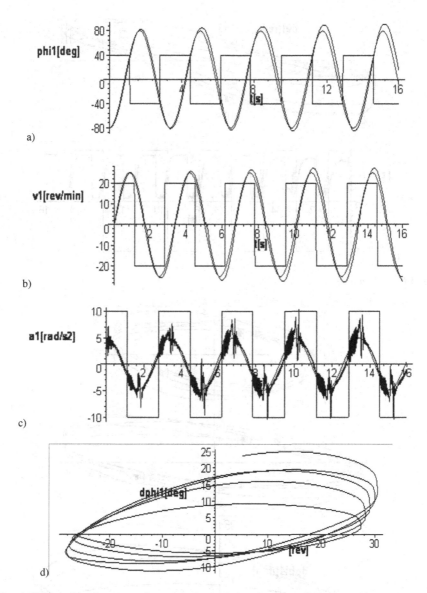

Fig. 4.67 Unstable character of pendulum regulation for $\varphi_0 = -80°$: a) position angle φ_1 b) angular velocity ω_1 c) angular acceleration a_1 d) position error ε_φ e) speed error ε_ω f) k_u factor

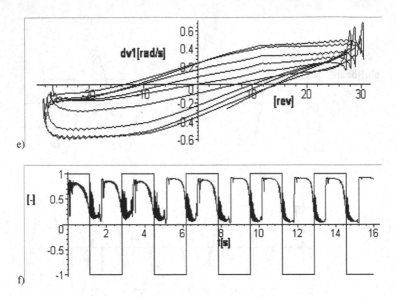

e)

f)

Fig. 4.67 (*continued*)

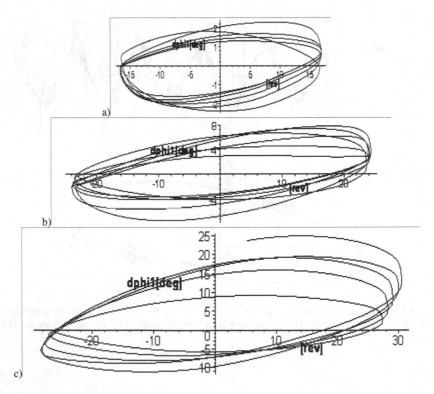

a)

b)

c)

Fig. 4.68 Comparison of position error ε_φ, for: a) $\varphi_0 = -50°$ b) $\varphi_0 = -70°$ c) $\varphi_0 = -80°$

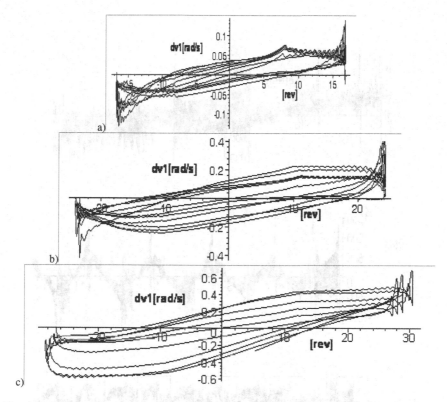

Fig. 4.69 Comparison of speed error ε_ω, for: a) $\varphi_0 = -50°$ b) $\varphi_0 = -70°$, c) $\varphi_0 = -80°$

The set of figures that follows, i.e. Figs. 4.70 - 4.74 illustrates solutions to the *Example 4.2b*, which presents start-up of a pendulum from an initial angle of $\varphi_0 = 10°$ to a swing with an amplitude of φ_a. The waveform illustrating the start-up in accordance with (4.84) is exponentially regulated with the time constant of T_a.

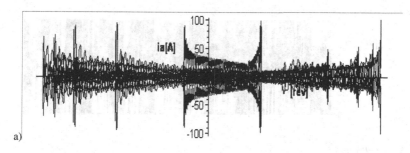

Fig. 4.70 Electromechanical variables for exponential starting of the pendulum swing from $\varphi_0 = -10°$ to $\varphi_a = -70°$, with PIDD regulator: a) armature currents b) electromagnetic torque $T_e = f(\omega)$ c) electromagnetic torque $T_e = f(t)$ d) k_u factor e) U_a/U motor voltage

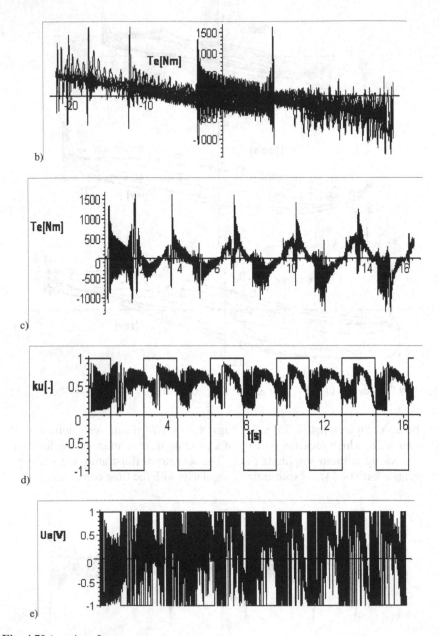

Fig. 4.70 (*continued*)

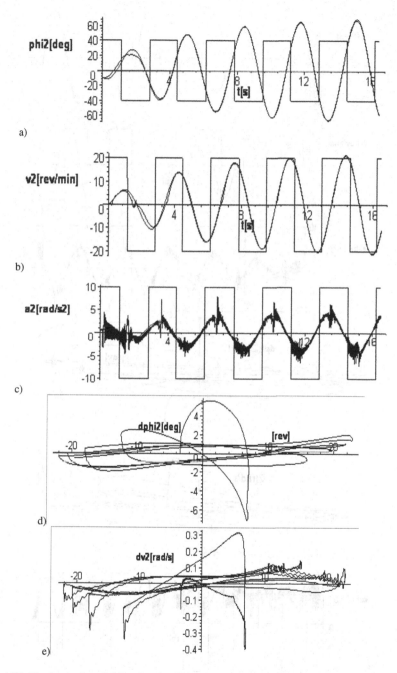

Fig. 4.71 Mechanical variables for the PIDD controlled pendulum swinging, like in Fig. 4.70: a) position angle φ_2 b) angular velocity ω_2, c) angular acceleration a_2 d) position error ε_φ e) speed error ε_ω

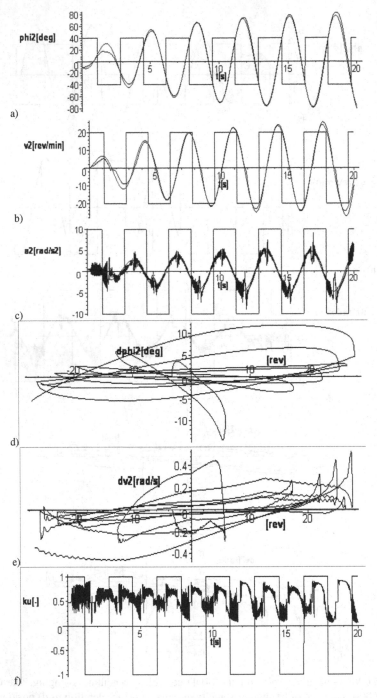

Fig. 4.72 Unstable starting movement of the pendulum for $\varphi_a = -80°$: a) position angle φ_2 b) angular velocity ω_2 c) angular acceleration d) position error ε_φ e) speed error ε_ω f) k_u factor

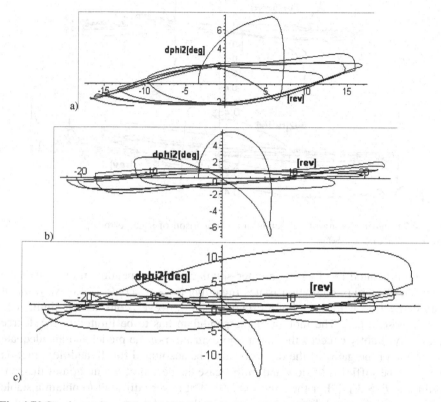

Fig. 4.73 Starting swing of the pendulum. Comparison of position error ε_φ, for: a) $\varphi_a = -50°$
b) $\varphi_a = -70°$ c) $\varphi_a = -80°$

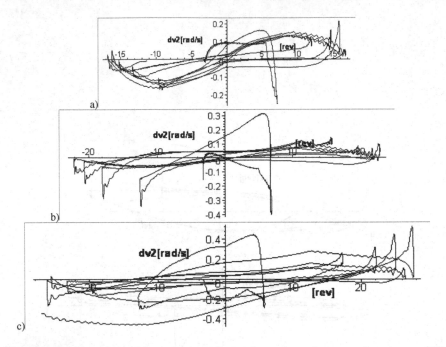

Fig. 4.74 Starting swing of the pendulum. Comparison of speed error ε_ω for: a) $\varphi_a = -50°$ b) $\varphi_a = -70°$ c) $\varphi_a = -80°$

The presented results of computer simulation of the pendulum drive indicate that such complex regulation issues regarding tracking control of drive reversal can be realized with the aid of a PID regulator; however, considerable limitations are imposed on it. The motion of the pendulum has to be limited to the degree that only slightly exceeds the linear approximation of the model and an adequate selection of the gains of the regulator must be accounted for. In addition, motion needs to be sufficiently slow and in this case the period of the motion of the pendulum is $T_p > 3$ [s]. For the angles $|\varphi_a| > 70°$ it is very difficult to obtain a stable operation of the pendulum and the errors of the regulations are becoming greater. However, we can completely confirm the applicability of BLDC motor for this type of drive.

4.5.4 Formal Linearization of BLDC Motor Drive

The transformed model of the motor (4.50) for variables:

$$\mathbf{q} = \left[\theta_r, \Omega_r, i_q, i_d\right]^T \tag{4.87}$$

can take the form:

$$\dot{\theta}_r = \Omega_r$$

$$\dot{\Omega}_r = (T_e - T_l - D\Omega_r)/J$$

$$i_q = \frac{1}{L_q}\left(u_q - R_q i_q + p\Omega_r\left(L_d i_d - \sqrt{\frac{3}{2}}\Psi_{rs}\right)\right) \qquad (4.88)$$

$$i_d = \frac{1}{L_d}\left(u_d - R_d i_d - p\Omega_r L_q i_q\right)$$

where:

$$T_e = pi_q\left(-3M_s i_d + \sqrt{\frac{3}{2}}\Psi_{rs}\right) = i_q(Ai_d + B_\Psi)$$

- denotes electromagnetic torque of the motor. As a result of the introduction to (4.88) of new variables : $z_1 = \theta_r$, $z_2 = \Omega_r$, $z_3 = T_e$, $z_4 = i_d$, we receive:

$$\dot{z}_1 = z_2$$

$$\dot{z}_2 = (z_3 - T_l - Dz_2)/J$$

$$\dot{z}_3 = v_q \qquad (4.89)$$

$$\dot{z}_4 = v_d$$

where:

$$v_q = i_q(Ai_d + B_\Psi) + i_q Ai_d$$

$$v_d = (u_d - R_d i_d - p\Omega_r L_q i_q)/L_d \qquad (4.90)$$

On the basis of the relation (4.90) and by the use of (4.88) it is possible to determine the space for such linearized control by determination of the transformed supply voltages u_q, u_d:

$$u_q = R_q i_q - p\Omega_r\left(L_d i_d - \sqrt{\frac{3}{2}}\Psi_{rs}\right) + \frac{L_q}{Ai_d + B_\Psi}(v_q - Ai_q v_d) \qquad (4.91)$$

$$u_d = v_d L_d + R_d i_d + p\Omega_r L_q i_q$$

The application of this linearized model of BLDC drive (4.89) can be the following: the control value v_q constitutes the derivative of the given electromagnetic torque T_{ed}:

$$v_q = \dot{T}_{ed} \qquad (4.92)$$

Concurrently, the control variable v_d constitutes the derivative of the given transient of the motor's current i_d

$$v_d = \dot{i}_d \qquad (4.93)$$

The design of a trajectory of the motion of the drive can easily apply these variables obviously in the admissible area of the control that is determined by voltages

u_d, u_q. Under the assumption that we aim to reduce reluctance torque of the motor that interferes with transients and concurrently we need to perform a trajectory of the motion, the course of action can be the following. Let us assume that approximately $i_d \approx 0$ and simultaneously on the basis of (4.67) under standard control conditions the relation $u_d \approx 0$ is fulfilled, it is possible to establish the following equality on the basis of (4.90):

$$v_d = -p\Omega_r \frac{L_q}{L_d} i_q = -p\Omega_r \frac{L_q}{L_d B_\Psi} T_{ed} \tag{4.94}$$

since $i_q = \dfrac{T_{ed}}{B_\Psi}$, when $i_d = 0$ and the reluctance torque is missing.

In the subsequent step, on the basis of (4.91) it is possible to derive voltage u_q:

$$u_q = R_q i_q + \Omega_r B_\Psi + \frac{L_q}{B_\Psi}\left(v_q + Ap\Omega_r \frac{L_q}{L_d}\left(\frac{T_{ed}}{B_\Psi}\right)^2\right) \tag{4.95}$$

$$u_d = 0$$

Under the assumption of a planned trajectory in the form of initial curves for the variables

$$\mathbf{P} = [i_d = 0, v_q = \dot{T}_{ed}, T_{ed}, \Omega_{rd}] \tag{4.96}$$

after the application of (4.95) we will obtain the function for desired control $u_{qd}(t)$. The control system developed in this manner has to ensure the execution of the planned regulation of $u_{qd}(t)$ and additionally has an adjustment system with feedback, which has been designed to ensure the desired precision of the control despite the occurrence of interference. Moreover, one can note that usually the value of A is small in comparison to B_Ψ, v_q since it is related to the variable term M_s of inductance and that could be premises for disregarding the second term in the parentheses of the expression (4.95). Hence, in most simplified form, the examined control can be reduced to:

$$u_q = R_q i_q + \Omega_r B_\Psi + \frac{L_q}{B_\Psi} v_q \tag{4.97}$$

A problem is associated with the fact that the calculated controls, for instance (4.95) and (4.97) have to be performed in an untransformed system, which means that there is a necessity to apply the inverse transformation to obtain a system of phase voltages of the armature. The proposed transients for $u_d = 0$ are relatively easy in execution with the aid of pulse width factor k_u (4.66). Another type of control designed to match the planned trajectory of motion without transformation of the variables will be presented in the section that follows.

4.5.5 *Regulation of BLDC Motor with Inverse Dynamics*

This type of regulation is effective with regard to a system in which case it is possible to plan the trajectory of motion, which involves the determination of the position, speed and acceleration of the mechanical variables. In the examined case of a brushless DC motor this means that the desired transients $\theta_{rd}, \Omega_{rd}, \dot{\Omega}_{rd} = a_{rd}$ need to be familiar. Subscript d denotes the desired value, i.e. the one that marks an ideal trajectory of motion. The actual trajectory $\theta_r, \Omega_r, \dot{\Omega}_r = a_r$ usually does and will diverge from the desired one as a result of the effect of a number of factors which are disregarded when stating initial assumptions that tend to simplify either the mathematical model of the drive, constant parameters of the supply or other factors that get in the way of the process of regulation. The difference between the transients forms the error of the regulation. It is further applied for improving the control signal in the additional component of the regulation system, i.e. the corrector.

$$\varepsilon_\theta = \theta_{rd} - \theta_r$$
$$\varepsilon_\Omega = \Omega_{rd} - \Omega_r \qquad (4.98)$$
$$\varepsilon_a = a_{rd} - a_r$$

On the basis of the designed trajectory θ_{rd}, Ω_{rd}, a_{rd} and mathematical model it is possible to derive the needed rotational torque for a drive

$$T_{ed} = f(\theta_{rd}, \Omega_{rd}, a_{rd}) \qquad (4.99)$$

that is subsequently performed by the regulator of the torque which involves the calculation and input of a signal, that has the value corresponding to the pulse

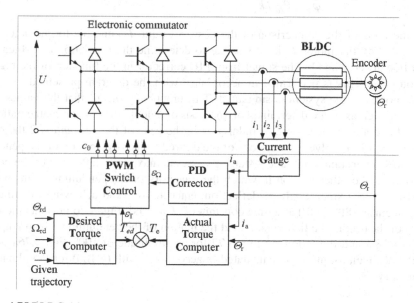

Fig. 4.75 BLDC drive control by the inverse dynamics method

width factor k_u, into the system of the electronic commutator. The diagram of the control system that applies the inverse dynamics method is presented in Fig. 4.75.

This method finds a common application in a number of drive systems whose task involves tracking of a given trajectory. The regulation system determines the basic control procedures on the basis of the computed required state while the PID corrector offer an input of an additional signal thus minimizing the control error. The benefits resulting from the application of this method include increase of the regulation speed, considerable reduction of errors since the predominant part of the control signal is given in advance and regardless of the error. Other advantages of this type of regulation are associated with effective control in non-linear systems since in this case the limitations regarding PID regulation are no longer in force. This method finds application in a number of industrial drives including the control of manipulators and robots. Below is a demonstration of the operating principle in practice with reference to an object from *Example 4.2* – control of a pendulum.

Example 4.3 This task involves the control of a large, massive pendulum to match a given trajectory, which is considerably diverges from its natural motion. The details are identical as in *Example 4.2*; besides, the illustration in Fig. 4.63 serves for the analysis. The control of the motion, which forms an example of tracking control can be based on the method of inverse dynamics. The trajectories are given by the functions (4.83), (4.84) for the variables of the pendulum in motion and not by the actual rotor movement. The determination of the desired torque in accordance with (4.81) offers the following result:

$$T_{ed}k_g = J_p\ddot{\varphi}_d + mgl\sin\varphi_d + (D_\varphi + D_r k_g^2)\dot{\varphi}_d$$
$$\varphi_d = \theta_{rd}/k_g \tag{4.100}$$

On the basis of the comparison of the desired torque T_{ed} and the torque actually generated by the motor T_e it is possible to determine the regulation k_u, which is additionally adjusted by the signal from PID corrector in the function of the regulation error. This is performed in accordance with the diagram presented in Fig. 4.75, however, the system is simulated. This practically means that the values θ_r, Ω_r, a_r as well as T_e are determined on the basis of calculations on the mathematical model and are not measured. Similarly as in the case of the *Example 4.2* the examination will involve two versions of the desired trajectory: motion of the pendulum with a constant amplitude – starting from the maximum deflection of the pendulum and the other version in which the trajectory of the pendulum begins with start-up from a small angle of deflection, equal to 10°, and achieving maximum displacement (80°, 100°) at a time constant of $T = 5$ [s]. Fig. 4.76 presents the results of the task in the first version and the motion of the pendulum with an amplitudes of 80° and the following Fig. 4.77 for an amplitude of 100°. Fig. 4.76a presents the given torque $T_{ed}k_g$ calculated in accordance with (4.100) for the planned trajectory.

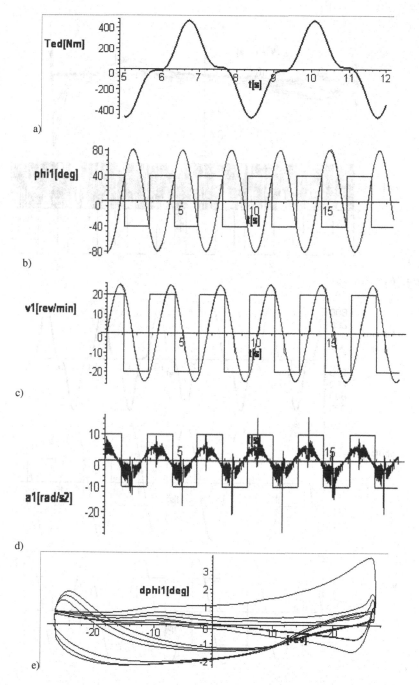

Fig. 4.76 Electromechanical variables for inverse dynamics control of swinging massive pendulum, for $\varphi_0 = -80°$: a) pre-computed desired torque $T_{ed}\,k_g$ b) position angle φ_1 c) angular speed, d) angular acceleration e) position error f) speed error ε_ω g) k_u factor

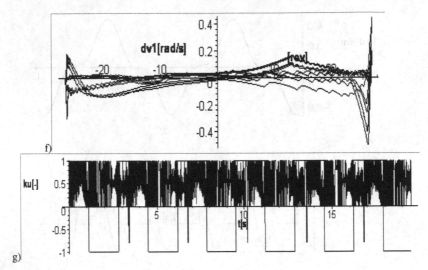

f)

g)

Fig. 4.76 (*continued*)

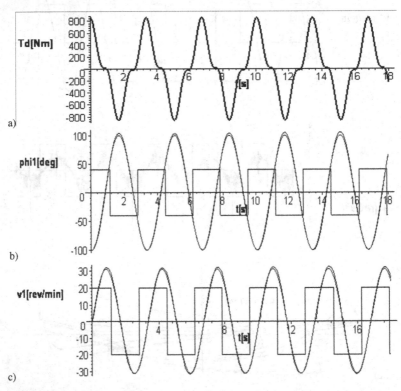

a)

b)

c)

Fig. 4.77 Electromechanical variables for inverse dynamics control of swinging massive pendulum, for $\varphi_0 = -100°$: a) desired torque $T_{ed}\,k_g$ b) position angle φ_1 c) angular speed d) angular acceleration e) position error ε_φ f) speed error ε_ω g) motor voltage h) electromagnetic torque $T_e\,k_g$

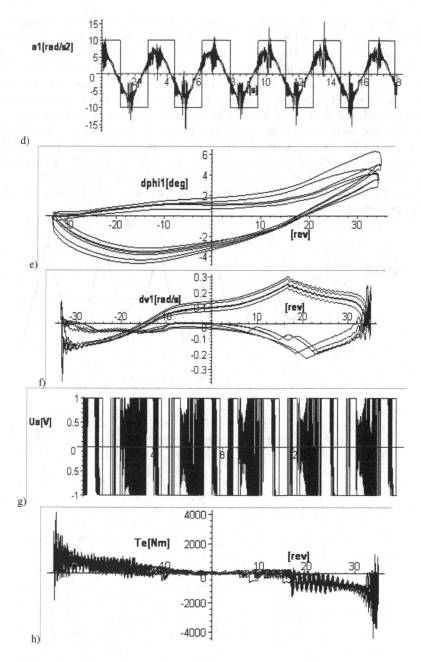

Fig. 4.77 (*continued*)

From the comparison of the transients in Fig. 4.65 with the ones in Fig. 4.76 (φ_0 = -70°) and the transients in Fig. 4.67 with the ones presented in Fig. 4.77

($\varphi_0 = -80°$ and $\varphi_0 = -100°$) leads to the conclusion that there is a fundamental dif-ference between the two. One can clearly conclude that the latter group i.e. results that are obtained by applying regulation with inverse dynamics, has considerable advantages over the ones presented earlier. The differences involve the stabiliza-tion of the transients to obtain large values of amplitudes of the swing of the pen-dulum and also very relevant improvement of the accuracy of the regulation as a result of applying inverse dynamics. The reduction of the positional error for sta-bilized transients is over 5 times, not to mention the possibility of steady operation for larger displacements. A set of figures that follows presents the results of the regulation for the alternative version of the task, i.e. for the start-up of the pendu-lum. Some samples of transients gained are presented in Fig. 4.79 for a rising am-plitude of the motion to reach the values $\varphi_a = 80°$ and $\varphi_a = 100°$ in the stable state of the pendulum.

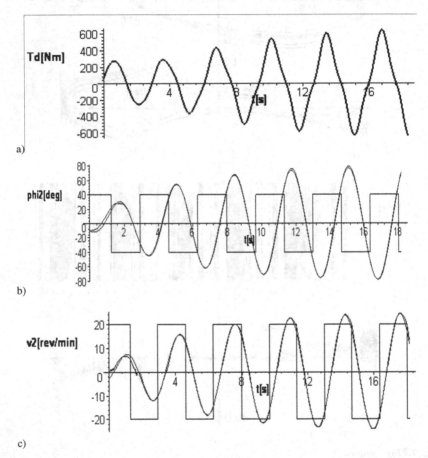

Fig. 4.78 Staring course of the pendulum for $\varphi_a = 80°$, with 'inverse dynamics' control: a) pre-computed desired torque T_{ed} b) position angle φ_2, c) angular speed d) angular accelera-tion e) position error ε_φ f) speed error ε_ω g) k_u factor

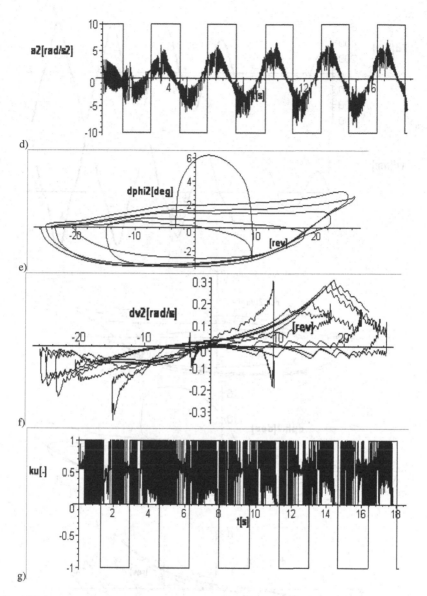

Fig. 4.78 (*continued*)

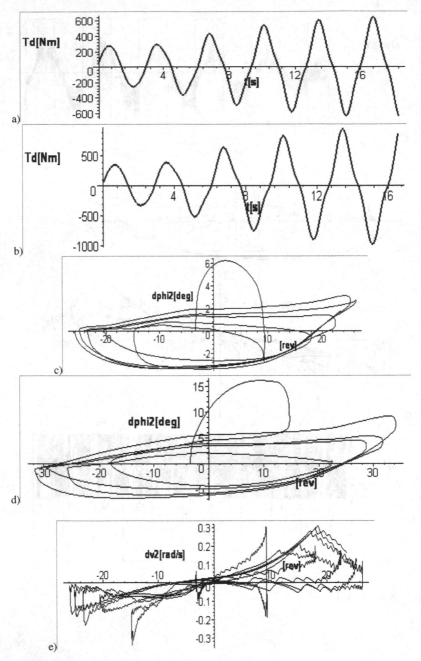

Fig. 4.79 Comparison of starting results of the pendulum for $\varphi_a = 80°$ versus $\varphi_a = 100°$, with inverse dynamics control: a) desired torque $T_{ed} k_g$ for $\varphi_a = 80°$ b) desired torque $T_{ed} k_g$ for $\varphi_a = 100°$ c) position error ε_φ for $\varphi_a = 80°$ d) position error ε_φ for $\varphi_a = 100°$ e) speed error ε_ω for $\varphi_a = 80°$ f) speed error ε_ω for $\varphi_a = 100°$ g) electromagnetic torque for $\varphi_a = 80°$ h) electromagnetic torque for $\varphi_a = 100°$

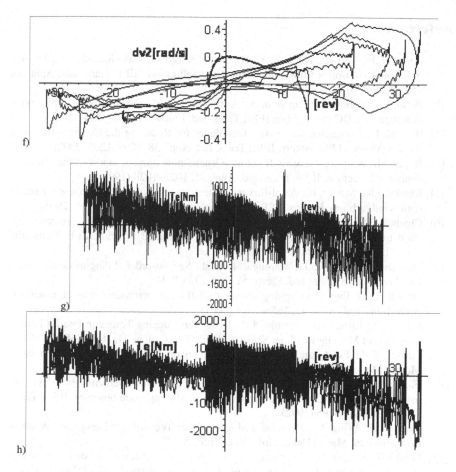

Fig. 4.79 (*continued*)

In the commentary of the results that were obtained one can compare the results presented in Fig. 4.72 with the ones in Fig. 4.78 since both of them refer to the same given trajectory, i.e. the start-up of a pendulum to reach $\varphi_a = 80°$ for various control procedures. For the case of PIDD regulation the start-up directly leads to non-stability (Fig. 4.72) and the error of the angular position increases and even exceeds already 10°. In contrast, in the case of the application of control using inverse dynamics (Fig. 4.78), the error ε_φ during the start - up tends to stabilize and reaches the range of ±2° for a minimum value of the speed error. The results of the subsequent Fig. 4.79 indicate that increasing the amplitude to $\varphi_a = 100°$ offers the possibility of effective regulation of the position in the tracking motion while the position error stabilizes in the range of ±5°. The examples presented here indicate that the method is very effective despite the fact that it has been applied without the precise selection of adequate regulation parameters.

References

[1] Bernal, F.F., Cerrada, A.G., Faure, R.: Model-Based Loss Minimization for DC and AC Vector-controlled Motors Including Core Saturation. IEEE Trans. Ind. Appl. 36, 755–763 (2000)

[2] Betin, F., Pinchon, D., Capolino, A.: A time-varying sliding surface for robust position control of DC motor drive. IEEE Trans. Ind. Electr. 49, 462–473 (2002)

[3] Bianchi, P., Bolognani, S.: Design Techniques for Reducing the Cogging Torque in Surface-Mounted PM Motors. IEEE Trans. Ind. Appl. 38, 1259–1265 (2002)

[4] Boger, M., Wallace, A., Spee, R., et al.: General pole number model of the brushless doubly-fed machine. IEEE Trans. Ind. Appl. 31, 1022–1028 (1995)

[5] Bumby, J.R., Martin, R.: Axial-flux permanent magnet air-cored generator for small-scale wind turbines. In: Proc. IEE, El. P. Appl., vol. 152, pp. 1065–1075 (2005)

[6] Cinchilla, M., Arnaltes, S., Burgos, J.C.: Control of permanent-magnet generators applied to variable-speed wind-energy systems connected to grid. IEEE Trans. En. Con. 21, 130–135 (2006)

[7] Damiano, A., Gatto, G.L., Marongiu, I., et al.: Second-order sliding-mode control of DC drives. IEEE Trans. Ind. Electr. 51, 364–373 (2004)

[8] Dosiek, L., Pillay, P.: Cogging torque reduction in permanent magnet machines. IEEE Trans. Ind. Appl. 43, 1656–1671 (2007)

[9] Filho, E.R., Lima, A.M., Araujo, T.S.: Reduction Cogging Torque in Interior Permanent Magnet Machines without Skewing. IEEE Trans. Mag. 34, 3652–3655 (1998)

[10] Furlani, E.P.: Field Analysis and Optimization on NdFeB Axial Field Permanent Magnet Motors. IEEE Trans. Mag. 33, 3883–3885 (1997)

[11] Furuhashi, T., Sangwongwanich, S., Okuma, S.: A position-velocity sensorless control for brushless DC motors using an adaptive sliding mode observer. IEEE Trans. Ind. Electr. 39, 89–95 (1992)

[12] Gieras, J.F., Wing, M.: Permanent Magnet Motor Technology: Design and Applications, 2nd edn. Marcel Dekker Inc., New York (2002)

[13] Glinka, T., Kulesz, B.: Comparison of rated power and efficiency of dc motor, induction motor and brushless motor with electronic commutator. In: Conf. Proc. 16th Int. Conf. on Electr. Mach. – ICEM 2004 (2004)

[14] Ha, I., Kang, C.: Explicit characterization of all feedback-linearizing controllers for a general type brushless DC motor. IEEE Trans. Aut. Contr. 39, 673–677 (1994)

[15] Hamdi, E.S.: Design of Small Electrical Machines. John Wiley & Sons, Chichester (1994)

[16] Hemati, N., Le, M.C.: Complete model characterization of brushless DC motors. IEEE Trans. Ind. Appl. 28, 172–180 (1992)

[17] Hendershot, J.R., Miller, T.J.E.: Design of Brushless Permanent Magnet Motors. Magna Physics and Clarendon Press, Oxford (1994)

[18] Holtz, J., Springob, L.: Identification and compensation of torque ripples in high-precision permanent motor drives. IEEE Trans. Ind. Electr. 34, 309–320 (1996)

[19] Hu, J., Dawson, D.M., Anderson, K.: Position control of a brushless DC motor without velocity measurements. IEE Proc.– El. P. Appl. 140, 113–122 (1993)

[20] Hung, J., Ding, Z.: Design of currents to reduce torque ripple in brushless DC motors. IEEE. Proc. El. P. Appl. 140, 260–266 (1993)

[21] Hwang, S.M., Lieu, D.K.: Reduction of torque ripple in brushless DC motors. IEEE Trans. Mag. 31, 3737–3739 (1995)

[22] Ito, M., Kawabata, K., Tajima, F., et al.: Coupled Magnetic Field Analysis with Circuit and Kinematics Modeling of Brushless Motors. IEEE Trans. Mag. 33, 1702–1705 (1996)

[23] Jabbar, M.A., Phyu, H.N., Liu, Z.: Analysis of the starting process of a disc drive spindle motor by time stepping finite element method. IEEE Trans. Mag. 40, 3204–3206 (2004)

[24] Jang, G.H., Kim, M.G.: Optimal Commutation of a BLDC Motor by Utilizing the Symmetric Terminal Voltage. IEEE Trans. Mag. 42, 3473–3475 (2006)

[25] Kabashima, T., Kawahara, A., Goto, T.: Force Calculation Using Magnetizing Currents. IEEE Trans. Mag. 24, 451–454 (1998)

[26] Kawase, Y., Hayashi, Y., Yamaguchi, T.: 3D finite element analysis of permanent magnet motor excited from square pulse voltage source. IEEE Trans. Mag. 32, 1537–1540 (2006)

[27] Kawase, Y., Yamaguchi, T., Hayashi, Y.: Analysis of cogging torque of permanent magnet motor by 3-d finite element method. IEEE Trans. Mag. 31, 2044–2047 (1995)

[28] Kenjo, T., Nagamori, S.: Permanent Magnet and Brushless DC Motors. Clarendon Press, Oxford (1985)

[29] Kim, D.K., Lee, K.W., Kwon, B.I.: Commutation Torque Ripple Reduction in a Position Sensorless Brushless DC Motor Drive. IEEE Trans. P. Electr. 21, 1762–1768 (2006)

[30] Kim, T.H., Ehsani, M.: Sensorless control of BLDC motors from near-zero to high speeds. IEEE Trans. P. Electr. 19, 1635–1645 (2004)

[31] Lateb, R., Takorabet, N., Meibody-Tabar, F.: Effect of magnet segmentation on the cogging torque in surface-mounted permanent magnet motors. IEEE Tran. Mag. 42, 442–445 (2006)

[32] Lederer, D.H., Kost, N.: Modeling of non-linear magnetic material using a complex effective reluctivity. IEEE Trans. Mag. 35, 3060–3063 (1998)

[33] Lee, W.J., Sul, S.K.: A New Starting Method of BLDC Motors without Position Sensors. IEEE Trans. Ind. Appl. 42, 1532–1538 (2006)

[34] Low, T., Tseng, K., Lee, T., et al.: Strategy for the instantaneous torque control of permanent magnet brushless DC drives. IEEE Proc. El. P. Appl. 137, 355–363 (1990)

[35] Łukaniszyn, M., Jagieła, M., Wróbel, R.: Optimization of permanent magnet shape for minimum cogging torque using a genetic algorithm. IEEE Trans. Mag. 40, 1228–1231 (2004)

[36] Łukaniszyn, M., Wrobel, R.: A study on the influence of permanent magnet dimensions and stator core structures on the torque of the disc-type brushless DC motor. El. Eng. 82, 164–169 (2000)

[37] Magnussen, F., Lendenmann, H.: Parasitic Effects in PM Machines with Concentrated Windings. IEEE Trans. Ind. Appl. 43, 1223–1232 (2007)

[38] Matsui, N.: Sensorless PM brushless DC motor drives. IEEE Trans. Ind. Electr. 43, 300–308 (1996)

[39] Melkote, H., Khorrami, F.: Nonlinear adaptive control of direct-drive brushless DC motors and applications to robotic manipulators. IEEE/ASME Trans. Mechatr. 4, 71–81 (1999)

[40] Shao, J., Nolan, D., Tessier, M., et al.: A novel microcontroller-based sensorless brushless (BLDC) motor drive for automotive fuel pumps. IEEE Trans. Appl. 39, 1734–1740 (2003)

[41] Wang, K., Atallah, K., Howe, D.: Optimal torque control of fault-tolerant permanent magnet brushless machines. IEEE Transactions on Mag. 39, 2962–2964 (2003)

[42] Zeroug, H., Boukakis, B., Sahraoui, H.: Modeling and Analysis of a Brushless DC Motor Drive. In: Conf. Proc. of 13th Int. Conf. El Mach. (ICEM 1998), pp. 1255–1260 (1998)

[43] Zhu, Z.Q., Howe, D., Chan, C.C.: An improved analytical model for predicting the magnetic field distribution in brushless permanent magnet motors. IEEE Trans. Mag. 38, 229–238 (2002)

[44] Zhu, Z.Q., Howe, D., Mitchell, J.K.: Magnetic Field Analysis and Inductances of Brushless DC Machines with Surface-Mounted Magnets and Non-Overlapping Stator Winding. IEEE Trans. Mag. 31, 2115–2118 (1995)

Chapter 5
Switched Reluctance Motor Drives

Abstract. The chapter is devoted to Switched Reluctance Motor (SRM) drives. Firstly nonlinear magnetizing curves of SRM are presented and their importance for motor's operation is discussed. The model presented in the chapter takes into account these nonlinear characteristics depending on phase current and rotor position angle, but ignores mutual magnetization of phases. The magnetization curves are regarded in a specific and original way as a product of nonlinear functions depending on magnetic saturation and rotor position angle. This approach seems to be useful as it enables one to analyze the influence of particular construction elements on characteristics of a motor. In consequence mutual inductances of phases are disregarded, but their actual influence is presented for two typical SRMs, and proved to be marginal. Several problems of SRM operation and control are presented based on mathematical models and results of computer simulations. Among others they are: determining a pulse sequence for starting, direct start up with current limitation, breaking and a comprehensive discussion of generator operation conditions. The problem of regulation parameters fitting is also presented, considered from a point of view of gaining possibly high efficiency and low torque ripple level. As far as control of SRM is concerned there is sliding mode control discussed as well as current control and DTC with an aim to minimize torque pulsation at various states of operation. Besides, there is the problem of a control with and without position/speed sensors presented and state observers application discussed that enable this kind of control.

5.1 Introduction

Switched reluctance motor (SRM) as an engineering solution to the design of the electric motor in rotational motion realizes one of the earliest ideas of operation principle for the electric motor, which originates from the first half of the 19th century. It employs the simple concept of an electromechanical system based on the attraction of a ferromagnetic element by an electromagnet. In order to make this idea viable for a rotational motor alternately energized coils are situated on Z_s stator teeth, while Z_r salient rotor teeth without windings are attracted by adequately energized stator windings. The art in their design as well as the fundamental technological problem is associated with adequate energizing and de-energizing stator windings in a proper phase sequence in order to ensure a smooth rotation of the rotor whose teeth are attracted by generated magnetic field. Despite simple operating principle the practical operation of SRM motor is associated with a need

to apply fast and efficient electronic switches in order to realize the switching sequence. Besides, sensors of rotor position with sufficient precision are needed to
secure commutation of the currents in the windings. For these reasons the development and wide application of the design started as late as 1980s when adequate
power electronic components were available to realize such switching functions
[24,32,33]. In order to ensure a starting torque and possibly smooth rotational motion it is necessary that rotor teeth are not aligned in respect to all corresponding
stator teeth concurrently. For a motor with a single pole pair $p = 1$, this simultaneous alignment occurs only for the opposite teeth, i.e. more generally at an angular
interval of π/p. For this reason the number of stator and rotor teeth is usually different. The most common solutions apply the following sequence of teeth numbers: $Z_s/Z_r = 6/4$ and $Z_s/Z_r = 8/6$. Self-evidently, other teeth number sequence is
encountered, e.g. $Z_s/Z_r = 4/2$ for a two-phase motor, or $Z_s/Z_r = 10/8$ for a motor
with five phase windings in the stator. The number of phases for motors with a
single pole pair is equal to $m = Z_s/2$. This results from the fact that the opposite
stator teeth are energized simultaneously with an equal current during the connection of the coils in series. But parallel connection of windings on opposite stator
teeth is also possible. As a result, two coils of opposite stator teeth form a single
phase winding of the motor. For higher pole pair numbers, the number of pairs of
stator and rotor teeth, whose axes overlap, is respectively higher and amounts to p
> 1 while the stator windings belonging to the same phase are energized in the
same sequence; hence, the number of phases is equal to $m = Z_s/(2p)$. This affects
the respectively higher electromagnetic torque of the motor. Fig 5.1 presents a
cross-section of a $Z_s/Z_r = 6/4$ SRM motor, i.e. a reluctance three-phase motor.

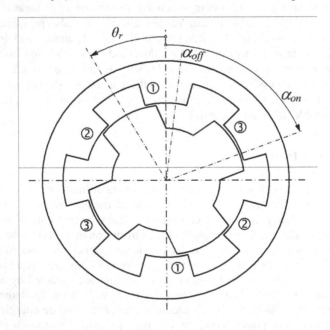

Fig. 5.1 Cross-section of a $Z_s/Z_r = 6/4$ SRM motor, with the indication of rotor position angle θ_r, and switch angles α_{on} and α_{off}

The application of SRM motor has opened new boundaries in the practice of the design of electric motor drives. In connection with this, it would be of value to characterize its basic parameters and define the scope of its application. In terms of the rated power SRM motors are manufactured in a range from very small units with the output of several watts to enormous drives with the power of several hundred kW and can even reach MW level. The comparison of the energy efficiency, start-up torques and torque overloads between the typical SRM motors and the corresponding induction motors leads to the conclusion that the SRMs offer more advantages; however, the advantage of the SRM motor is not significant. A comparative analysis of a SRM motor with a corresponding induction motor accounting also for noise emission level is based on the example of [5]. One of the characteristics of SRM motor includes the possibility of gaining high rotational speeds, as high as 10,000 [rev/min] without special engineering changes in the motor. At the same time, SMR motors whose special design makes them high-speed can reach as high as 100,000 [rev/min]. One has to note that such machines have a smooth cylindrical rotor while the effect of variable reluctance is obtained as a result of the application of materials that vary in terms of magnetic permeability along the circumference of the rotor. Another special feature of SRM motor involves its mechanical characteristic $\Omega_r = f(T_l)$ whose waveform is similar to the series wound DC motors. This means that during an increased load this motor considerably slows down it rotation and when the load is reduced it accelerates. The negative characteristics of SRM motors include high torque ripple and higher noise level in comparison to e.g. induction motors [5,8,16,17,38]. The counter-measures include the proper magnetic circuit construction and application of adequate control systems thus reducing torque pulses generated by the motor. As far as the applications of SRM motors is concerned, they are similar to the uses of induction motors and series wound DC motors. In particular, they find application in traction drives and car drives [23,37] due to flexible mechanical characteristics, large torque overload capability, simple construction and high level of reliability. SRM motors can be successfully applied in servomotors and actuators. SRM machines can also play the role of generators; however, due to the passive role of the rotor the magnetic excitation has to occur as a result of current passing through stator windings, which is associated with specific requirements regarding control and affects the efficiency of the machine as a generator. Bibliography devoted to this problem is numerous [18,19,23] also in the context of the construction of wind power stations [34,36] and this area is the subject in chapter 5.4.3. In terms of the investment a SRM motor drive is cheaper in comparison to an induction motor drive and so is converter as a result of the much more simplified construction of the stator winding, lack of rotor windings and less complicated system needed to supply the machine. At present, the lower popularity of the motor and smaller offer on the part of manufacturers result in the fact that SRM motors have not yet been able to demonstrate all the advantages they have over induction motors.

5.2 Operating Principle and Supply Systems of SRM Motors

The most fundamental issue in the control of SRM motor drives is associated with adequate sequential voltage feeding to and disconnecting from motor's windings. Fig. 5.2 presents the cross-section of a motor with $Z_s/Z_r = 8/6$ teeth and angular position α_{on} and α_{off} of the rotor for which respective phase windings are supplied and subsequently disconnected from an external source. An intuitive understanding of the operating principle of such a motor suggests that the positive torque $(T_e > 0)$, i.e. one in accordance with direction of rotation of the rotor is encountered when the rotor transfers from the position of highest reluctance to the position with smallest reluctance with respect to the teeth of the stator whose winding is supplied. The characteristic positions of the rotor are denoted as unaligned position and aligned position, the latter of which refers to the overlapping of the axis of stator teeth and the one on rotor's teeth. The winding of a given phase should be supplied with a current of an adequate value in the range of this rotation angle, i.e. $\alpha_{on} < \theta_r \leq \alpha_{off}$. This means that the supply should be switched on as defined by angle α_{on} slightly before the unaligned position, while the supply is disconnected (as described by angle α_{off}) slightly in advance in relation to the instant when a tooth reaches an aligned position. The difference between the angle of the switching on of a power supply to the phase winding and the angle when it is disconnected is known as conduction angle, which is equal to:

$$\alpha_z = \alpha_{on} - \alpha_{off} \tag{5.1}$$

As one can conclude, the above mentioned advance of switching on and off of the voltage for a specific phase of the motor results from the dynamic characteristics of current increase in the phase winding following an instant the supply is switched on and subsequent decay of the current after the supply is disconnected. Moreover, it is relative to the rotor's speed, inductance of the windings as well as the control of converter switches. For small rotational speeds the current in the winding increases relatively fast with regard to the entire conduction angle α_z while the value of the current is controlled as a result of using PWM method. In contrast, for high rotational speeds the increase after switching on and decay of the current after disconnecting the supply comes relatively slowly in the time period which is determined by the conduction angle α_z since this angular range of the rotor motion is covered in a very short period of time. Additionally, relatively large back EMF is induced in the windings and the increase of the current is enforced only by the difference in voltage between the supply voltage u for a given phase and the back EMF e_b. As one can see, the control of the switch on angle α_{on} and the conduction angle α_z forms the basic method applied in the control and adaptation of characteristics of SRM motor, beside the possibility of controlling supply voltage usually achieved with the aid of PWM method. As a consequence of analyzing the operation of SRM motor one can determine the theoretical conduction angle for a single phase ε - denoted as stroke angle, which results from the number of phases and teeth of the rotor under the assumption of a separate conduction of the windings. In a typical SRM the difference in numbers of a stator and rotor teeth per pole pair is 2, i.e. $Z_s - Z_r = 2p$. Hence the stroke angle ε, as the smallest angular distance between closest stator and rotor teeth, in a state of alignment occurring for some other pair of teeth, is

$$\varepsilon = \frac{2\pi}{p}\left(\frac{1}{Z_r} - \frac{1}{Z_s}\right) = \frac{2\pi}{p}\frac{2p}{Z_sZ_r} = \frac{2\pi}{mZ_r} \tag{5.2}$$

For a motor with $Z_s/Z_r = 6/4$ teeth this gives an angle $\varepsilon = 360°/(3\cdot4) = 30°$, while for one with $Z_s/Z_r = 8/6$ teeth stroke angle is equal to $\varepsilon = 360°/(4\cdot6) = 15°$. The actual value of the conduction angles α_z is always greater than the stroke angle ε as a result of the processes of current increase and decay in the phase windings and, hence, during the operation of the SRM motor there are periods when 2 or even 3 phase windings are in conducting ('ON') state. Since usually the aim is to gain high values of electromagnetic torque, the conduction period is extended within the range of strong attraction of the rotor's tooth by the electromagnet made up by a pair of stator teeth so that the switch off angle α_{off} only slightly precedes the aligned position. For this reason the process of current decay in a given phase is accelerated as much as possible to avoid negative torque values. This occurs after the rotor reaches the position determined by the angle α_{off} as a result of energizing this winding with is reverse voltage u that supplies the phase and, thus, causing the energy return to the source. Such a capability has to be secured through the commutation system of the phases of SRM motor.

The basic system of the power supply and commutation of a single phase winding of SRM involves an asymmetric transistor /diode H bridge shown in Fig. 5.3. Since SRM is a reluctance motor and the direction of the torque is not relative to

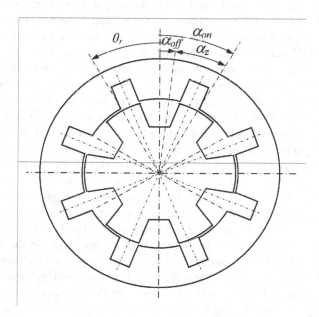

Fig. 5.2 Cross-section of a $Z_s/Z_r = 8/6$ SRM motor, with the indication of rotor position angle θ_r, switching angles α_{on} and α_{off}, as well as conduction angle α_z

the direction of the current flow through the phase winding, the commutator bridge does not need to facilitate current flow through the winding in both directions and it is sufficient to apply two power transistors and two diodes to ensure energy supply and energy return to the source.

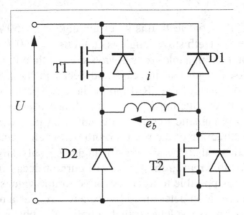

Fig. 5.3 A typical commutation H-bridge circuit for switching current of a single phase of a SRM

Over the period when the winding is in supply state from the source both transistors T1 and T2 are in the ON state. This occurs for the angle of the rotation in the range from α_{on} to α_{off} and when simultaneously there is a increase of the current in the cycle of the PWM regulation of the voltage. However, in the section of PWM cycle when phase current decreases, only one transistor and one diode in H bridge are in ON state. This could be transistor T2 and diode D2 of the bridge and in this case we have to do with a decay of the current in the circuit in which these elements short a phase winding. In contrast, after the rotor reaches position α_{off}, both transistors are switched off and the current in the winding is closed in the circuit formed by two diodes D1 and D2 and the power source. This direction of the current flow determined by the diodes results in the return of the energy stored in the electromagnetic field into the source accompanied with a rapid decay of the current in the winding. We can also consider an option of mechanical energy conversion over this period and it is only relative to the actual sense of electromagnetic torque and its value. However, for the purposes of rough explanation of the operating principle of the motor we can assume that in the vicinity of the aligned position of the rotor the electromagnetic torque is small and, as a result, the return of the energy consists only in the return of the energy stored in the electromagnetic field. The following figures (Fig 5.4 – Fig 5.6) contain an illustration of the operation of the commutation system for a single phase winding, for the respective low rotational speed, i.e. for $n = 600$ [rev/min], for the mid speed ranges, i.e. 1600 [rev/min] (Fig. 5.5) and for higher speeds, e.g. for $n = 3000$ [rev/min] (Fig. 5.6). The variable deciding on the switching sequence is the value of the current in this winding and it is responsible for voltage switching between the values of u and 0 - i.e. performing so-called 'soft-chopping'.

After the completion of the switching cycle in a given phase, i.e. after the rotor angle exceeds position α_{off}, both transistors are switched off and the current in the winding decays quickly and recuperation of energy, due to the supply with $-u$ voltage, takes place.

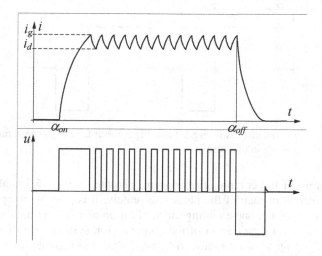

Fig. 5.4 Current commutation during a single conduction cycle of a phase winding with a current limitation, for a low speed range of SRM

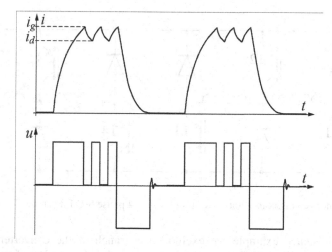

Fig. 5.5 Similar commutation as presented in Fig 5.4, but for a medium speed range of SRM

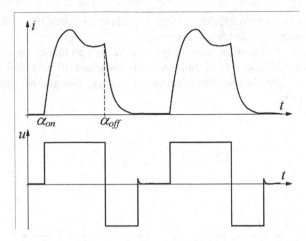

Fig. 5.6 Current commutation as in Fig 5.4 and Fig 5.5, but for a high speed range. Current regulation is not possible in this case

The diagram of the commutation system presented in Fig. 5.3 enables one to control the current in each of the phases separately. It is possible to apply a more economical system, i.e. one enabling the application of a smaller number of power electronic switches in the design of the commutation system. One of such examples is found in Fig. 5.7 for a motor with $m = 4$ phases or a greater even number of phases. The control of the switching sequence of the phases is based on an assumption that during the operation of the motor we don't have to do with simultaneous conduction in more than two phase windings.

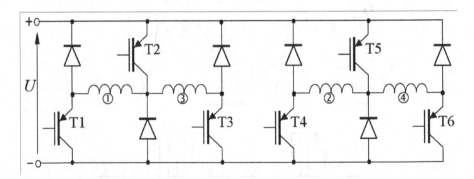

Fig. 5.7 Branch-saving commutation system for $m = 4$ phase SRM machine

In the presented example we exclude one branch of the converter per four branches in two complete H bridges, which means that the number of components decreases by 25%.

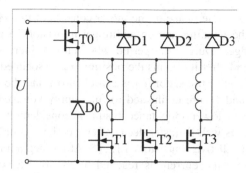

Fig. 5.8 Switch saving commutation scheme with one general T0 switch and single switches for each of phase windings

A system that goes even further in terms of the economical use of electronic components is presented in Fig. 5.8. Each of the phases apply a single transistor and there is one transistor T0 that is common to them all. This system offers the possibility of reducing commutation losses; however, practically it does not permit recuperation of energy of the phase that is being in the switching off state, because it requires the transistor T0 to be switched off. In this case we have to do with the decay of the current in the closed circuit of this phase winding across phase diode and transistor T0. Considerable opportunities in terms of the improvement of the operation in the range of high rotational speed is offered by the system [47], in which phase windings containing coils situated in the opposite stator teeth have available clamps to supply each coil separately. It means that the phase winding can be effectively divided into two equal parts. In this case it is possible to apply an alternative, in-series or parallel supply of the two parts of the phase winding and, as a result, considerably accelerate the increase and decay of the current in the winding and, additionally, increase the range of the speeds for which it is possible to control the current. Such system configuration involving division of a winding and enabling series power supply for lower speeds and, concurrently, parallel connection at higher speeds is presented in Fig. 5.9.

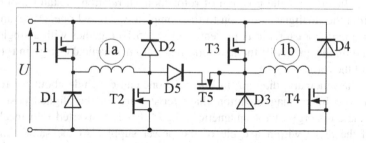

Fig. 5.9 Commutation scheme of a divided SRM phase winding for changeable supply configuration of both parts

This system for a single phase of the motor consists of two H bridges that supply the halves of the winding and an additional transistor T5 and diode D5 that connects these bridges. During the period when current increases after the rotor gains α_{on} position and after it obtains the position α_{off} associated with the termination of the phase supply, transistor T5 is switched off and the two halves of the phase winding 1b and 1b are connected to full supply voltage u. This fact combined with two times lower inductance result in considerably faster current increase after it is fed as well as faster decay after the phase is disconnected as a result of switching off all five transistors T1...T5. At the beginning of a cycle, after the required value of the current is reached, transistor T5 connecting the two bridges is switched on and the winding halves are connected in a series to form a single phase winding. Within a single cycle of the supply the switching of the part phases to parallel and, subsequently, to series supply can be performed several times thus increasing the range in which it is possible to control the current in the motor, as in Fig. 5.5. For the case when after the switching from the parallel to series connection the currents in the two halves are not equal the currents in them have to be balanced and the surplus of the current in one section of the winding returns through one of the diodes – D2 or D4. As it was indicated by computer simulations and the operation of such experimental set-up [47], the application of series-parallel switching of the winding halves in a two-phase motor ($m = 2$) has led to an increase of the rotational speed by 80% under rated loading and the rated current has not been exceeded.

5.3   Magnetization Characteristics and Torque Producing in SRM Motor

In terms of the construction the capability of a motor to transform energy and produce a torque is determined by the magnetization characteristics, whose waveforms result from the engineering details and properties of the ferromagnetic materials applied. What is meant here is the family of magnetization characteristics in the function of the position of rotor tooth in relation to stator tooth in the range from the unaligned position to the completely aligned one. An example of magnetization characteristic is presented in Fig. 5.10 together with a single cycle of converting the energy of the magnetic field into mechanical energy in rotational motion of the rotor.

From the schematic diagram in Fig. 5.10 one can conclude about the relation: the more non-linear magnetization characteristic are for the aligned position the greater the co-energy of the magnetic field T' that is converted into mechanical work of the drive within a cycle of the power supply. At the same time, less energy of the magnetic field T_f is returned to the source during the power diode

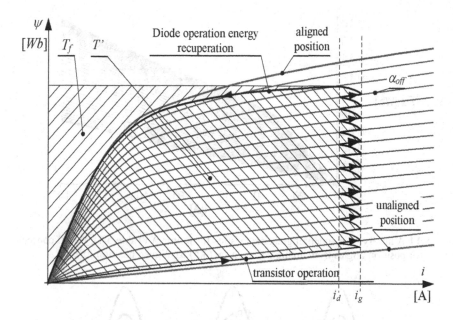

Fig. 5.10 A single cycle of energy conversion in SRM. $T \checkmark$ is co-energy conversed from magnetic to mechanical form, T_f is magnetic energy recuperated to a source in a diode conduction period of the current decay

conduction duty period. On the basis of the cycle of energy conversion presented in Fig. 5.10 it is possible to assess the motor's torque:

$$T_e = \frac{\text{mechanical power/cycle}}{\text{rotation angle/cycle}} = \frac{P_m}{\theta_r} = \frac{mZ_rT'}{2\pi} \quad [Nm] \qquad (5.3)$$

The relation presented here (5.3) and the interpretation of the cycle of energy conversion in Fig. 5.10 are relevant for a simplified case when the particular machine duty cycles are separate and there isn't a period of a common conduction.

In the further part of this chapter the examples and illustrations will be based on two typical layouts of SRM motors:

- motor A with rated values: Z_s/Z_r =8/6; P_n = 900 [W]; U_n = 32 [V]
- motor B, for which: Z_s/Z_r =6/4; P_n = 900 [W]; U_n = 310 [V].

Details of the two machines are summarized in Table 5.1 in Chapter 5.4. Below is a summary of the characteristics of magnetization and several other characteristics associated with torque generation for motor A.

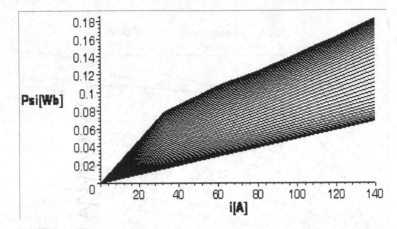

Fig. 5.11 A family of magnetizing characteristics for motor A (Table 5.1), for 50 consecutive rotor position angles from unaligned position (30°) to aligned position (0°)

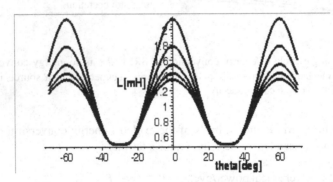

Fig. 5.12 Saturated inductance curves of motor A for increasing stator phase current values $i =$ 40,60,80,100,120 [A]

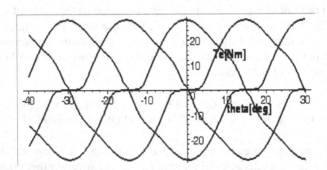

Fig. 5.13 Electromagnetic torque curves, for an individual supply of consecutive phase windings of motor A with $i = 120$ [A] as a function of a rotor position θ_r

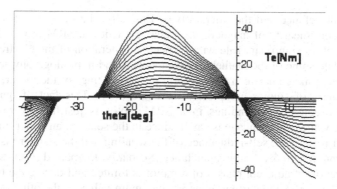

Fig. 5.14 Electromagnetic torque curves during a single phase supply with increasing current $i = 8,16,24,\ldots160$ [A] as a function of a rotor position θ_r

5.4 Mathematical Model of SRM Motor

5.4.1 Foundations and Assumptions of the Mathematical Model

The mathematical model forms the basic tool for conducting simulations of dynamic courses and finding characteristics of the motor as well as a whole drive. It is also indispensable in the research of drive control systems. The degree of complication and the precision of a given mathematical model is relative to the practical application of a model, i.e. the number, type and relevance range of the characteristics that are determined by its use. The form of the mathematical model is clearly relative to the simplifying assumptions adopted during the statement of the model. It is also important to note the source of information serving for the purposes of developing a model, i.e. whether the source originates in engineering data or data gained on the basis of measuring and testing existing objects, or data has been gained in some other way. The models based on measurements can either be deterministic in nature or originate on the basis of artificial intelligence methods. The latter, however, have a limited scope of application since they serve in order to examine objects the information of which is often approximated [21,22,29,49].

In the bibliography in this subject there is a considerable number of mathematical models of this kind, which, however, differ in details. For instance, mathematical models of SRM, which apply engineering details and linkage between the windings are presented in [13], and the ones without linkage are found in [45]. In turn, another model designed for dynamic calculations and simulations of waveforms is presented in [19] under the assumption of familiarity of functions of windings inductance and their derivatives. Basically, the majority of models account for non-linearity of magnetization characteristics in SRM motor. This is so because it plays significant role with respect to generation of the electromagnetic torque. While some of the mathematical models found in bibliography account for the magnetic linkage between the adjacent phase windings of the motor, a considerable number of studies completely disregard them. This fact of ignoring the linkage between phase windings in a SRM motor is justified by two details. Firstly, the value of the linkage is small, since in the sense of an upper limit it does not exceed 10% of the self-inductance of the winding and the standard is that their value is from 5% to 6% of this inductance. Secondly, the period of the simultane·ous conduction of adjacent phases of a motor is limited and during the time when in one of the phases the current gains its maximum value, in the other one it is already decaying. The presented mathematical model is designed to conduct swift and multiple dynamic calculations of the waveforms of SRM; hence, it is considerably simplified. It accounts for non-linearity of the characteristics of magnetization of the motor, which is indispensable, but disregards magnetic linkage between the windings. It leads to the simplification of the model concurrently not causing relevant errors as a result of the application of the model. This is confirmed by the characteristics and waveforms gained on the basis of measurements and, in particular, this pertains to characteristics gained for the case of reversing the current in the adjacent phase windings during comparative analysis and measurement of inductance in standard SRM motors. The inconsiderable differences between the characteristics gained on the basis of such measurements indicate that the linkage between phases can be disregarded without the deterioration of the precision of the results collected by the application of this model. The mathematical model of the motor is derived on the basis of Lagrange's equations for an electromechanical system while preserving the notations used in Fig. 5.2. The number of the degrees of freedom in the system is equal to:

$$s = m + 1 \qquad (5.4)$$

while m denotes the number of the electric degrees of freedom of the system and the electric charges associated with phase windings form the respective variables

$$q_1 = Q_1 \qquad q_2 = Q_2 \quad \dots \quad q_m = Q_m$$

and their time derivatives have the meaning of phase currents of the motor:

$$\dot{Q}_k = i_k \qquad k = 1 \dots m \qquad (5.5)$$

Concurrently, the remaining degree of freedom is reserved for the mechanical variable of the system

$$q_{m+1} = \theta_r$$

and denotes the angle of rotation of the rotor. Under the assumption of the lack of magnetic linkage between phase windings, Lagrange's function for this system takes the form:

$$L = \sum_{k=1}^{m} \int_{0}^{i_k} \Psi_k\left(\theta_r, \tilde{i}_k\right) d\tilde{i}_k + \frac{1}{2} J \dot{\theta}_r^2 \qquad (5.6)$$

The virtual work of the system corresponding to the exchange of energy between the system and the surrounding environment is equal to:

$$\delta A = \sum_{k=1}^{m} \left(u_k - R_k i_k\right) \delta Q_k + \left(-T_l + D\dot{\theta}_r\right) \delta\theta_r \qquad (5.7)$$

in which the particular terms refer to:

$\Psi_k\left(\theta_r, i_k\right) = M_{kk}\left(\theta_r, i_k\right) i_k$ - magnetic flux associated with k-th phase winding,
R_k - resistance associated with current flow through k-th phase winding, accounting for resistance of electronic elements and resistance of the supply source
u_k - voltage applied to k-th phase winding,
J - moment of inertia associated with motor's shaft,.
D - viscous damping coefficient associated with damping of the motion,
T_l - shaft load torque.

5.4.2 Equations of Motion for the Motor

The generalized form of equations of motion in accordance with Lagrange's model (2.51) is in the form

$$\frac{d}{dt} \frac{\partial L}{\partial \dot{q}_j} - \frac{\partial L}{\partial q_j} = P_j \qquad j = 1 \ldots m+1$$

where generalized force (2.70)

$$P_j = \frac{\partial(\delta A)}{\partial(\delta q_j)} \qquad (5.8)$$

is the force acting along its j-th generalized coordinate calculated as the respective partial derivative of virtual work δA (5.7). For the electric circuits of the motor in accordance with the assumptions of disregarding mutual phase linkages, we obtain the total of m equations in the form:

$$\frac{d}{dt}\left(M_{kk}\left(\theta_r, i_k\right) i_k\right) = u_k - R_k i_k \qquad k = 1 \ldots m \qquad (5.9)$$

After calculation of the time derivative, it is:

$$\frac{di_k}{dt}\left(M_{kk}\left(\theta_r, i_k\right) + i_k \frac{\partial M_{kk}\left(\theta_r, i_k\right)}{\partial i_k}\right) = u_k - \underbrace{i_k \Omega_r \frac{\partial M_{kk}\left(\theta_r, i_k\right)}{\partial \theta_r}}_{e_k(i_k, \theta_r, \Omega_r)} - R_k i_k \qquad (5.10)$$

where $\Omega_r = \dot{\theta}_r$ - is the speed of the rotational motion of the rotor. The term e_k in equation (5.10) denotes back EMF of rotation:

$$e_k = i_k \Omega_r \frac{\partial M_{kk}(\theta_r, i_k)}{\partial \theta_r} \tag{5.11}$$

which is proportional to the current in the winding and the angular speed of the rotor. This term determines the similarity between SRM machine and series wound DC motor since the phase current i_k plays here the same role as the excitation current in the DC machine. The equations (5.10), in the consideration of the lack of linkages, can be arranged in the standard form:

$$\frac{di_k}{dt} = \left(u_k - R_k i_k - e_k(i_k, \theta_r, \Omega_r)\right) / \left(M_{kk}(i_k, \theta_r) + i_k \frac{\partial M_{kk}(\theta_r, i_k)}{\partial \theta_r} \right) \tag{5.12}$$

The equation of motion for the mechanical variable is the following:

$$\frac{d}{dt}\left(\frac{\partial L}{\partial \dot{\theta}_r}\right) - \frac{\partial L}{\partial \theta_r} = -T_l - D\dot{\theta}_r \tag{5.13}$$

Assuming a constant value J for the inertia, as a result we obtain:

$$J\ddot{\theta}_r = T_e - T_l - D\dot{\theta}_r \tag{5.14}$$

The electromagnetic torque in this equation is equal to:

$$T_e = \frac{\partial L}{\partial \theta_r} = \sum_{k=1}^{m} \int_0^{i_k} \frac{\partial M_{kk}(\theta_r, \tilde{i}_k)}{\partial \theta_r} \tilde{i}_k d\tilde{i}_k \tag{5.15}$$

Using (5.11), the formula for the torque can be restated as:

$$T_e = \frac{1}{\Omega_r} \sum_{k=1}^{m} \int_0^{i_k} e_k(\tilde{i}_k, \theta_r, \Omega_r) d\tilde{i}_k \tag{5.16}$$

5.4.3 Function of Winding Inductance

A more detailed insight into the mathematical model, given by equations (5.12)...(5.16) is associated with the need of noting the functions of windings inductances, which play a key role in this model. In the examined model the following form of the inductance function has been provided [41,42]

$$M_{kk}(\theta_r, i_k) = M(\vartheta_k)\lambda(\vartheta_k, i_k) \tag{5.17}$$

Such inductance function notation in the form of a product may seem complicated; however, it has a number of advantages. The term $M(\vartheta_k)$ presents (Fig 5.15) the unsaturated inductance of the winding in the form of the function ϑ_k, which is the rotor's angular position reduced to the pitch of the teeth $\tau_r = 2\pi/Z_r$:

$$\vartheta_k = \tau_r\, frac\left[(\theta_r - (k-1)\varepsilon)/\tau_r\right] \tag{5.18}$$

where the term $frac(x)$ denotes the fractional part of argument x. Concurrently, the second term in (5.17), presented in the graphical form in Fig. 5.16, is responsible for the magnetic saturation and introduces the adequate functional relation from rotor's angular position and current in the winding. Such a presentation of inductance coefficient makes it possible to study the effect of saturation as well as engineering changes on inductance as well as estimate the parameters of the motor on the basis of measured characteristics. The respective partial derivatives of the inductance (5.17) take the form:

$$\frac{\partial M_{kk}(\theta_r, i_k)}{\partial \theta_r} = \frac{\partial M(\vartheta_k)}{\partial \vartheta_k}\lambda(\vartheta_k, i_k) + M(\vartheta_k)\frac{\partial \lambda(\vartheta_k, i_k)}{\partial \vartheta_k}$$

$$\frac{\partial M_{kk}(\theta_r, i_k)}{\partial i_k} = M(\vartheta_k)\frac{\partial \lambda(\vartheta_k, i_k)}{\partial i_k} \tag{5.19}$$

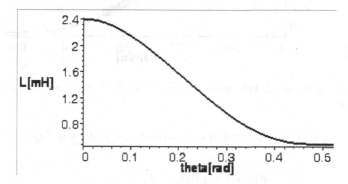

Fig. 5.15 Unsaturated inductance coefficient of a phase winding for SRM motor A in a function rotor position angle θ_r

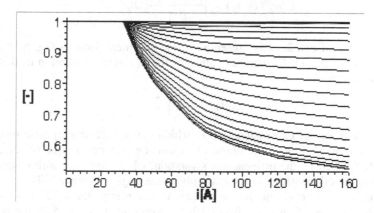

Fig. 5.16 Saturation factor $\lambda(\theta_r, i)$ of a phase winding inductance, for position angle values $\theta_r = 0°, 1.5°, 3°, \ldots 30°$, as a function of winding's current

This way of representing the inductance of the winding and its derivatives (5.19) affects the detailed form of the expression of electromagnetic torque (5.15), which takes the following form

$$T_e = \sum_{k=1}^{m} \left(\frac{\partial M(\vartheta_k)}{\partial \vartheta_k} cla(\vartheta_k, i_k) + M(\vartheta_k) cdlafi(\vartheta_k, i_k) \right) \qquad (5.20)$$

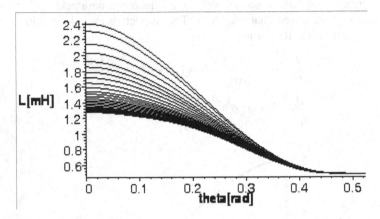

Fig. 5.17 Phase winding's inductance coefficient, for $i = 5, 10, 15, \ldots 150$ [A], as a rotor position function

There are used two integrals with respect to current i_k in the winding

$$cla(\vartheta_k, i_k) = \int_0^{i_k} \lambda(\vartheta_k, \tilde{i}_k) d\tilde{i}_k$$

$$cdlafi(\vartheta_k, i_k) = \int_0^{i_k} \frac{\partial \lambda(\vartheta_k, \tilde{i}_k)}{\partial \vartheta_k} \tilde{i}_k d\tilde{i}_k \qquad (5.21)$$

The first integral $cla(\vartheta_k, i_k)$ is presented in the graphical form in Fig. 5.18 in the function of the current for various values of the ϑ_k angle. Evaluation of this integral indicates that

$$cla(\vartheta_k, i_k) < \tfrac{1}{2} i_k^2 \qquad (5.22)$$

The first term of the expression (5.20), which defines electromagnetic torque of the motor, reminds one of the classical expression denoting torque of reluctance origin diminished by the influence of saturation, which contrasts with unsaturated inductance function, which is relative only to the angle of rotation. This term denotes the basic component of the torque. Concurrently, the other term of the torque in the form of an integral $clafi(\vartheta_k, i_k)$ is presented in Fig. 5.19 for positive values of the machine's angle of rotation. Since it is an odd function and the motor's mode of operation takes place for negative values of angle ϑ_k, the second term in the expression (5.20) determines decrease of the basic torque computed by use of the first term of (5.20). However, one can note that the change of the torque associated with non-linearity of magnetization characteristics plays a more important role for large values of the current. For the case of motor A, whose characteristics are presented here, it demonstrates for $i > 60$ [A], i.e. the value of the current that is higher than the rated current.

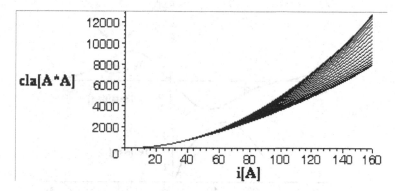

Fig. 5.18 The integral $cla(\vartheta_k, i_k)$ according to (5.21) – as a function of phase current, for a position angle $\vartheta_k = 30°, 28.5°, 27°, \dots 0°$, (top – down)

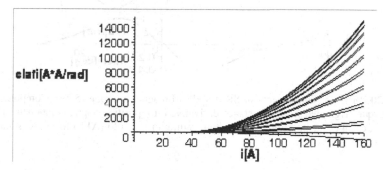

Fig. 5.19 The integral $clafi(\vartheta_k, i_k)$ according to (5.21) – as a function of phase current, for a position angle $= 30°, 28.5°, 27°, \dots 0°$, (top – down)

This is illustrated by Fig. 5.20, which presents both components of the torque with respect to the angle of rotation, i.e. the basic term relative to the derivative of the inductance – as component I and the other term that is relative to the change of the saturation– as component II. This decomposition of the electromechanical torque of a motor is presented for high value of the phase current in the motor $i = 120$ [A] in Fig. 5.20a and for the current $i = 50$ [A] in Fig. 5.20b, i.e. for small magnetic saturation. As one can see from the illustrations, the less important component II of the torque is the smaller with the smaller saturation of the

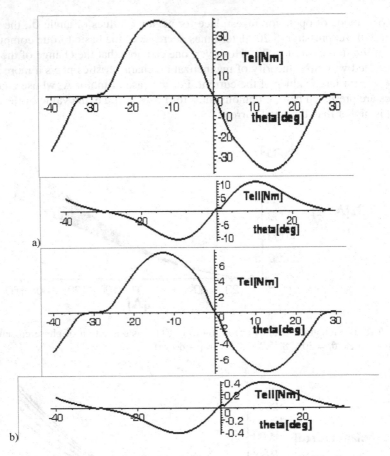

Fig. 5.20 Particular components of SRM's electromagnetic torque (5.20). Component I – basic torque originated from inductance derivative on a position angle; component II – reflecting a change in saturation: a) for a high saturation $i = 120$ [A] b) for a low saturation $i = 50$ [A]

magnetic circuit. For the phase current $i = 120$ [A] it is equal to around 30% of the value of the basic torque, while for $i = 50$ [A] it corresponds only to 5% of this torque.

5.5 Dynamic Characteristics of SRM Drives

5.5.1 Exemplary Motors for Simulation and Tests

For the purposes of illustrating the characteristics and dynamic courses of SRM motor drives a selection of two low power motors was made. The two motors are found in the catalogues and were the subject of the research and measurements in laboratory conditions [41]. A summary of the data is found in Table 5.1.

Table 5.1 Selected data for two typical SRM

Parameters	Motor A	Motor B
Z_s - stator's teeth number	8	6
Z_r - rotor's teeth number	6	4
P_n - rated power [kW]	0.9	0.9
n_n - rotor speed [rev/min]	3200	3600
U_n - rated voltage [V]	32	310
I_n - rated source current [A]	35.0	3.0
T_n - rated torque value [Nm]	2.6	2.4
R_{ph} - phase winding's resistance [Ω]	0.045	1.9
L_{max} - aligned inductance [mH]	2.2	140
L_{min} - unaligned inductance [mH]	0.46	17
$M_{1-2;2-3,3-4}$ - mutual inductance [% L_{max}]	3.5 – 7.0	2.5 – 8.5
$M_{1-3;2-4}$ - mutual inductance [% L_{max}]	0.9	-
η - motor's efficiency [%]	84	86
α_{on}; α_{off} - switching angles [deg]	38° ; 10°	51° ; 15°

5.5.2   Starting of SRM Drive

A considerable problem is encountered during the start-up of SRM motors since the torque is relative to the initial position of the rotor, which is unfamiliar, as a rule. Incremental encoders are used in order to control the motor and determine an instantaneous position of the rotor. However, the latter do not provide information regarding the position when it is stalled. Another problem with the start-up, especially with regard to a motor with a small number of teeth e.g. Z_s/Z_r = 4/2 but also Z_s/Z_r = 6/4 to a certain extent, is associated with the fact that they do not develop required start-up torque in every position of the rotor for both directions of rotation. Therefore, for the motors, which for engineering purposes are incapable of starting in every initial position of the rotor, the start-up process either occurs for a small load or the rotor is positioned prior to the starting procedure. Concurrently, the problem associated with determining the initial position of the rotor before starting the motor can be solved by:

- application of absolute encoders, which provide a reading of the initial position,

- application of resolvers, which as externally energized inductance devices that secure a precise measurement of the position in every situation. The two solutions are, however, rather expensive and they are not applied in commonly used drives. Some other possibilities of preparing the drive to the perform the start-up include:

- positioning as a result of forcing adequately strong current flow through selected windings, i.e. its alignment prior to the start-up,

- determination of the rotor's position and selection of a starting sequence using test pulses applied to the windings prior to its starting, when the rotor is stalled.

The latter option will be examined in more detail below.

5.5.2.1 Start-Up Control for Switched Reluctance Motor by Pulse Sequence

The application of starting pulse sequence to determine the position of the rotor will be presented with regard to both motors from Table 5.1 (motor A and B). Fig. 5.21 contains a summary of the test result for motor B ($Z_s/Z_r = 6/4$) within the angular range of the rotation of the rotor 0°-60°.

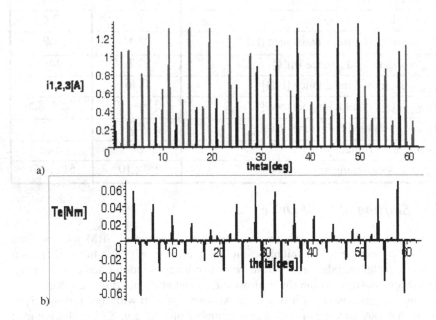

Fig. 5.21 General picture of: a) test current pulses and b) respective torque response for the $Z_s/Z_r = 6/4$ motor, in the rotor position range 0°-60°

On the basis of Fig. 5.21 one can conclude that the best conditions for the start-up of this motor are encountered in the range from $\theta_r = 0°...6°$, while the least optimum ones are for the angles close to $\theta_r = 15°$. This is so because in this position the teeth on the rotor are displaced in relation to the stator teeth by ±15°,±45°, i.e. the position of the rotor in which the derivative of the rotor winding's inductance assumes a small value. In contrast, for $\theta_r = 0°$ the remaining teeth on the rotor are in the position ±30° from the axes of the phases except for the pair in the aligned position. This, in turn, offers optimum conditions for the start-up. The next figure (Fig. 5.22) presents a more detailed response of the motor to testing pulse for selected rotor angles.

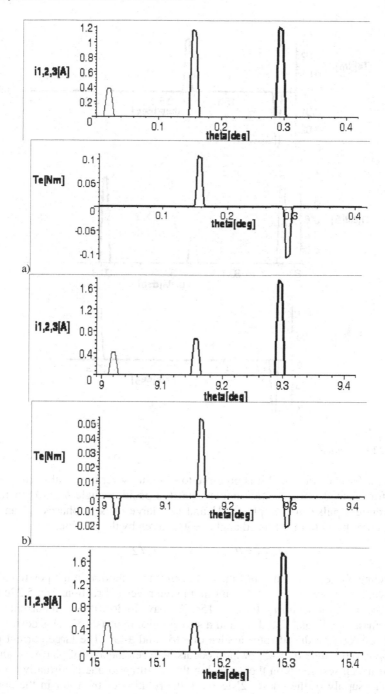

Fig. 5.22 SRM (motor B, $m = 3$) response to test voltage pulses in various rotor positions: a) $\theta_r = 0°$ b) $\theta_r = 9°$ c) $\theta_r = 15°$ d) $\theta_r = 30°$

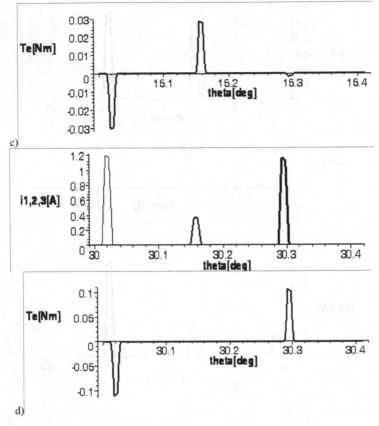

Fig. 5.22 (*continued*)

On the basis of Fig. 5.22 it is possible to examine several typical start-up situations for selected rotor position angles. For the position angle $\theta_r = 0°$ there is a small current pulse in the phase '1' and two large ones in phases '2' and '3', which correspond to the reduced angles ±30°, given by the relation:

$$\vartheta_{1,2,3} = \theta_r \pm \varepsilon \qquad \vartheta_r \leq \tau_r / 2 \tag{5.23}$$

Energizing phase '2' during start-up will result in the motion in the positive direction, while energizing phase '3' in motion in the reverse direction. Fig. 5.22c illustrates the situation occurring for $\theta_r = 15°$. We have to do with small current pulses in the phase windings '1' and '2' and a strong pulse in phase '3'. This corresponds to the respective reduced rotor angles of ±15° and ±45°. The large current pulse corresponds to angle $\theta_r = 45°$, which occurs between the axis of phase '3' and the axis of the closest tooth on the rotor, but the resulting torque is virtually nonexistent. The supply of the phase '2' leads to the rotation of the rotor in the positive direction, while of the phase '1' in the negative one. In these cases the torque is, however, three times smaller than for the angle $\theta_r = 0°$ and, hence, the start-up can be impeded. The most problematic starting conditions are encountered for $\theta_r = 9°$, i.e. for $\vartheta_{1,2,3} = 9°,-21°,39°$ - in Fig. 5.22b, while as phases '1' or '3' are energized the negative torque is very small as it reaches around 20% of the value for angle θ_r = 0°. This results from the fact that the position angle 9° is quite small in relation to the half of the pitch of the tooth $\tau_r/2$ = 45°, it is too close to the aligned position. Concurrently, for angle 39° the current is already considerably large since inductance is small but it is too close to the limit of 45°, when we have to do with a change of the torque sense. So in consequence there is not a good option for the negative direction start-up. The final illustration in Fig 5.22d presents the effects of a cyclic power supply – the situation is such like for θ_r = 0°, only phase '2' replaces phase '1', while phase '3' replaces phase '2' and, in turn phase '1' is in the place of phase '3'.

The more general conclusion from the test is that in order to conduct the start-up one should energize the phase winding for which there is a current pulse with the mean value or a winding for which the pulse response is as close to the mean value as possible. During energizing of the winding, whose pulse precedes the phase in which there was a pulse with the highest value, the start-up torque is positive and if we supply the phase that follows the one with the highest pulse response, the start-up torque is negative. One has to bear in mind that all responses in the form of current pulses are positive as we apply positive voltage pulses, which in the presented examples are equal to 30% of the rated voltage and are 0.5 [ms] in duration. The negative value of a current in a phase winding does not change the sense of the reluctance torque (see 5.20 -5.22). The following figures illustrate the situations associated with pulse determination of the rotor's position and setting the start-up sequence of motor A (Z_s/Z_r = 8/6) in the range of the rotation angle 0°...15°.

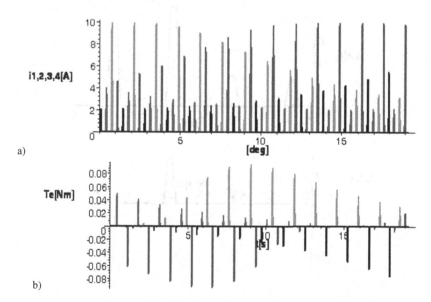

Fig. 5.23 Response for voltage test pulses in SRM motor A (m = 4) for rotor position θ_r = 0°...20°: a) current peaks b) corresponding torque jerks

From Fig. 5.23 it is possible to conclude that the best start-up conditions are encountered for the rotor angle $\theta_r = 6°...9°$. However, for 4 phase windings in every position it is possible to conduct the start-up even under considerable motor load. The following Fig. 5.24 illustrates the results of the pulse start-up test for a number of selected rotor positions.

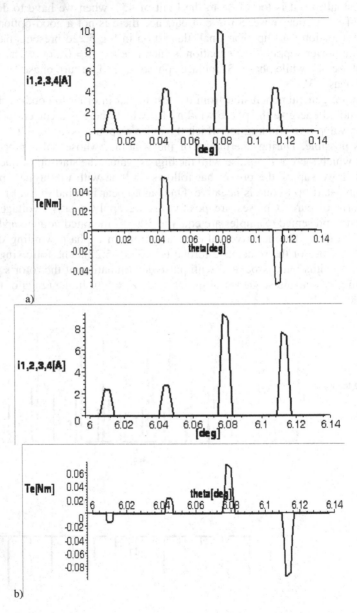

Fig. 5.24 SRM response (motor A, $m = 4$) for test voltage pulses applied to the consecutive windings in several rotor positions: a) $\theta_r = 0°$ b) $\theta_r = 6°$ c) $\theta_r = 12°$ d) $\theta_r = 15°$

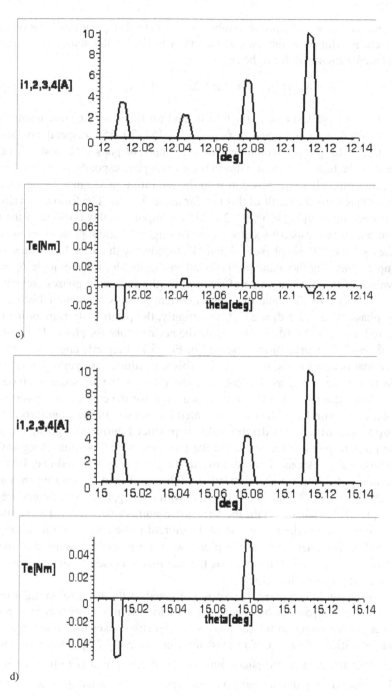

Fig. 5.24 *(continued)*

For the motor with the phase number $m = 4$, the subsequent positions of the rotor's teeth in relation to the axes of the stator teeth, i.e. the subsequent values of the reduced angle result from the relation:

$$\vartheta_{1,2,3,4} = \theta_r \pm \varepsilon \pm 2\varepsilon \qquad \vartheta_r \le \tau_r / 2 \qquad\qquad (5.24)$$

Fig. 5.24a contains test results for $\theta_r = 0°$ and for this case the subsequent values of the reduced angle are equal to $\vartheta_{1,2,3,4} = 0°,-15°,-30°,15°$, respectively, and the large start-up torque is encountered for the supply of phases '2' and '4', i.e. for $\vartheta_r = \pm 15°$. The highest current pulse in the winding corresponds to angle $\vartheta_r = \pm 30°$ since it corresponds to the position with the minimum inductance in the winding. Fig. 5.24b presents the result of this test for angle $\theta_r = 6°$. The following values of the reduced angle: $\vartheta_{1,2,3,4} = 6°,-9°,-24°,21°$ correspond to this position of the rotor teeth in relation to respective stator teeth. The highest current pulses correspond to the angles $°,-24°,21°$ for phases '3' and '4', together with the highest values of the start-up torque. Another example is found in Fig. 5.24c, for the angle $\theta_r = 12°$. The values of the reduced angle of rotation for the following phases are equal to: $\vartheta_{1,2,3,4} = 12°,-3°,-18°,27°$, and the highest current pulse is encountered while energizing phase '4', i.e. for $\vartheta_4 = 27°$. Concurrently, the positive start-up torque is encountered for phase '3' ($\vartheta_3 = -18°$) while the negative one for phase '1', i.e. for the angle $\vartheta_1 = 12°$. The situation presented in Fig. 5.24d regards angle $\theta_r = 15°$, i.e. the one that is equal to the stroke angle. This well illustrates cyclic characteristics of the test pulse since, as one can see, the roles of the subsequent phases are shifted. The rules regarding the start-up sequence for the case of the motor with $m = 4$ phases are similar to the ones presented previously. In the examined case the start-up torque with a considerable value is produced during energizing the phase whose pulse response is the second of the four in terms of its value along with the one whose pulse response is most close to the previously selected one. In the example in Fig. 5.24c these are, respectively, phases '3' and '1', and for the case in Fig. 5.24b – phases '3' and '4'. The torque with the negative value is produced for the supply of the winding with the successive number (modulo m) in relation to the phase with the highest response of the current pulse and the positive torque is generated during energizing of the phase with the preceding number of the one with the largest pulse if it qualifies as the one that is closest to the previously selected one in terms of the value.

This may sound complex; thus, it is presented in the form of an algorithm in Fig. 5.25. This algorithm has been prepared for the motor with $m = 4$ phases, however, for the motor with three phases the algorithm takes the same form except for the limitation of the set of phases' numbers to $z \in \{1,2,3\}$. This algorithm involves determination of the phase number $z \in \{1,2,3,4\}$ that is to be energized in the first order during motor start-up depending on the selected direction of the rotational speed: $n > 0$ $(T_e > 0) \rightarrow z^+$ or $n < 0$ $(T_e < 0) \rightarrow z^-$ and on the basis of the

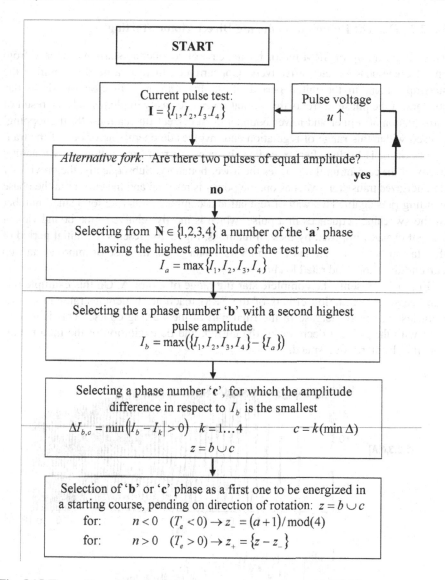

Fig. 5.25 The outline of the algorithm for a starting sequence of SRM based on test impulses before starting

pulse test for all 4 phases: $\mathbf{I} = \{I_1, I_2, I_3, I_4\}$. One should note that in the algorithm the pulse '**c**' determined as the one that is closest in terms of the amplitude to the second highest called '**b**' could be the one with the largest value , i.e. the one described as '**a**'.

5.5.2.2 Current Delimitation during Direct Motor Starting

The direct start-up of SRM motor for the case of conducting starting sequence from
an adequate phase occurs effectively. Concurrently, there is a need of limiting the
start-up current in the initial period of starting sequence. In case the drive has
sensors of the phase currents this limitation can be easily implemented as a result of
introduction of upper and lower boundaries of current fluctuations. 'Soft-chopping'
is used within this range of regulation and involves de-energizing of one of the transistors of the H bridge after exceeding the upper current limit followed by a natural
decay of the current until it reaches the lower boundary. Subsequently, the previously
de-energized transistor switches on, the power is restored and the current in the phase
winding rises again. This way of regulation accomplishes its role for a small number
of the switching sequences in a pulse, which is mostly relative to the boundaries of
current changes imposed by these limitations. Fig. 5.26 presents the initial period of
the start-up for motor A, and Fig. 5.28 for motor B, for the case of imposed start-up
current limitations and rated loading.

Fig. 5.27 presents the complete starting range of motor A. On this example one
can see series excitation effects of the SRM machine. It demonstrates in a slow
long lasting speed increase in the final part of the starting course. This is associated with the gradual decrease of the current and de-excitation of the motor, thus
causing the increasing speed.

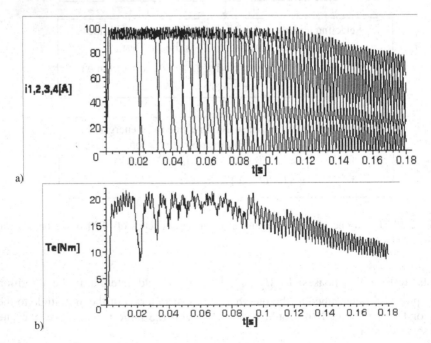

Fig. 5.26 Initial part of starting current of SRM motor A, with current limitation within the
range of 100 [A]...90 [A]: a) phase currents b) electromagnetic torque

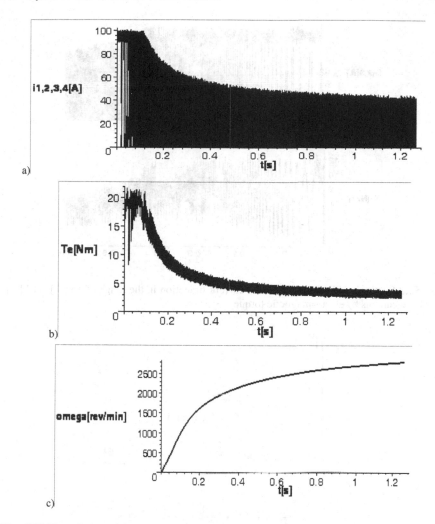

Fig. 5.27 The whole starting course of SRM motor A with a current limitation and a nominal load: a) phase currents b) electromagnetic torque c) rotational speed

5.5.3 Braking and Generating by SRM

Generator regime of operation of SRM machine is not provided for either in the engineering structure of the machine itself nor due to the structure of the semi-controlled H bridge from which it is controlled (Fig. 5.3). The rotor of the reluctance SRM motor is not energized and the excitation flux of the machine comes from the current in the stator windings. Hence, there is a lack of separately regulated excitation current that is typical for generators or permanent magnets that offer an excitation flux of the machine practically regardless of the machine load.

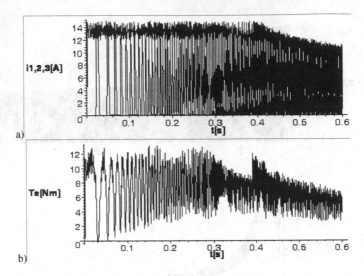

a)

b)

Fig. 5.28 Starting of SRM motor B with current limitation in the range of 15 [A]...12 [A]: a) phase currents b) electromagnetic torque

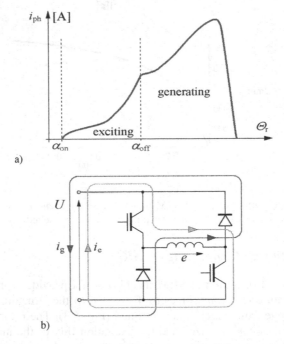

a)

b)

Fig. 5.29 Excitation period and generating period of SRG machine during one switching cycle: a) current cycle b) current flow in a H bridge for one phase winging; i_e - braking current flow, i_g - generating current flow

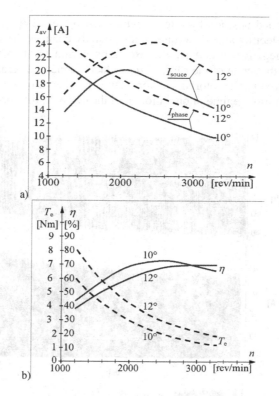

Fig. 5.30 Steady-state characteristics for a generating mode of SRM operation (motor A) as a function of rotor speed, for $u = U_n$, $\alpha_{on} = 16°$, $\alpha_{off} = -10°, -12°$: a) phase and source currents b) torque and overall efficiency

The typical transistor-diode H bridge that energizes the machine's windings from the DC source makes it possible to control the machine's current only for the case of the motor regime of operation. The return of power from phase windings into the source can only occur after switching off both transistors and takes place in an uncontrolled manner until the magnetic field associated with the winding decays. Moreover, the generation and braking of SRM machine is not steady, which results from the curves of the static characteristics presented in Fig. 5.30 and is confirmed by the waveforms presenting the unsteady operation in the vicinity of the equilibrium point in the system without feedback, as shown in Fig. 5.31.

However, for adequate control using angles α_{on}, α_{off} and in the system with feedback for the regulation of the output power the SRM machine is capable of performing the duties of the generator [19,28,34,36,40,43]. In the generation regime of operation each pulse of the machine's current consists of two parts. In the first part, for two transistor in the ON state there is an excitation of the SR generator, sometimes denoted as SRG, since the machine draws current from the source and it operates as a brake or a motor depending on the sense of the torque produced in this period. The transformation of the mechanical power in this period is inconsiderable since the angle α_{on} for which the machine operates as a generator precedes the aligned position by a few degrees and the excitation occurs for the small values of electromagnetic torque which actually changes the sense from the positive to the negative one. After the excitation period the two transistors are switched off for the angle of the rotor of α_{off}, which happens several degrees after the aligned position of the rotor and stator teeth is reached. Following we have to do with the generation regime of operation until the decay of the current in the

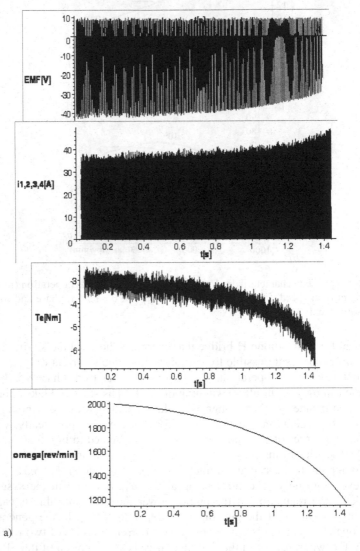

Fig. 5.31 Unstable operation of SR machine as a generator: EMF, phase currents, torque and rotor's speed courses: a) under an equilibrium point of balance b) above an equilibrium point of balance

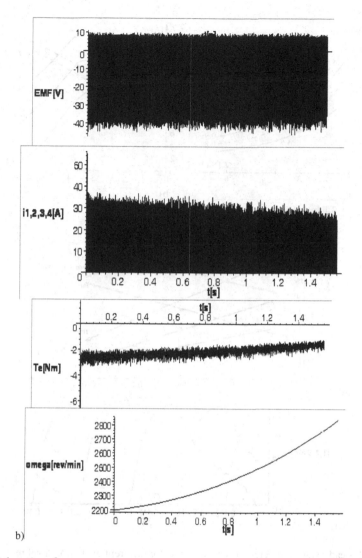

Fig. 5.31 (*continued*)

winding, which in this case closes through two diodes of the H bridge and thus energy returns into the source. In this range of operation the energy conversion is performed at the expense of mechanical energy, since the electromagnetic torque is negative and to some extent due to magnetic field energy associated with winding's current. The illustration of the excitation and subsequently, generator regime operation of SRM is presented in Fig. 5.29.

For adequately selected control angles α_{on}, α_{off} the mean value of the generated current is considerably higher than the mean value of excitation current and the

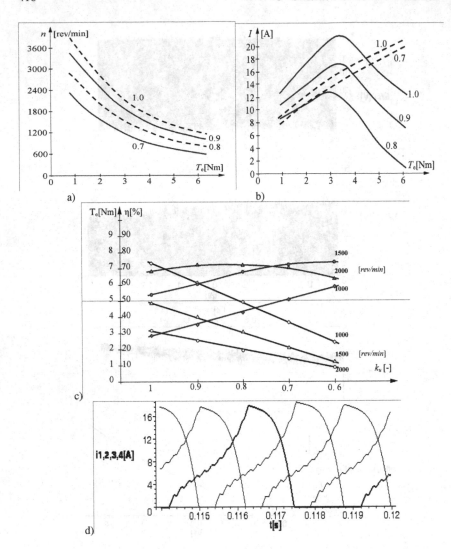

Fig. 5.32 Steady state characteristics of SRG (A) for different excitation level realized by k_u factor: a) rotational speed b) phase and source currents as a torque functions c) torque and efficiency as k_u functions d) shape of current pulses

machine operates as a generator with a decent energy efficiency (Fig. 5.30), which is, however, lower than for its motor operation.

From the characteristics in Fig. 5.30 one can see that the electromagnetic torque and phase currents decrease along with an increase of the rotor speed, which fore-casts an unsteady characteristics of the machine's operation within this range. This is confirmed by its transients, which present the behavior of the generator after the balance is disturbed (Fig. 5.31).

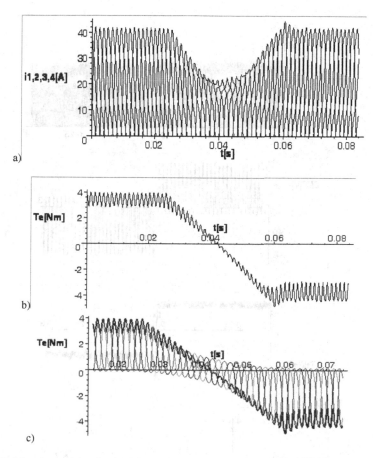

a)

b)

c)

Fig. 5.33 Linear transition from motor to generator regime of work of SRM (A), after linear change of: $\alpha_{on} = 35°\rightarrow13°$ and $\alpha_{off} = 10°\rightarrow-12°$, $n=$ 1850 [rev/min]: a) phase currents b) electromagnetic torque T_e c) partial torques constituting total T_e torque

The unstable equilibrium in the conditions presented in Fig. 5.31 occur for the speed $n = 2100$ [rev/min] and the load $T_l = -3.0$[Nm]. The generation regime can be stabilized as a result of including adequate feedback relative to the speed and acting upon the pulse width modulation coefficient k_u defined for the PWM control, which in case of constant source voltage U regulates a level of an excitation. This is so because the PWM voltage control by the k_u factor is acting during transistor operation of the H bridge and this way it effects the excitation level. As a consequence, k_u factor regulation is a main tool to control indirectly electromagnetic torque during the generator operation with a constant source voltage. The characteristics that indicate this possibility for motor A are presented in Fig. 5.32.

A smooth transfer from the motor to generator regime of operation is illustrated in Fig. 5.33. In this case we have to do with linear change of the control angles $\alpha_{on} = 35°\rightarrow13°$ and $\alpha_{off} = 10°\rightarrow-12°$ as well as a change of the torque on the

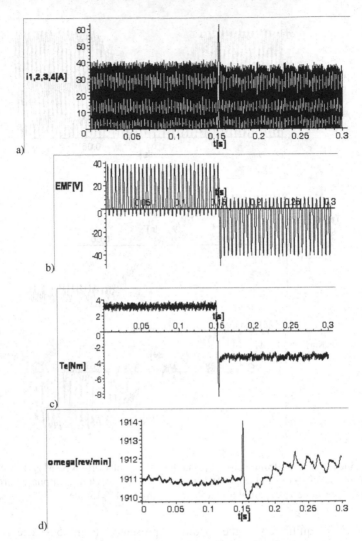

Fig. 5.34 Fast change from motor to generator operation for: $\alpha_{on} = 35° \rightarrow 16°$, $\alpha_{off} = 10° \rightarrow -10°$, $T_l = 3.0 \rightarrow -3.3$ [Nm]: a) phase currents b) EMF in a winding c) electromagnetic torque d) rotational speed

machine shaft from the load of $T_l = 3.2$ [Nm] to the torque driving the generator $T_l = -3.8$ [Nm]. The energy efficiency for this state of the generator regime is equal to $\eta_g = 68\%$, in contrast to $\eta_s = 82.5\%$ for the motor regime preceding the change of the operating regime.

The transfer from the motor to generator regime can also occur fast and does not pose any problems to the stability of the drive. An example of such fast change is illustrated in Fig. 5.34. In the presented example we have to do with a prompt

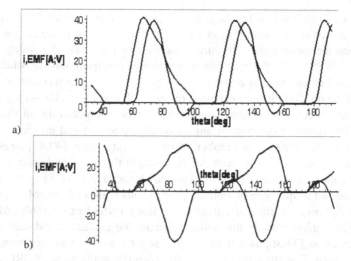

Fig. 5.35 Detail shape of a phase current and induced EMF in a phase winding for: a) motor b) generator mode of SRM operation, control parameters like in Fig 5.34

switching of the control: $\alpha_{on} = 35° \rightarrow 16°$ and $\alpha_{off} = 10° \rightarrow -10°$ and for the load torque of $T_l = 3.0$ [Nm] $\rightarrow -3.3$ [Nm].

Fig. 5.35 presents the detailed time waveform of the phase current and induced EMF in the SRM machine for motor and generator regime in the conditions defined in the illustration in Fig. 5.34. One can note the change in the shape of the waveform for phase current and EMF, which reminds of a reverse rotation of the machine for the motor regime.

The effectiveness of operation and energy efficiency of the SRM machine during generator regime are considerably relative to the control angles α_{on}, α_{off} and k_u factor controlling the excitation level. Under the assumption that energy is supplied in an optimum way during excitation most of it can be returned into the source during generator regime and, thus, the efficiency is quite high. Under the assumption of a constant rotational speed and constant source voltage U of the drive this efficiency can be expressed by the relation:

$$\eta_g = \frac{P_2}{P_1} = \frac{P_{exc} - P_{gen}}{P_{m,av}} = \frac{U(I_{exc,av} - I_{gen,av})}{T_l \, \Omega_{r,av}} \qquad (5.25)$$

where the particular symbols denote:

P_1 - mechanical power output of the drive
P_2 - electric power used by the drive
P_{exc} - electric power drawn form the source during excitation part of the cycle
P_{gen} - electric power returned into the source during generation part of the cycle
T_l - load torque
$I_{exc,av}$ - mean value of source current during excitation part of a cycle
$I_{gen,av}$ - mean value of source current during generation part of a cycle
$\Omega_{r,av}$ - mean rotational speed averaging the effect of pulsation.

The efficiency defined in this way, which is positive for generator regime, has a negative numerator since the mean current in the generator cycle is considerably higher than the mean excitation current as well as a negative denominator of the expression (5.25) since the machine is actually driven from outside, which means that the load torque is also negative. Examples of the characteristics of the efficiency of the machine A are presented in Fig. 5.30 and Fig. 5.32c – they illustrate clearly the effect of the control angle and k_u factor respectively on their waveforms. The effects of the control angles on the operation and efficiency of SRG generator are presented in a number of bibliography items [40,43] stressing the complexity of the issue. This complexity is due to the fact that during the generator regime the same point on the mechanical characteristics of SR generators (i.e. rotor speed and torque) can be obtained for various values of control variables α_{on}, α_{off} and k_u however, the current and efficiency differ considerably. Moreover, some bibliography items in this subject discuss the excitation and self-excitation of SR generators [34,36] as well as generation for a high and low rotational speeds of the machine. The latter results from the potential application of SRGs in wind power stations in the engineering models involving mechanical gear and ones without it.

5.6 Characteristics of SRM Machines

5.6.1 Control Signals and Typical Steady-State Characteristics

Although SRM motor is a reluctance machine and has a completely passive rotor, there are three control quantities deciding about the characteristics of the operation of the drive. They are: supply voltage u, initial angle of energizing the phase winding – switch on angle α_{on} and initial angle of de-energizing power supply from the winding – switch off angle α_{off}. Alternative to the switch off angle, the set of the control signals can apply the conduction angle α_z (5.1). The above control angles α_{on}, α_{off} (Fig. 5.1, Fig. 5.2) are meant to be the angles that precede the aligned position of stator and rotor teeth. For a typical supply of the SRM drive the following condition is fulfilled: the conduction angle $\alpha_z > \varepsilon$, where ε is the stroke angle (5.2). Usually in order to apply the possibilities of driving the motor the control is performed in the range:

$$\varepsilon \le \alpha_z < 2\varepsilon \tag{5.26}$$

At the same time, switching on occurs for

$$\alpha_{on} = \alpha_z + \alpha_{off} \approx 1.5\varepsilon\ldots3.0\varepsilon \tag{5.27}$$

which is aimed at obtaining a large value of the current in the adequate range of the reduced rotation angle ϑ_r (5.18). The higher values of α_{on} concern SR motors with higher number m of phase windings. Since the number of steady-state characteristics that can be presented for these control variables is large, the presentation here will focus only on selected characteristics calculated for motor B ($Z_s/Z_r = 6/4$). The first group of characteristics (Fig. 5.36) is performed in the function of the load torque T_l for three values of the supply voltage: $U = 1.0, 0.75, 0.5\ U_n$.

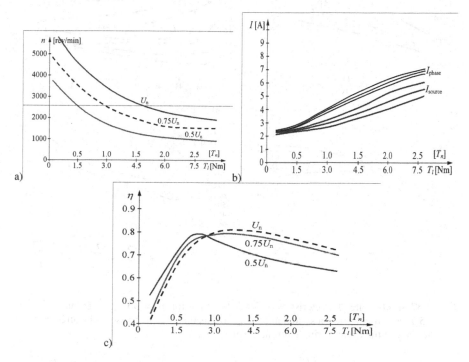

Fig. 5.36 Steady-state characteristics of 6/4 SRM drive for $U = 1.0, 0.75, 0.5\ U_n$ and $\alpha_{on} = 51°$, $\alpha_z = 36°$, as a function of a load torque: a) speed curves b) phase and source currents c) efficiency of the drive

The presented characteristics in the function of the angle α_{on} indicate that there is an optimum selection of the advance angle for energizing winding which occurs for the examined machine somewhere in the range from $\alpha_{on} = 50°...52°$, i.e. for $\alpha_{on} \approx 1.7\varepsilon$ in this case. Concurrently, the smallest pulsations of the torque are encountered for $\alpha_{on} = 42°...46°$, i.e. for $\alpha_{on} \approx 1.4\varepsilon$. Hence, it results that the late beginning of the conduction process, e.g. for $\alpha_{on} \approx 1.4\varepsilon$, leads to an uninterrupted current flow in the windings and reduces pulsations. However, the disadvantage thereof is associated with negative torque components originated from current flow in particular phase windings of the motor, which are manifested after the aligned position is exceeded. The presented characteristics do not illustrate the effect of the conduction angle α_z on the characteristics, which for the cases in Fig. 5.36, Fig. 5.37 is equal to 36°, i.e. $\alpha_z \approx 1.2\varepsilon$, which corresponds to a standard value.

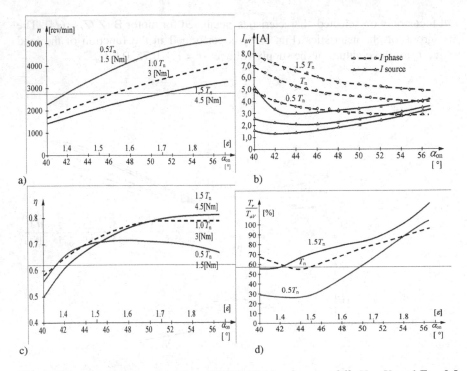

Fig. 5.37 Steady-state characteristics of 6/4 SRM drive for, $\alpha_z = 36°$, $U = U_n$ and $T_l = 0.5$, 1.0, 1.5 T_n as a function of a switch angle α_{on}: a) speed curves b) phase and source currents c) efficiency of the drive d) ripple torque level T_{rip}/T_{av}

5.6.2 Efficiency and Torque Ripple Level of SRM

Having three control parameters α_{on}, α_{off}, u it is possible to determine the same point of operation of the drive along the mechanical characteristic of the motor n, T_l for a series of various control parameters. Thus, at the same operating point it is possible to transform energy for various efficiencies and for various levels of torque ripple - T_{rip}. Such research has been presented in [41,42], and the general conclusion is that the quasi-optimal selection of control parameters of SRM motors is technically possible. The control values during quasi-optimal operations vary along with rotational speed. For small speeds the control occurs as a result of changing u, for a constant values of α_{on}, α_z. In the intermediate range of the rotational speeds the value of the voltage u remains constant, while the switch on angle α_{on} and the conduction angle α_z increase. Within the range of the high speeds only the angle α_{on} increases, while the remaining parameters of control remain constant. The curves for control variables for a quasi-optimal control of SRM motor are presented in Fig. 5.38.

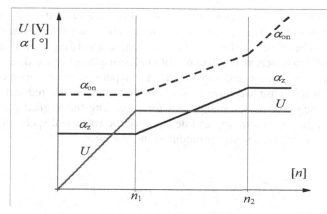

Fig. 5.38 Change of control variables for a maximum efficiency of SRM drive in respect to change of rotational speed

   The preservation of the control in accordance with the principle presented in Fig. 5.38 makes it possible to secure the operation of the drive with maximum efficiency. Concurrently, the ripple component of the torque is at a minimum for a given load torque of a motor from the threshold speed n_1 as well. The threshold values n_1, n_2, for which there should be a change in the control manner are relative to the motor's load torque T_l and they are derived on the basis of the minimum current. For the case of the mathematical model of SRM motor (5.12) discussed here, it is possible to present electromagnetic torque produced by the motor as the sum of the component torques resulting from the flow of particular phase currents. This is so because in accordance with the assumptions made during the development of this model, the phase windings are not magnetically linked and the current coming from each of the phases generates a magnetic flux for a single pair of stator teeth regardless of the currents in the remaining phase windings.
   In particular, this offers a possibility of graphical presentation of how the control angles α_{on}, α_{off} affect the history of electromagnetic torque and what values of the control angle are beneficial for the reduction of torque ripple. Fig. 5.39 presents the development of torque in motor B ($Z_s/Z_r = 6/4$) for the rated values of the supply and load. Subsequently, Fig. 5.40 presents the torque of the motor for the control angles selected in a manner in which the pulsations of the torque are the smallest. This occurs for $\alpha_{on} = 43°$ and $\alpha_{off} = 9°$, i.e. for the conduction of the phase across $\alpha_z = 34°$. In the first of the cases, the pulse component of the torque is equal to 70% of the mean value of the torque, and in the latter case (presented in Fig. 5.40) for $T_{rip} = 0.72$ [Nm], which corresponds to around 27% of the mean torque. This is done at the expense of the reduction o the system's efficiency by 4 per cent points. If the control angles were to be selected at the values $\alpha_{on} = 44°$ and $\alpha_{off} = 10°$, the pulsation level would be only equal to 33% , and the efficiency loss would be two times lower, which means 2 per cent points (Fig. 5.41). As a result, a compromise with regard to the selection of control parameters is possible with a considerable benefit to the quality of the drive's operation, which is generally characterized by the quasi-optimal curves of the control parameters in Fig. 5.38.

The illustrations of the curves in Figs. 5.39 - Fig 5.41 present why for a motor with three phases there is a certain loss of energy efficiency during limiting pulsation. This is so because the flat waveform of torque according to time and reduction of the pulsation occurs for the control angles displaced in the direction of the aligned position of stator and rotor teeth in comparison to the operation in the rated state. In this case we have to do with two phenomena reducing the efficiency: large negative torque component for exceeding the aligned position with the current in the given winding and decrease of the rotational speed of the rotor, which results in the smaller power output of the machine.

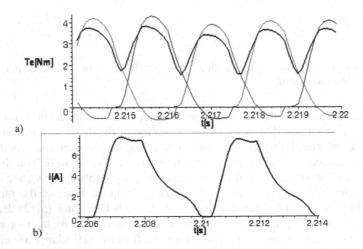

Fig. 5.39 Electromagnetic torque T_e of 3-phase SRM as a sum of partial torques originated from single phase currents for $\alpha_{on} = 51°$, $\alpha_{off} = 15°$: a) torque-ripple curves with torque ripple level $T_{rip}/T_{av} = 70\%$ b) phase currents

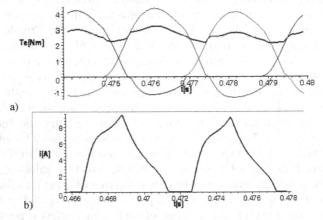

Fig. 5.40 Electromagnetic torque T_e of 3-phase SRM as a sum of partial torques for $\alpha_{on} = 43°$, $\alpha_{off} = 9°$: a) torque-ripple curves with torque ripple level $T_{rip}/T_{av} = 27\%$ b) phase currents

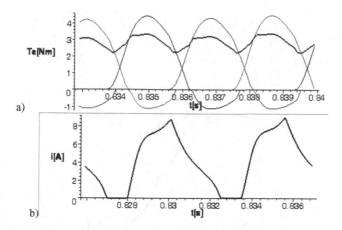

Fig. 5.41 Electromagnetic torque T_e of 3-phase SRM as a sum of partial torques for $\alpha_{on} = 44°$, $\alpha_{off} = 10°$, as a compromise between efficiency and torque –ripple level: a) torque-ripple curves with $T_{rip}/T_{av} = 33\%$ b) phase currents

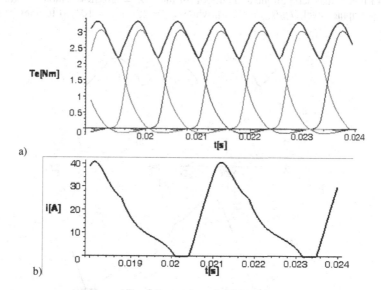

Fig. 5.42 Electromagnetic torque and current for the $Z_s/Z_r = 8/6$ SRM - motor A, for the nominal state: $\alpha_{on} = 38°$, $\alpha_{off} = 10°$: a) torques b) phase current

The data given above and Figs. 5.39 - 5.41 concern motor B whereas for motor A ($Z_s/Z_r = 8/6$) the effect of the parameters on the level of pulsation is relatively smaller. For this motor the level of pulsation is close to the minimum $T_{rip}/T_{av} = 32\%...40\%$ within a wide range of the control angles and it is difficult to obtain a level of pulsation below 30%. It is possible to exceed this boundary; however, this can only occur for the loads of the motor that are greater than the rated load

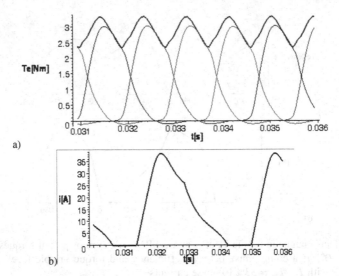

a)

b)

Fig. 5.43 Electromagnetic torque and current for the Z_s/Z_r = 8/6 SRM - motor A, for minimal torque-ripple level (T_{rip}/T_{av} = 32%), while α_{on} = 35°, α_{off} = 10°: a) torques b) phase current

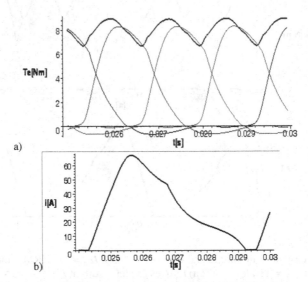

a)

b)

Fig. 5.44 Electromagnetic torque and current for the Z_s/Z_r = 8/6 SRM - motor A, for a high load of the motor T_l = 7.8 [Nm] and α_{on} = 38°, α_{off} = 10°: a) torques, while torque ripple level is T_{rip}/T_{av} = 26% , n =1905 [rev/min] b) phase current

(Fig 5.44). Concurrently, for the control corresponding to the rated state the waveforms are presented in Fig. 5.42, for α_{on} = 38°, α_{off} = 10°. In this case, the respective values of the efficiency and torque ripple level are the following: η = 83.5%, T_{rip}/T_{av} = 39%. The lowest level of torque ripple, which is equal to T_{rip}/T_{av} = 32%

takes place for the control: $\alpha_{on} = 35°$, $\alpha_{off} = 10°$, and the efficiency is even higher, as it is equal to $\eta = 84.8\%$. The two latest states differ in terms of the value of the power output of the motor due to the definitely different values of rotational speed, which are respectively equal to 3220 [rev/min] and 2780 [rev/min].

As one can see, in SRM motors there is a possibility of reducing the level of pulsation as a result of adequate selection of the control angles α_{on}, α_{off}. This, however, is possible within a limited range and may lead to a slight decrease of the efficiency, in particular for low rotational speeds [30,41,42].

5.6.3 Shapes of Current Waves of SRS

The shapes of phase currents reflect the mode of the control of SRM motor and assume specific waveforms depending on the control angles and rotational speed of the rotor. It is also possible to distinguish the generator operation of the machine from the motor regime on the basis of its waveform (Fig. 5.29a). Fig. 5.45 presents the shapes of the phase current of motor A which differ in terms of the

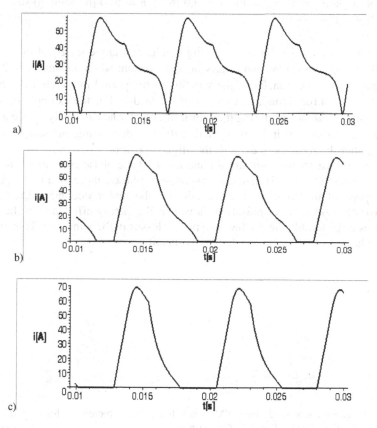

Fig. 5.45 Phase current of the (A) SRM for $\alpha_{on} = 35°$, $T_l = 5.0$ [Nm] and different α_{off} values: a) $\alpha_{off} = 5°$, b) $\alpha_{off} = 10°$ c) $\alpha_{off} = 15°$

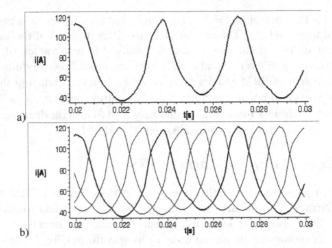

Fig. 5.46 Phase current of the (A) SRM for a continuous conduction resulting from late switch-off angle $\alpha_{off} = 5°$, $\alpha_{on} = 40°$, $T_l = 3.0$ [Nm], $n = 2900$ [rev/min]: a) single phase current b) all 4 phase currents

switch off angle α_{off}, i.e. for a decreasing conduction angle α_z, equal to $\alpha_z = 30°$, 25°, 20°, respectively. One can clearly note the instant when the supply voltage is disconnected and the transfer of the winding to the period in which it returns the energy to the source through the diodes of the bridge. For the example presented in Fig. 5.45a for $\alpha_z = 30°$, during the return of energy one can easily notice a bulge on the waveform which is associated with the decreasing inductance of the winding after the rotor tooth exceeds the aligned position.

Fig. 5.46 presents the continuous conduction of the phase currents of the SRM motor, which occurs for the late de-energizing of phases, large load and high rotational speed of the rotor. In these conditions the motor operates correctly and demonstrate a low level of pulsation; however, the energy efficiency of the motor decreases considerably due to the large power losses in the windings. This, in turn, brings a hazard of motor failure due to overheating.

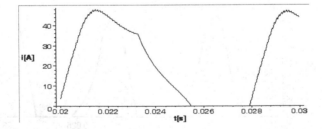

Fig. 5.47 Phase current shape by PWM controlled voltage (motor A), for $k_u = 0.7$, $\alpha_{on} = 33°$, $\alpha_{off} = 8°$, $T_l = 3.0$ [Nm], $n = 1200$ [rev/min]

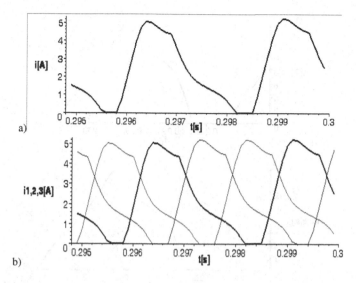

a)

b)

Fig. 5.48 Phase current shape in case of high rotor speed n = 5400 [rev/min], for motor B, while: α_{on} = 51°, α_{off} = 15°, T_l = 0.5 [Nm]: a) single phase current b) all 3-phase currents

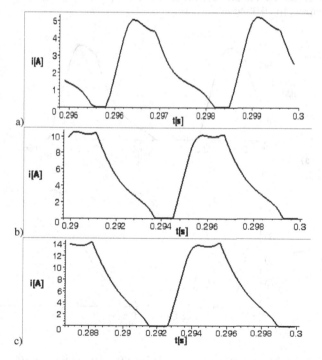

a)

b)

c)

Fig. 5.49 An influence of a load on current shape (motor B), α_{on} = 51°, α_{off} = 15°: a) T_l = 0.5 [Nm], n = 5400 [rev/min] b) T_l = 4.5 [Nm] n = 2700 [rev/min] c) T_l = 7.5 [Nm], n = 2030 [rev/min]

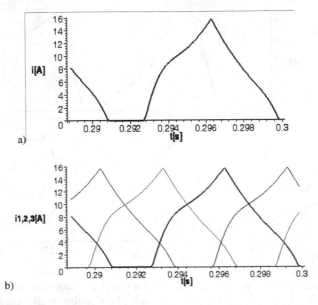

a)

b)

Fig. 5.50 Late switch on and off of SRM (motor B), $\alpha_{on} = 40°$, $\alpha_{off} = 4°$, $T_l = 3.0$[Nm]: a) single phase current b) all 3 phase currents

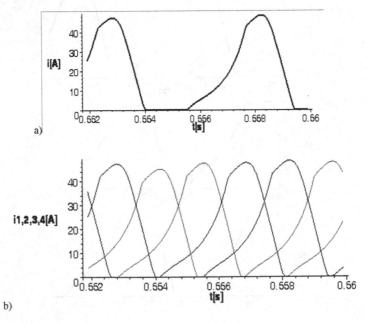

a)

b)

Fig. 5.51 Current shape for a generator operation of SR (motor A): $\alpha_{on} = 10°$, $\alpha_{off} = -18°$: a) single phase current b) all 4 phase currents

In Fig 5.49 one can see that the load does not have a significant effect on the waveform of the current, which is similar for both small (T_l = 0.5 [Nm]) as well as large (T_l = 7.5 [Nm]) load torque of the motor. However, the rotational speed of the drive differs significantly in these two cases, which can be also concluded from static characteristics – Fig 5.36.

5.7 Control of SRM Drives

5.7.1 Variable Structure – Sliding Mode Control of SRM

Sliding mode control is widely used in a up-to-date electric drives. This results from the fact that switching of constant supply voltage to electric motor windings formally constitutes of one of the possible control modes with variable structure. This, in turn, was made possible as a result of the major development in the field of power electronics.

Sliding mode control involves the control of the motion along a sliding surface $\sigma(\mathbf{q},t)$ =0 [25,44]. This group of methods involves control applying PWM technique (Pulse Width Modulation), as well as other more advanced control methods, including DTC (Direct Torque Control), with regard to induction motor drives for instance, as described in Chapter 3.4. The natural stability of the system together with high frequency of the switching make it possible in majority of cases for the trajectory of the drive's motion to follow in a close vicinity of a sliding edge even without application of special efforts and precise selection of the parameters. Sometimes engineers involved in its practical application do not see it necessary to bother themselves with proving stability of a drive. In such cases the experience resulting from laboratory tests and a narrow range of requirements regarding the control of the drive make it possible to design control on the basis of one's experience. However, in a wide range of other cases and, in particular, in actuators realizing complex and variable trajectories of motion the selection of control parameters tends to be more formal and most often it is based on the direct Lyapunov method for the analysis of the stability of the system [25,4/1,16/1]. It finds application in servomechanisms with stepper motors as well as BLDC drives [53/3]. Similarly, in SRM drives it is possible to realize the given trajectory of the motion by a proper switching the supply voltage in the range ±u, or ±u, 0 with an adequately high frequency. The sliding control of SRM motor that is applied here has certain limitations resulting from the nature of the motor as well as the adopted assumptions. One of them is that the control occurs simultaneously only for a single phase winding of the motor, as a result of which it involves just one dimension. The second restriction concerns the fact that the control occurs only for the conduction of transistors, i.e. for the flow of energy from the source. In contrast, during the diode conduction and return of the energy into the source the motor's current and torque are uncontrolled for the duration of this stage of operation. Obviously, there is a possibility of application of transistor-diode control mode, i.e. −u, 0 during the periodic switching of the winding and some kind of the effect of current control is obtained in this way; however, it works only in the direction of increasing the time needed for the decay of the current and it will not be applied here. The presentation further on will concentrate on the application of sliding control with regard to SRM machine realizing given trajectory of motion and limiting torque ripple, i.e. overcoming one of the drawbacks of this drive in terms of vibration and noise production. The presentation will include DTC control and current control of the motor in accordance with the given sliding curve realizing these targets. It is possible to plan other control techniques involving sliding, all of which realize the given trajectories of the motion of the drive. The presentation of practical examples of such systems will be the subject of the following sections.

5.7.2 Current Control of SRM Drive

The current control used here involves the sliding control of SRM drive in which the sliding surface is defined by the given function of the currents of the phase windings of a machine

$$\sigma\left(\sum_k i_k, \theta_r\right) = 0 \qquad \sigma_k(i_k, \theta_r) = 0 \tag{5.28}$$

and the control itself is defined in the standard way:

$$u_k(i_k, \theta_r) = \begin{cases} +u_k\left(\sum_k i_k, \theta_r\right) & \text{for } \sigma\left(\sum_k i_k, \theta_r\right) > 0 \\[2mm] -u_k\left(\sum_k i_k, \theta_r\right) & \text{for } \sigma\left(\sum_k i_k, \theta_r\right) < 0 \end{cases} \tag{5.29}$$

The sliding control described by the formulae (5.28) and (5.29), which is further called current control due to the fact that the sliding surface is designed on the basis of the values of the phase currents, is not a typical one. This comes as a consequence of the fact that the analytical expression (5.28) of sliding surface does not involve time t in an explicit way but the control is relative to other variable, i.e. the angle of the rotation of the rotor θ_r, thus, it is a phase surface. Secondly, as it will be presented later, the sliding surface is determined in the function of the sum of phase currents. In the practice of SRM motor control this means the dependence of the sliding surface on a single phase current that is drawn from the source, which is controllable– since it is supplied through the transistors, and possibly on a single or more phase currents that are in the phase of decay. This, in turn, means that the control is single-dimensional dependent on the sum of the currents in windings, while only one of the currents is controllable, i.e. the one that is energized from the source. The selection of the sliding surface has two roles to play: it minimizes the pulsations of the torque and executes the given curve of the electromagnetic torque of the motor that is defined to adequately reflect the required trajectory of the motion of the drive. As one can see, the task set in this

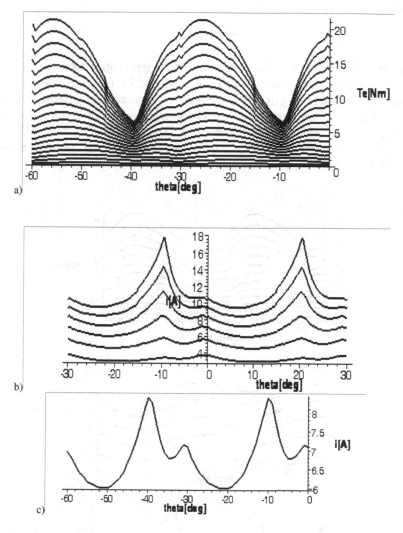

Fig. 5.52 Characteristic curves for $Z_s/Z_r = 6/4$ SRM as a function of rotor position angle: a) electromagnetic torque T_e for two neighboring phase windings supplied by $i = 1.0, 2.0, 3.0, \ldots, 20$ [A] b) Σi_k of currents in two neighboring phase windings required for $T_e = 1.0, 2.0, \ldots, 6.0$ [Nm] c) Σi_k of currents required for $T_e = 3.0$ [Nm] in detail form

way requires that the problem that is inverse to the motor torque function has to be solved, i.e.

$$i_k = f(T_e) \qquad (5.30)$$

Since this problem is non-linear and concurrently the current is normally conducted through more than a single phase winding, the above should be restated as follows

$$\sum_k i_k = f(T_e) \tag{5.31}$$

In practice, this problem is associated with the determination of the sum of two currents being conducted through adjacent windings, i.e. one that is supplied and one that returns the energy, in a way that enables generation of a desired instantaneous value of electromagnetic torque. The solution of this task is the reverse to the relation presented in Fig. 5.52a and Fig. 5.53a respectively for the SRM motor with the teeth number $Z_s/Z_r = 6/4$ and $Z_s/Z_r = 8/6$. These are the machines B and A in Table 5.1.

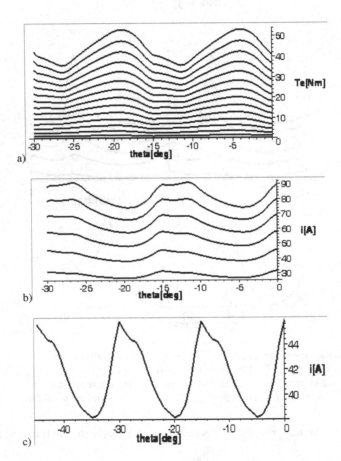

Fig. 5.53 Characteristic curves for for $Z_s/Z_r = 8/6$ SRM as a function of rotor position angle: a) electromagnetic torque T_e for two neighboring phase windings supplied by $i = 10.0$, 20.0, 30.0, ...,150 [A] b) Σi_k current in two neighboring phase windings required for $T_e = 3.0, 6.0, 9.0, ...,18.0$ [Nm] c) Σi_k current required for $T_e = 6.0$ [Nm] in detailed form

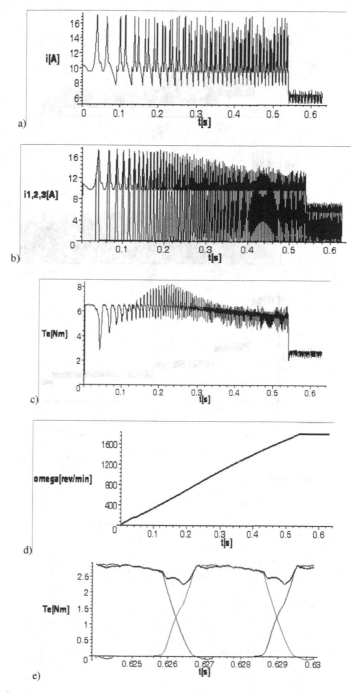

Fig. 5.54 Starting of motor B ($Z_s/Z_r = 6/4$) with current controller set for segmented constant torque values: a) required Σi_k current b) phase currents c) electromagnetic torque d) rotor speed e) partial and resultant torque for a final speed

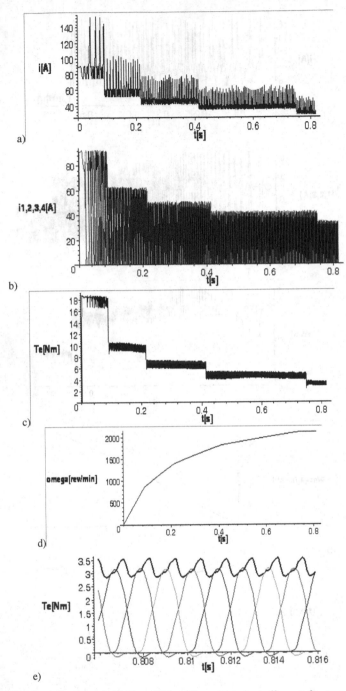

Fig. 5.55 Starting of motor A (Z_s/Z_r = 8/6) with current controller set for segmented constant torque values: a) required Σi_k current b) phase currents c) electromagnetic torque d) rotor speed e) partial and resultant torque for a final speed

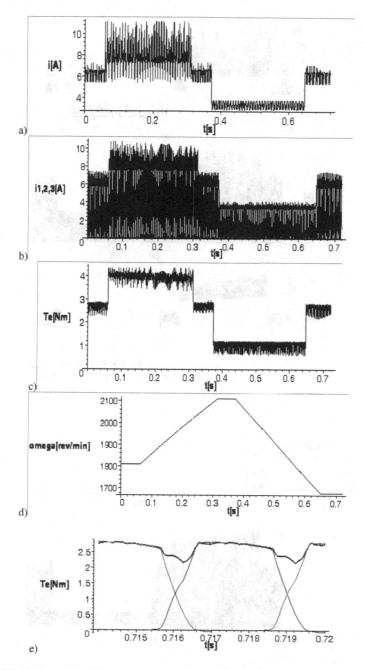

Fig. 5.56 Trajectory formed in the B SRM machine under a current controller: a) required current b) phase currents c) electromagnetic torque d) rotational speed e) partial and resultant torques for $\Omega_r = 180$ [rad/min]

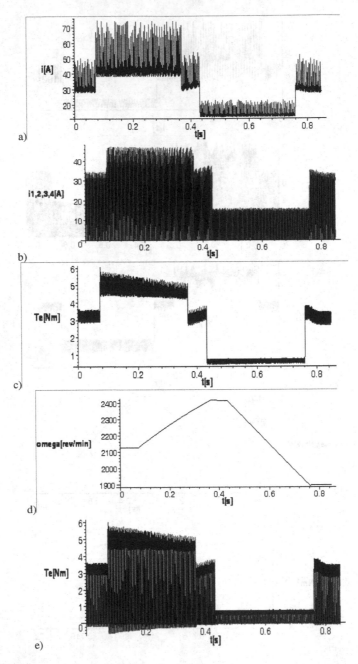

Fig. 5.57 Trajectory formed in A SRM machine under a current control rule: a) required current b) phase currents c) electromagnetic torque d) rotational speed e) partial and resultant torques for $\Omega_r = 200$ [rad/min]

From the results in Fig. 5.52 and Fig. 5.53 it stems that gaining high value of the torque for motor B (Z_s/Z_r = 6/4) poses a much more difficult task than for the motor with 4 phase windings, i.e. motor A (Z_s/Z_r = 8/6) as a result of the larger distances between the teeth for the first of them. However, in both cases it is possible to perform the current control by the presented method, which will be demonstrated on the basis of several examples. First, we will discuss the start-up of motors with the application of current control. For motor B it is presented in Fig. 5.54.

Current control, as presented can be effectively used to form the trajectory of the motion of a drive as a result of applying a given waveform of torque produced by a motor. This torque for the application of the current control needs to be subsequently transformed (5.31) this gaining the required current waveform necessary to perform the task given by Σi_k. This is presented on the illustrations of the operation of the drive for both motors A and B.

The presented examples of the application of current control (Fig. 5.54... Fig. 5.57) prove that the presented method is effective with regard to the both motors considered as exemplary ones, i.e. for the motors with 3 and 4 phase windings. This allows one to form the waveforms of rotational speed, torque and current, the latter of which is in this method the quantity that is directly regulated, as well as enables one to limit the pulsations of the torque. However, this method can be effective only within the range in which there is an adequate surplus of the regulation, which in this case means a sufficient surplus of the supply voltage, that will enable one to perform the planned current control. This limited surplus of the control is the reason that in the start-up of motors (Fig. 5.54, Fig. 5.55) is carried out with a torque decreasing by stepwise sections along with increasing speed. The forming of the trajectory of the motion occurs regarding the rotational speed that is permitted by the supply voltage in order to ensure that the given current shape resulting from assumed trajectory of the motion were possible to perform by the control system.

5.7.3 Direct Torque Control (DTC) for SRM Drive

DTC control with regard to SRM motor also forms an application of sliding method for drive regulation since it occurs as a result of rapid switching of the voltage applied to the windings of a machine's stator in a way that ensures that the given waveform of electromagnetic torque is realized. This method in its practical application is similar to current control, which has already been the focus of presentation earlier in the chapter. The specific characteristics of DTC control involve the fact that the sliding surface is constructed on the basis of the desired waveform of the torque:

$$\sigma\big(T_e(\mathbf{q}),\theta_r\big)=0 \tag{5.32}$$

This can be restated more directly as:

$$\sigma:\quad T_r(\theta_r,\Omega_r,t)-T_e(\mathbf{q})=0$$

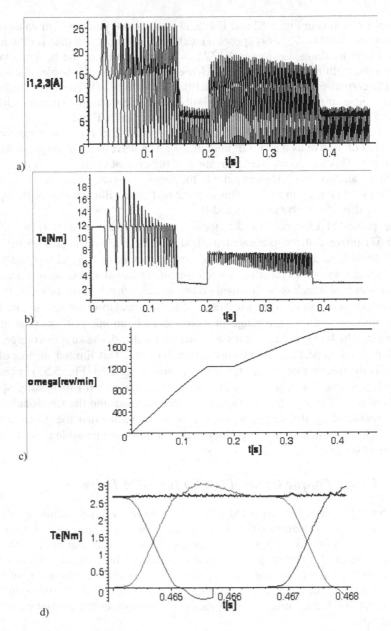

Fig. 5.58 Starting of B SRM machine by use of DTC control method with required torque, respectively T_e = 12.0, 2.5, 7.0, 2.7 [Nm]: a) phase currents b) electromagnetic torque c) rotational speed d) partial and resultant torques for T_e = 2.7 [Nm]

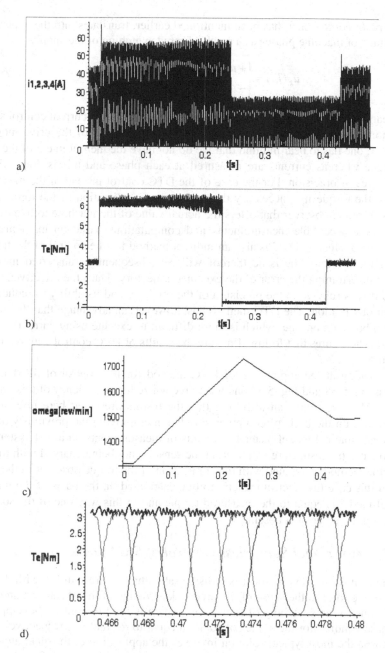

Fig. 5.59 Starting of A SRM machine by use of DTC control method with required torque respectively T_e = 3.1, 6.2, 1.0, 3.2 [Nm]: a) phase currents b) electromagnetic torque c) rotational speed d) partial and resultant torques for T_e = 3.2 [Nm]

This type of control, as it has been mentioned earlier, translates into the control of the voltage of machine phases as a result of rapid switching of the supply

$$u_k(T_e) = \begin{cases} +u_k & \text{for} \quad \sigma > 0 \\ -u_k & \text{for} \quad \sigma \le 0 \end{cases} \tag{3.33}$$

It appears that this method should offer more advantages than current control since it is more direct with regard to realizing a given trajectory of the drive motion. However, one has to bear in mind that we do not have the measurement of the motor torque whereas currents are measured at each phase and this is done with a high degree of precision. For the case of the DTC control instead of the measurement of the torque it is necessary that we apply an estimation of the torque, i.e. usually a torque observer that offers the actual value of the machine torque on the basis of the accessible measurements and computations based on mathematical model. In conclusion, DTC is also an indirect method for the control of the trajectory of the motion. The issue thereof will be subsequently transferred into the stage of determining the error of the executed trajectory. Thus, the effectiveness of this control is relative to the precision of the observer and its ability to reduce the error of observation. The DTC method has, however, an advantage that the trajectory can be given on-line, which is more difficult to execute using current control method. The figures that follow illustrate the results of DTC control with regard to SRM motor.

By looking at the application of DTC method for the control of SRM motor drive in Fig. 5.58 and Fig. 5.59 one has to recognize high efficiency of this type of control. However, one can also note that the results presented here refer to the ones gained on the basis of computer simulation employing the previously developed mathematical model instead of results of measurements on a real system. In consequence the results are idealized in the sense of not being charged with the error of the method associated with the application of the torque observer in the control. In this case the electromagnetic torque calculated on the basis of the mathematical model is equal to the measured torque and in this way one of the sources of the significant error is absent.

5.7.4 Sensor- and Sensorless Control of SRM Drive

For the control of SRM motor it is indispensable that we are familiar with the position of the rotor in the sense of the precise knowledge of the of rotation angle θ_r. This is due to the switching of transistors, which is used to control the supply of the phase winding for the angles of rotation equal to α_{on} and α_{off}, respectively. For this reason the most typical solution involves the application of the quadrature encoder in the control system, whose signals are transformed into information regarding the position of the rotor, its rotational speed and direction of the rotational motion. A block diagram of the control using the signal from position sensor is presented in Fig. 5.60.

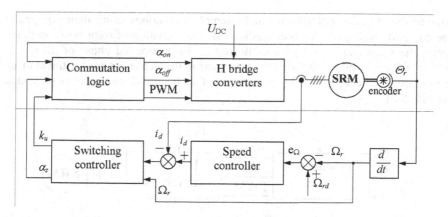

Fig. 5.60 Block diagram of control of SRM drive applying encoder sensor signal

An alternative to the control system using encoder involves sensorless control [12,14,15,31,48], which in brief means that the sensor is absent from the system. Such a solution is made possible as a result of applying a position estimator or observer of motor state, thus, leads to savings in terms of the investment, reduction of mass and space occupied by the system and increase in efficiency. This improvement in terms of reliability is connected with the lack of an additional mechanical device on the shaft that is also common to the rotor and the connection leading from it to the control system. It is not always possible to apply sensorless control, and in particular, it may not be possible to use it in systems in which it is necessary to have a very high degree of precision of regulation. The observer itself will be the subject of discussion later and now we will focus on the earlier concept regarding position estimator. It is formed by a complex measurement and calculation unit of the control system with the previously prepared characteristics of magnetization or characteristics of windings' inductance. Here we apply the relations that are reverse to the magnetization characteristics, that is:

$$\vartheta_k = f\left(\Psi_k, i_k\right) \qquad \vartheta_k = f\left(L_k, i_k\right) \tag{5.34}$$

where subscript k denotes the number of a phase winding, Ψ_k - magnetic flux coupled with this winding, and ϑ_k - angle of rotation reduced to the pitch of the teeth for the k-th winding. In order to use the relation (5.34) it is necessary to measure phase currents and voltages supplying phase windings so that the instantaneous value of the flux linkage associated with the k-th winding is familiar:

$$\Psi_k = \int_t \left(u_k - R i_k\right) d\tau \tag{5.35}$$

While we have the value of the flux linkage and current available, it is possible to precisely determine the value of the angle of rotation in terms of the rotor's tooth position in respect to the axes of the given winding on the basis of look-up tables based on the reverse characteristics of magnetization (5.34). It is self-evident that a useful device for such a control is a signal processor (DSP) and the particular manufacturers offer publications and exemplary solutions using their equipment beside part catalogues. Another method for the estimation of rotor position θ_r applies the technique of test pulses injected to the unsupplied phase of SRM machine, which is quite similar to the one used in the determination of the start-up sequence – see chapter 5.4.2. An example of such sensorless control is presented in Fig. 5.61.

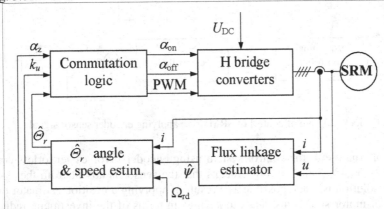

Fig. 5.61 General block diagram of a sensorless control of SRM with a position estimator based on flux linkage

5.7.5 State Observer Application for Sensorless Control of SRM

Sensorless control denotes here, just as in the previous examples, the lack of a position/speed sensor, such as encoder or resolver in a system. Concurrently, the system has to contain sensors of phase current, which are applied commonly and do not pose any technical problem. Their use allows for the application of adequate emergency devices and various diagnostic methods regarding the state the drive. In this manner, they lead to an increase of reliability of the system at a low cost and can be applied in the operation of the state observer. The role of the observer consists in on-line determination of estimates $\hat{\mathbf{q}}$ of variables $\mathbf{q}$ and reduction of the observation error

$$\mathbf{e} = \mathbf{q} - \hat{\mathbf{q}} \qquad (5.36)$$

to zero within a given time. Taking after [25] the non-linear model of the dynamic state in the form

$$\dot{\mathbf{q}} = f(\mathbf{q}) + g(\mathbf{q}, \mathbf{u}) \qquad (5.37)$$

the vector of the observer in the form:

$$\mathbf{y} = h(\mathbf{q}, \mathbf{u}) \qquad (5.38)$$

and the difference from the estimation of the state vector:

$$r = y - \hat{y} = y - h(\hat{q}, u) \tag{5.39}$$

we can define the following equation for the observer:

$$\dot{\hat{q}} = f(\hat{q}) + g(\hat{q}, u) + \kappa(y - h(\hat{q}, u)) \tag{5.40}$$

In accordance with (5.40) this observer constitutes the dynamic model of the system (5.37) accounting for the estimated vector of variables $\hat{q}$ plus a non-linear function κ, whose role is to reduce the error of the system (5.39) to zero in the steady state. For the SRM motor, whose mathematical model is presented by equations (5.12) – (5.15), the possible state observer is presented by the equations:

$$\frac{d\hat{i}_k}{dt} = \left(u_k - R_k \hat{i}_k - e_k(\hat{i}_k, \hat{\theta}_r, \hat{\Omega}_r)\right) \Big/ \left(M_{kk}(\hat{i}_k, \hat{\theta}_r) + \hat{i}_k \frac{\partial M_{kk}(\hat{i}_k, \hat{\theta}_r)}{\partial \hat{\theta}_r}\right)$$

$$\dot{\hat{\Omega}}_r = \left(T_e(\hat{i}_k, \hat{\theta}_r) - T_l - D\hat{\Omega}_r\right) \Big/ J + \kappa_1(i_k - \hat{i}_k) \tag{5.41}$$

$$\dot{\hat{\theta}}_r = \hat{\Omega}_r + \kappa_2(i_k - \hat{i}_k)$$

The equations of the observer (5.41) form a direct repetition of the dynamic equations to which undetermined correction functions κ_1, κ_2 are supplemented and applied with regard to the equations whose variables are not observed. A problem that is widely discussed in the literature involves a method of finding correction functions for a non-linear system. Thus, the lack of a general method leads to a number of specific solutions, which are applied on the basis of analogy to similar systems. In these circumstances it is important to select an appropriate method of testing whether the estimation error decays in time for the experimentally selected correlation functions κ. This is possible with the aid of the generalized Lyapunov method [25,16/1,21/1,23/2] after the selection of positively determined candidate function V in an given area. This function needs to be positively determined and relative to the estimation errors. In order to secure the asymptotic error decay the first derivative of the function has to be negative in that area, in accordance with the Laypunov theorem. In the examined case of estimation of the position and rotational speed of SRM motor, the candidate function can be assumed in the form:

$$V = \varepsilon_\theta^2 + \varepsilon_\Omega^2 = \left(\theta_r - \hat{\theta}_r\right)^2 + \left(\Omega_r - \hat{\Omega}_r\right)^2 \tag{5.42}$$

This function is self-evidently positively determined in the entire area of the occurrence of the estimation error. Concurrently, the requirement of the asymptotic decay error comes down the inequality in the form

$$\dot{V} = 2\dot{\varepsilon}_\theta \varepsilon_\theta + 2\dot{\varepsilon}_\Omega \varepsilon_\Omega < 0 \tag{5.43}$$

Hence, it is necessary to study two inequalities

$$\dot{\varepsilon}_\theta \varepsilon_\theta < 0 \quad \text{and} \quad \dot{\varepsilon}_\Omega \varepsilon_\Omega < 0 \tag{5.44}$$

for the presented model of the observer (5.41) and selected correlation functions κ_1, κ_2. When it comes to the selection of these functions, the most extreme solution [25] involves the use of a sliding mode observer, which switches a constant function depending on the sign of the observation error. Thus, it imposes the function to remain in the vicinity of the observed value. In the examined case the application of the sliding observer with regard to (5.41) means that:

$$\kappa_1(i_k - \hat{i}_k) = K_1 sign(i_k - \hat{i}_k)$$
$$\kappa_2(i_k - \hat{i}_k) = K_2 sign(i_k - \hat{i}_k)$$

(5.45)

K_1, K_2 denote here the gains which can assume positive or negative values depending on the sign of estimation error. By looking at the conditions of estimation error decay (5.44) one can imagine how the sign changes of correlation functions (5.45) lead to the negative value required in these conditions. After testing and selecting adequate gain factors K_1, K_2 this practically enables one to apply such an observer in the control of SRM motor without the application of the position sensor.

References

[1] Alrifai, M., Zribi, M., Krishnan, R., et al.: Nonlinear Speed control of Switched Reluctance Motor Drives Taking into Account Mutual Inductance. J. of Contr. Sc. Eng., ID 491625, Hindawi (2008)

[2] Astrom, K.J., Wittenmark, B.: Adaptive Control. Addison-Wesley, New York (1989)

[3] Bartoszewicz, A.: Time-varying sliding modes for second order systems. IEE Proc. Contr. Th. Appl. 143, 455–462 (1996)

[4] Becerra, R.C., Ehsani, M., Miller, T.J.E.: Commutation of SR Motors. IEEE Trans. P. Electr. 8, 257–263 (1993)

[5] Binder, A.: Switched reluctance Drive and Inverter-fed Induction Machine – a comparison of Design Parameters and Drive Performance. El. Eng. 82, 239–248 (2000)

[6] Bjaaberg, F., Kjaer, P.C., Rasmussen, P.O.: Improved digital Current Control in Switched Reluctance Motor Drives. IEEE Trans. P. Electr. 14, 666–669 (1999)

[7] Blanke, M.: Diagnosis and Fault-Tolerant Control. Springer, Berlin (2006)

[8] Cameron, D., Lang, J., Umans, S.: The origin and reduction of acoustic noise in doubly salient variable-reluctance motors. IEEE Trans. Ind. Appl. 28, 1250–1255 (1992)

[9] Chang, Y.: On the design of power circuit and control scheme for switched reluctance generator. IEEE Trans. Ind. Electr. 55, 445–454 (2008)

[10] Cheok, A.D., Zhang, Z.F.: Fuzzy logic rotor position estimation based switched reluctance motor dsp drive with accuracy enhancement. IEEE Trans. P. Electr. 20, 908–921 (2005)

[11] Chi, H.P., Lin, R.L., Chen, J.F.: Simplified flux-linkage model for switched reluctance motors. IEE Proc. El. P. Appl. 152, 577–583 (2005)

[12] Deger, M.W., Lorenz, R.D.: Using Multiple Saliences for the Estimation of Flux Position and Velocity. IEEE Trans. Ind. Appl. 34, 1097–1104 (1998)

[13] Deihimi, A., Farhangi, S., Henneberger, G.: A general nonlinear model of switched reluctance motor with mutual coupling and multiphase excitation. E. Eng. 84 (2002)

[14] Ehsani, M., Husain, I., Mahajan, S., et al.: Modulation Encoding Techniques for Indirect Rotor Position Sensing in Switched Reluctance Motors. IEEE Trans. Ind. Appl. 30, 85–91 (1994)

[15] Gallegos-Lopez, G., Kjaer, P.C., Miller, J.E.: A New Sensorless Method for Switched Reluctance Motor Drives. IEEE Trans. Ind. Appl. 34, 832–840 (1998)

[16] Hong, J.P.: Stator Pole and Yoke Design for Vibration Reduction of Switched Reluctance Motor. IEEE Trans. Mag. 38, 929–932 (2002)

[17] Hussain, I.: Minimization of Torque Ripple in SRM Drives. IEEE Trans. Ind. Electr. 49 (2002)

[18] Husain, I., Radun, A.: Fault analysis and Excitation Requirements for Switched Reluctance Generators. IEEE Trans. E. Conv. 17, 67–72 (2002)

[19] Ichinikura, O., Kikuchi, T., Nakamura, K.: Dynamic Simulation Model of Switched Reluctance Generator. IEEE Trans. Mag. 39, 3253–3255 (2003)

[20] Jack, A.G., Mecrow, B.C., Haylock, J.A.: A comparative study of permanent magnet and switched reluctance motors for high-performance fault-tolerant operation. IEEE Trans. Ind. Appl. 32, 889–895 (1996)

[21] Kamper, M.J., Rasmeni, S.W., Wang, R.J.: Finite element time step simulation of the switched reluctance machine drive under single pulse mode operation. IEEE Trans. Mag. 43, 3202–3281 (2007)

[22] Khalil, A., Husain, I.: A Fourier series generalized geometry based analytical model of switched reluctance machines. IEEE Trans. Ind. Appl. 43, 673–684 (2007)

[23] Kjaer, P.C., Gribble, J.J., Miller, T.J.E.: Dynamic testing of switched reluctance motors for high-band actuator applications. IEEE/ASME Trans. Mechatr. 2, 123–135 (1997)

[24] Krishnan, R.: Switched Reluctance Motor Drives. CRC Press, Cambridge (2001)

[25] Levine, W.S. (ed.): The Control Handbook. CRS Press / IEEE Press (1996)

[26] Lin, F.C., Yang, S.: An approach to producing controlled radial force in switched reluctance motor. IEEE Trans. Ind. Electr. 54, 2137–2146 (2007)

[27] Lin, F.C., Yang, S.M., Lee, J.: Instantaneous Shaft Radial Force Control with Sinusoidal Excitations for Switched Reluctance Motors. IEEE Trans. En. Conv. 22, 629–636 (2007)

[28] Liptak, M., Rafajdus, P., Hrabcova, V.: Optimal Excitation of a Single-Phase SR Generator. In: Conference Proceedings of 16th Int. Conf. El. Mach. ICEM 2004 (2004)

[29] Lu, W.Z., Keyhani, A., Fardoun, A.: Neural network based modeling and parameter identification of switched reluctance motors. IEEE Trans. En. Conv. 18, 284–290 (2003)

[30] Mademlis, C., Kioskederis, I.: Performance optimization in switched reluctance motor drives with online commutation angle control. IEEE Trans. En. Conv. 18, 448–457 (2003)

[31] McCann, R.A., Islam, M.S., Hussain, I.: Application of a Sliding Mode Observer for Position and Speed Estimation in Switched Reluctance Motor Drives. IEEE Trans. Ind. Appl. 37, 51–58 (2001)

[32] Miller, T.J.E.: Switched Reluctance Motors and Their Control. Magna Physics Publishing / Oxford University Press, Hillsboro (1993)

[33] Miller, T.J.E.: Optimal Design of Switched Reluctance Motors. IEEE Trans. Ind. Electr. 49 (2002)

[34] Nasserdine, M., Rizk, J., Nagrial, M.: Switched Reluctance Generator for Wind Power Applications. World Acad. Sc. Eng. Techn. 41, 126–130 (2008)

[35] Parreira, B., Rafael, S., Pires, A.J.: Obtaining the magnetic characteristics of an 8/6 switched reluctance machine: from fem analysis to the experimental tests. IEEE Trans. Ind. Electr. 52, 48–53 (2005)

[36] Radimov, N., Ben-Hail, N., Rabinivici, R.: Switched Reluctance Machines as Three-Phase AC Autonomous Generator. IEEE Tran. Mag. 42, 3760–3764 (2006)

[37] Rahman, K.M., Schultz, S.E.: High-Performance Fully Digital Switched Reluctance Motor Controller for Vehicle Propulsion. IEEE Trans. Ind. Appl. 38, 1062–1071 (2002)

[38] Russa, K., Husain, I., Elbumk, M.E.: Torque-Ripple Minimization in Switched Reluctance Motors over a Wide Speed Range. IEEE Trans. Ind. Appl. 34, 1105–1112 (1998)

[39] Sobczyk, T.J., Drozdowski, P.: Inductances of electrical machine winding with non-uniform airgap. Arch. f. Elektr. 76, 213–218 (1993)

[40] Sozer, Y., Torrey, D.A.: Closed Loop Control of Excitation Parameters for High Speed Switched Reluctance Generator. IEEE Trans. P. Electr. 19 (2004)

[41] Tomczewski, K.: Optimization of switched reluctance motors control. Dissertation, Politechnika Opolska (2002)

[42] Tomczewski, K., Wach, P.: Control Characteristics for quasi-optimal operation of switched reluctance motors. El. Eng. 85, 275–281 (2003)

[43] Torrey, D.A.: Switched Reluctance Generators and Their Control. IEEE Trans. Ind. Electr. 49, 3–14 (2002)

[44] Utkin, V.I.: Sliding Mode Control Design Principles and Applications to Electric Drives. IEEE Trans. Ind. Electr. 40, 23–36 (1993)

[45] Vujičić, V., Vukosavić, S.: A simple nonlinear model of switched reluctance motor. IEEE Trans. En. Conv. 15 (2000)

[46] Wallace, R., Taylor, D.: A balanced commutator for switched reluctance motor to reduce torque ripple. IEEE Trans. P. Electr. 7, 31–36 (1992)

[47] Witkowski, A., Tomczewski, K.: SRM supply system with switched phase winding configuration. In: Proc. Symp. El Mach., SME, vol. 41(2), pp. 660–667 (2005)

[48] Xu, L., Wang, C.: Accurate Rotor Position Detection and Sensorless Control of SRM for Super-High Speed Operation. IEEE Trans. P. Electr. 17, 757–763 (2002)

[49] Xue, X.D., Cheng, K.W.E., Ho, S.L.: Simulation of switched reluctance motor drives using two dimensional cubic spline. IEEE Trans. En. Conv. 17, 471–477 (2002)

[50] Yoo, B., Ham, W.: Adaptive fuzzy sliding control of nonlinear system. IEEE Trans. Fuz. Sys. 6, 315–319 (1998)

[51] Zhang, J.H., Radun, A.V.: A new method to measure the switched reluctance motor's flux. IEEE Trans. Ind. Appl. 42, 1171–1176 (2006)

Index